AF594975

Multimedia Cartography

Multimedia Cartography

Editors

Beata Medyńska-Gulij
David Forrest
Paweł Cybulski

MDPI • Basel • Beijing • Wuhan • Barcelona • Belgrade • Manchester • Tokyo • Cluj • Tianjin

Editors

Beata Medyńska-Gulij
Department of Cartography and Geomatics
Adam Mickiewicz University
Poznan
Poland

David Forrest
School of Geographical Earth Sciences
University of Glasgow
Glasgow
United Kingdom

Paweł Cybulski
Department of Cartography and Geomatics
Adam Mickiewicz University
Poznan
Poland

Editorial Office
MDPI
St. Alban-Anlage 66
4052 Basel, Switzerland

This is a reprint of articles from the Special Issue published online in the open access journal *ISPRS International Journal of Geo-Information* (ISSN 2220-9964) (available at: www.mdpi.com/journal/ijgi/special_issues/Multimedia_Cartography).

For citation purposes, cite each article independently as indicated on the article page online and as indicated below:

LastName, A.A.; LastName, B.B.; LastName, C.C. Article Title. *Journal Name* **Year**, *Volume Number*, Page Range.

ISBN 978-3-0365-3680-4 (Hbk)
ISBN 978-3-0365-3679-8 (PDF)

Contents

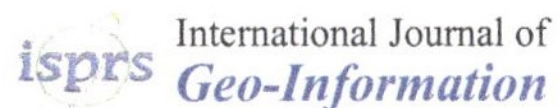
International Journal of
Geo-Information

Editorial

Modern Cartographic Forms of Expression: The Renaissance of Multimedia Cartography

Beata Medyńska-Gulij [1], David Forrest [2] and Paweł Cybulski [1,*]

1 Department of Cartography and Geomatics, Adam Mickiewicz University, 61-680 Poznan, Poland; bmg@amu.edu.pl
2 School of Geographical & Earth Sciences, University of Glasgow, Glasgow G12 8QQ, UK; David.Forrest@glasgow.ac.uk
* Correspondence: p.cybulski@amu.edu.pl

Abstract: This article summarizes the Special Issue of "Multimedia Cartography". We present three main research fields in which multimedia cartography and the study of the effectiveness of multimedia maps are currently taking place. In each of these fields, we describe how published research is embedded in the broader context of map design and user studies. The research refers to contemporary technological trends such as web HTML5 standards, virtual reality, eye tracking, or 3D printing. Efficiency, performance, and usability studies of multimedia maps were also included. The research published in this issue is interdisciplinary. They combine traditional mapping methods with new technologies. They are searching for new places for cartography in, e.g., the environment of computer games. They combine the design of the map with its perception by users.

Keywords: multimedia; cartography; animation; spatial visualization; multimedia cartographic product; medium efficiency; medium attractiveness

Citation: Medyńska-Gulij, B.; Forrest, D.; Cybulski, P. Modern Cartographic Forms of Expression: The Renaissance of Multimedia Cartography. *ISPRS Int. J. Geo-Inf.* **2021**, *10*, 484. https://doi.org/10.3390/ijgi10070484

Received: 6 July 2021
Accepted: 9 July 2021
Published: 14 July 2021

1. Introduction

The growing volume of data that can be presented by cartographic visualization requires map makers to use different means of expression. In modern cartographic communication, they are often related to multimedia. They allow integration the map with video, sound, or animation in an interactive environment. Multimedia cartography is a research area in cartography that focuses on the utilization of various multimedia for the effective and efficient visualization and communication of spatiotemporal data. The multidimensionality of geographical data creates certain demands for mapping products, and the answer to these demands is multimedia cartography. Assessing multimedia effectiveness, efficiency, and usability is possible thanks to user studies. In this approach, actual users are involved in the overall process of map design.

2. Dynamic Spatiotemporal Data

This Special Issue focuses on the crucial features of modern maps, which are effectiveness, efficiency, and attractiveness of information communication. The issues of multimedia cartography include, among others, the study of dynamic maps. Due to its visual attractiveness, animation is eagerly used for data that change over time. It is often presented in global news websites, as well as education, entertainment, weather forecasts on the web or on television. For this reason, it is crucial to conduct research on effectiveness of animated map communication. Popular time-series animation causes some perceptual difficulties for map users. Therefore, Traun et al. [1] conducted empirical research on the effect of local outlier preserving value generalization of animated choropleth maps on the ability to detect general trends in spatiotemporal data. Such an approach results from searching for solutions to reduce the cognitive load that animated map users experience. The research experiment with the participation of users presented in this study assumed the identification of outliers and the recognition of spatial trends. Study results show that, based on the spatial generalization of choropleth map animation, participants had higher outlier detection performance. However, it turned out that there was no significant improvement of participants' ability to correctly recognize the temporal trend. Moreover, the trend detection performance decreased with participants' age.

The effectiveness of presentation of spatiotemporal data was the main issue of the paper from Medyńska-Gulij et al. [2]. In this study, the main focus was on presentation of moveable and stationary objects in dynamic cartographic presentation. Three types of animated maps were proposed to represent people gathering at mass events. The study adopted experts' opinion and objective effectiveness evaluation. Therefore, various mapping techniques were presented of how to visualize people at a mass event. Methods proposed combined experts subjective opinion and a user study on effectiveness. The results show that the most difficult was to assess the dynamics of spatial relations (e.g., crowdedness, the number of participants). The study results showed the great importance of visually highlighting stationary objects and the need to minimize the number of map elements such as scale bar and north arrow.

Another study in this Special Issue related to cognitive aspects of spatiotemporal data was conducted by Cybulski [3]. This study takes into account two map backgrounds, which are a road map (abstract image) and a satellite map (realistic image). On these two cartographic backgrounds, participants were presented with a set of animated routes of varying difficulty (dependent on length and number of turns). The main goal of the aforementioned study is to assess the effectiveness and efficiency of route memorization based on the cartographic image. Additionally, the study adopts an eye tracking methodology to examine and explain the visual strategy that follows effective route memorization in different map-based tasks conditions. The main result shows that the effectiveness and efficiency of route memorization depends on the route difficulty. The research shows that, in this type of map-based task, the cartographic background has no impact on the memorization process nor visual strategy. The study reveals that, although the starting point of animation was the most fixated location, most errors in route memorization were related to the starting position. Additionally, the study results present the effect of subjective spatial abilities on visual strategy.

The abovementioned studies are in line with the map perception field of cartographic research. They result from the need for an objective assessment of multimedia cartographic products by the users themselves. Such results provide necessary feedback cartographers and modern map makers [4,5].

3. Maps and Interactions

Interactivity is an crucial feature of modern multimedia cartographic products [6]. Research conducted by Lorek and Horbiński [7] analyze old cartographic material in the context of usefulness in designing an interactive multimedia map of spatial development of Gliwice. The spatial development was based on vectorization spatial features from six maps: Urmesstischblätter map, Messtischblätter map, three topographical maps from different time periods, and the OpenStreetMap (OSM) database. A modern spatial database together with archival cartographic material contributed to the design of an interactive map of European freeway junction A1/A4 development. This study presented how to use old cartographic materials with a spatial database and implementation of a JavaScript Library–Leaflet. Research results focus on describing the utility of interactions such as mouse over or mouse click and implementation of W3C standards for new cartographic products.

An important part of interactive multimedia maps is the Graphical User Interface (GUI). The GUI often consists of several interactive buttons which that different functionality of a web mapping service. However, each global mapping service (e.g., Google Maps, Baidu Maps, OSM) has its unique set button graphics and arrangement. Therefore, users experience interaction differently in each of these services. This user experience (UX) while performing interactive map-based tasks was a main focus in the study from Cybulski and

Horbiński [8]. The study examined subjects' experience of interaction through the several different GUI and registered eye movement during the task. The study resulted in some valuable recommendations for map designers. It turned out that it is better for effective interaction to group buttons with similar functionality in the corners of the screen.

Another vital issue related to the interface and interaction is the maps' ability to adapt to the device on which it is displayed. In this context Horbiński et al. [9] performed a study on the responsiveness of GUI and its effectiveness in information communication. The design process of the multimedia map draws from modern JavaScript libraries, external plugins and HTML5 standards. In this study interactions and responsiveness of GUI were based on Google Maps, OpenStreetMap, and additional button arrangement based on the study of Horbiński et al. [10]. Analysis of mobile and desktop map-based tasks in a user testing research experiment with recording of eye movements was possible thanks to the use of an online questionnaire and Tobii X2-60 eye tracker. Study participants used three main functions which were: geolocation, spatial search, and route find. The study presented a novel GUI effectiveness index based on three parameters: the time to the first fixation on a target button, time of identification, and the time to the first mouse click. The study results reveal that the most effective way of performing all the interactive tasks was when the GUI was designed first on the desktop application and adapted for the mobile device. This was highly related to participants preferences of using web mapping services on the desktop computer. However, the most effective way of spatial search was when the buttons were in the lower part of the mobile screen, which was closest to the thumb.

The ease of receiving spatial information can prompt users to visit popular sightseeing websites. Therefore, Kato and Yamamoto [11] proposed the development a sightseeing spot recommendation system that could be useful and efficient for tourists. In the study, they an integrated social networking service, a web-geographic information system, and a recommendation system for interactive maps. The research experiment involved users' assessment of the system according to the viewing function, submitting functions, and recommendation functions of sightseeing spots. The study urges designers to take into account users' preferences in the map design process. The significant aspect of this study is that designed system could adopt the knowledge-based recommendations with collaborative recommendations according to users' preferences.

Multimedia presentations can combine text, image, videos, charts and maps, and through the interactive dashboard enables selecting, searching, and filtering the spatial data [12]. Zuo et al. [13] designed a novel map-based interactive dashboard supporting users' spatiotemporal knowledge acquisition and analysis. In the design of the dashboard, they applied interactive maps. Initially, the experiment consisted of free exploration in which participants could interact freely with the dashboard. Subsequently, there was a tasks-solving stage in which participants explored the dashboard to check authenticity of statements being displayed. The experimental study adopted a Gazepoint GP3 eye tracker with 60 Hz frequency. For several tasks, the study revealed the visual strategy that was most common for knowledge acquisition. This resulted in significant suggestions for designers of such dashboards. First of all, the study mentions the important role of the font size and proper labelling. Secondly, the panel arrangement in the interactive dashboard should follow a logical order (e.g., panels with similar content should be placed together).

In summary, interactive maps are the basic element of interest in multimedia cartography [14]. The GUI is especially important because global mapping services such as Google Maps, Bing Maps or Baidu Maps use an interactive interface as well as StreetView photos, videos and images. Studying the user experience during human–map interaction is a source of knowledge for effective and efficient mapping solutions. It also gives you the opportunity to optimize interactions for specific mapping products.

4. New Visualization Methods

The growing number of spatial and temporal features which can be presented through a cartographic product require searching for new visualization methods. Wielebski et al. [15]

proposed a set of complementary visualizations, interactive and static, to examine issues of people's spatial behavior. The aim of this study was to examine the usefulness for data analysis and interpretation of the following track features of human movement in urban space: trajectory, correctness, length and visibility. It also considered additional aspects such as walking time, walking speed, stops, spatial context, and motivation for getting to the finish point. The study involved recording the movement of people with the use of GPS and a questionnaire about pedestrian motivations. A space-time cube (STC) was designed and modified so that trajectories of movement could be presented in pseudo 3D view along with other mentioned features. Visualization was enhanced with buildings in City GML standard. However, the route graph, proposed by Andrienko and Andrienko [16] was assessed as the most useful for differentiation of pedestrian tracks, visualizing track length, walking time, motivation of route choice and pace/stops/tempo. The STC was the most useful for visualizing spatial context, and location in geographic space.

Other hew ways of using multimedia in cartography involves virtual tours and virtual reality (VR). Some of these virtual images uses official data (e.g., topographic databases Halik [17]) and are used by leisure industries to attract tourists [18]. Lai et al. [19] performed a research about processing panoramic images for a VR environment integrated in a multimedia platform. This study contributed especially to improving panoramic photography and image stitching quality, proposed a strategy for producing high quality panoramic images, combination of information technologies, and assessing teaching effectiveness through the multimedia platform. The results show that RMSE of stitched images was below five pixels using a full-frame single-lens reflex camera equipped with an ultra-wide-angle zoom lens on a panoramic instrument (GigaPan EPIC Pro V).

User experience in VR could be different based on previous experience with this technology. Medyńska-Gulij and Zagata [20] performed an experimental study in which they asked experts and gamers to take an immersive experience in a virtual stronghold of Ostrów Lednicki (Poland). Participants had to take a virtual walk around the stronghold, visit the palatium with a chapel, then go to the viewing point. The VR environment was built in the Unity game engine. The building geometry was based on a point cloud acquired from LiDAR. The gamers group needed less time than the experts group to complete the tour. However, expert users made more comments, and their observations made references to their expertise from academic publications. It turned out that gamers were more homogenous in their comments and opinions. Based on this, the study concludes that cartographic materials could be a fundamental source of information in immersive VR for cultural objects and reconstructions.

Another project related to VR technology and cartographic context is provided by Zagata et al. [21]. They studied the effectiveness of a mini-map for navigation in a reconstructed stronghold. The research adopted eye tracking technology in order to capture the total time users gaze on the mini-map. Participants were asked to find and collect Mieszko I's coins within the stronghold, highlighted on the mini-map. One group of participants used walking, and the second used teleportation for movement in space. Study participants were gamers, playing computer games a minimum 10 h per week. The study was supported by several software packages, e.g., SketchUp, CloudCompare, Photoshop, and Unity game engine. The main hardware was HTC Vive Pro Eye googles with an in-built eye tracking device. The study reveals that participants who examined the mini-map longer needed more time to finish collecting all the coins. There was no significant difference between people walking and teleporting in mini-map examination time. However, participants who used teleportation finished the task faster.

New visualization methods draw from various data gathering methods. Another study analyzed land use mapping techniques according to pedestrian movement registration with unmanned aerial vehicle (UAV) [22]. The research aimed to develop a methodological approach for thematic maps production based on low-level aerial images with recording of pedestrian movement. The study proposed point-to-polygon transformation of pedestrian visualization. The outcome of the research was several thematic maps of pedestrian density and land use occupied by pedestrians. The results show that formal land use classification could be changed due to the pedestrian movement behavior. The usage of point-to-polygon transformation allowed qualitative data to be obtained, therefore providing complementary information to actual land use visualization.

Visualization of spatial data is not only related to digital cartography. Harding et al. [23] proposed a web application enabling 3D printing of terrain models. This type of terrain visualization is something opposite to the VR environment, and the models that are produced are static in nature. Three-dimensional models provide the aspect of tangibility of cartographic product. Printed physical terrain models are commonly used in geography and geology teaching. The TouchTerrain application uses Google Earth Engine and can process digital 3D terrain models according to given coordinates. Additionally, it provides the possibility to import vector data such as GPS (GPS Exchange Format). In the study, Google Analytics were used to determine TouchTerrain users' characteristics. It turned out that most users are returning users, and only 21% of all users are new users (the results are based on the session between 1 July 2019 and 26 December 2020). The primary location of users, based on IP geolocation, suggests US and Western Europe.

5. Conclusions

This Special Issue on multimedia cartography highlights detailed research in specific research fields. However, the field of interest in multimedia cartography is much wider. Particularly noteworthy is the importance of the cartographic symbol and the map itself as the core of the multimedia visualization. The leading topic at present is the effectiveness of individual media in conveying information, in particular the effectiveness and efficiency of cartographic communication in multimedia presentation. In the context of changing technology, it is interesting to note the importance of new media such as VR, AR, holography and their significance for cartography. Among the research included in this Special Issue, one can also find questions about the place of maps in new technologies.

As the presented research shows, cartography plays an crucial role in the information communication process. The map is becoming an inseparable element of navigation in space, especially on mobile devices. In addition to the real world, it is part of virtual tours and allows finding specific locations in virtual space. Cartographers search for mapping methods that will be most useful and attractive for users by researching effectiveness, efficiency, and user experience implementing new research methods and technologies. Multimedia maps are used in education, storytelling, navigation, visualization and, above all, in everyday activities.

Author Contributions: Conceptualization, Beata Medyńska-Gulij, David Forrest, and Paweł Cybulski.; methodology, Beata Medyńska-Gulij, David Forrest, and Paweł Cybulski; writing—original draft preparation, Beata Medyńska-Gulij, David Forrest, and Paweł Cybulski. All authors have read and agreed to the published version of the manuscript.

Funding: This research received no external funding.

Institutional Review Board Statement: Not applicable.

Informed Consent Statement: Not applicable.

Data Availability Statement: Not applicable.

Conflicts of Interest: The authors declare no conflict of interest.

References

1. Traun, C.; Schreyer, M.L.; Wallentin, G. Empirical Insight from a Study on Outlier Preserving Generalization in Animated Choropleth Maps. *ISPRS Int. J. Geo-Inf.* **2021**, *10*, 208. [CrossRef]
2. Medyńśka-Gulij, B.; Wielebski, Ł.; Halik, Ł.; Smaczyński, M. Complexity Level of People Gathering Presentation on an Animated Maps–Objective Effectiveness Versus Expert Opinion. *ISPRS Int. J. Geo-Inf.* **2020**, *9*, 117. [CrossRef]
3. Cybulski, P. Effectiveness of Memorizing an Animated Route–Comparing Satellite and Road Map Differences in the Eye-Tracikng Study. *ISPRS Int. J. Geo-Inf.* **2021**, *10*, 159. [CrossRef]

4. Griffin, A.L. Cartography, visual perception and cognitive psychology. In *The Routledge Handbook of Mapping and Cartography*; Kent, A.J., Vujakovic, P., Eds.; Routledge: New York, NY, USA, 2017; pp. 44–54.
5. Medyńska-Gulij, B. Educating tomorrow's cartographers. In *The Routledge Handbook of Mapping and Cartography*; Kent, A.J., Vujakovic, P., Eds.; Routledge: New York, NY, USA, 2017; pp. 561–568.
6. Roth, R.E. Interactive maps: What we know and what we need to know? *J. Spat. Inform. Sci.* **2013**, *6*, 59–115. [CrossRef]
7. Lorek, D.; Horbiński, T. Interactive Web-Map of the European Freeway Junction A1/A4 Development with the Use of Archival Cartographic Sources. *ISPRS Int. J. Geo-Inf.* **2020**, *9*, 438. [CrossRef]
8. Cybulski, P.; Horbiński, T. User Experience in Using Graphical User Interfaces of Web Maps. *ISPRS Int. J. Geo-Inf.* **2020**, *9*, 412. [CrossRef]
9. Horbiński, T.; Cybulski, P.; Medyńska-Gulij, B. Web Map Effectiveness in the Responsive Context of the Graphical User Interface. *ISPRS Int. J. Geo-Inf.* **2021**, *10*, 134. [CrossRef]
10. Horbiński, T.; Cybulski, P.; Medyńska-Gulij, B. Graphic design and placement of buttons in mobile map application. *Cart. J.* **2020**, *57*, 196–208. [CrossRef]
11. Kato, Y.; Yamamoto, K. A Sightseeing Spot Recommendation System That Takes into Account the Visiting Frequency of Users. *ISPRS Int. J. Geo-Inf.* **2020**, *9*, 411. [CrossRef]
12. Few, S. *Information Dashboard Design: The Effective Visual Communication of Data*; O'Reilly Media, Inc.: Newton, MA, USA, 2006.
13. Zuo, C.; Ding, L.; Meng, L. A Feasibility Study of Map-Based Dashboard for Spatiotemporal Knowledge Acquisition and Analysis. *ISPRS Int. J. Geo-Inf.* **2020**, *9*, 636. [CrossRef]
14. Cartwright, W.; Peterson, M.P. Multimedia Cartography. In *Multimedia Cartography*; Cartwright, W., Peterson, M.P., Gartner, G., Eds.; Springer: Berlin/Heidelberg, Germany, 1999; pp. 1–10.
15. Wielebski, Ł.; Medyńska-Gulij, B.; Halik, Ł.; Dickmann, F. Time, Spatial, and Descriptive Features of Pedestrian Tracks on Set of Visualizations. *ISPRS Int. J. Geo-Inf.* **2020**, *9*, 348. [CrossRef]
16. Andrienko, N.; Andrienko, G. Visual analytics of movement: An overview of methods, tools and procedures. *Inf. Vis.* **2003**, *12*, 3–24. [CrossRef]
17. Halik, Ł. Challenges in converting Polish topographic database of built-up areas into 3D virtual reality geovisualization. *Cart. J.* **2018**, *55*, 391–399. [CrossRef]
18. Wu, Z.-H.; Zhu, Z.-T.; Chang, M. Issues of designing educational multimedia metadata set based on educational features and needs. *J. Internet Technol.* **2011**, *12*, 685–698.
19. Lai, J.-S.; Peng, Y.-C.; Huang, J.-Y. Panoramic Mapping with Information Technologies for Supporting Engineering Education: A Preliminary Exploration. *ISPRS Int. J. Geo-Inf.* **2020**, *9*, 689. [CrossRef]
20. Medyńska-Gulij, B.; Zagata, K. ExSperts and Gamers on Immersion into Reconstructed Strongholds. *ISPRS Int. J. Geo-Inf.* **2020**, *9*, 655. [CrossRef]
21. Zagata, K.; Gulij, J.; Halik, Ł.; Medyńska-Gulij, B. Mini-Map for Gamers Who Walk and Teleport in a Virtual Stronghold. *ISPRS Int. J. Geo-Inf.* **2021**, *10*, 96. [CrossRef]
22. Smaczyński, M.; Medyńska-Gulij, B.; Halik, Ł. The Land Use Mapping Techniques (Including the Areas Used by Pedestrians) Based on Low-Aerial Imagery. *ISPRS Int. J. Geo-Inf.* **2020**, *9*, 754. [CrossRef]
23. Harding, C.; Hasiuk, F.; Wood, A. TouchTerrain–3D Printable Terrain Models. *ISPRS Int. J. Geo-Inf.* **2021**, *10*, 108. [CrossRef]

International Journal of
Geo-Information

Article

Empirical Insights from a Study on Outlier Preserving Value Generalization in Animated Choropleth Maps

Christoph Traun [1,*], Manuela Larissa Schreyer [2] and Gudrun Wallentin [1]

[1] Interfaculty Department of Geoinformatics-ZGIS, University of Salzburg, 5020 Salzburg, Austria; gudrun.wallentin@sbg.ac.at
[2] AMAG Austria Metall AG, 5282 Ranshofen, Austria; manuelalarissa@freenet.de
* Correspondence: christoph.traun@sbg.ac.at

Abstract: Time series animation of choropleth maps easily exceeds our perceptual limits. In this empirical research, we investigate the effect of local outlier preserving value generalization of animated choropleth maps on the ability to detect general trends and local deviations thereof. Comparing generalization in space, in time, and in a combination of both dimensions, value smoothing based on a first order spatial neighborhood facilitated the detection of local outliers best, followed by the spatiotemporal and temporal generalization variants. We did not find any evidence that value generalization helps in detecting global trends.

Keywords: choropleth map animation; local outliers; generalization

Citation: Traun, C.; Schreyer, M.L.; Wallentin, G. Empirical Insights from a Study on Outlier Preserving Value Generalization in Animated Choropleth Maps. *ISPRS Int. J. Geo-Inf.* **2021**, *10*, 208. https://doi.org/10.3390/ijgi10040208

Academic Editors: Beata Medynska-Gulij and Wolfgang Kainz

Received: 11 February 2021
Accepted: 16 March 2021
Published: 1 April 2021

1. Introduction

Temporal animation of choropleth maps is a popular way to depict time-series data aggregated to enumeration units. Although movies from snapshots of choropleth maps are simple to understand conceptually, they often exceed our perceptual limits [1].

When viewing the sequence of time slices, users can easily miss important changes in the map during a saccade [2] or due to weak change signals from unattended areas of peripheral vision [3,4]. However, even when people apparently sense the often simultaneous changes in the quick succession of maps, they are easily overwhelmed by the sheer amount of transient information and often fail to derive appropriate mental models of the mapped process [1]. According to cognitive load theory [5,6], the cognitive bottleneck is the visual working memory, not able to store more than four new objects simultaneously [7]. Visual information, like objects changing their color in the map, needs to be shuttled into long-term memory to understand and further build upon it. If new information arrives during this process, the exchange between working memory and long-term memory is either cancelled by switching attention to new visual objects ("proactive inhibition"), or the new objects cannot be perceived due to blocked working memory resources by prior objects ("retroactive inhibition") [1]. The resulting "cognitive overflow" is further exacerbated in situations of split attention, when, for example, the user tries to simultaneously grasp the changing map contents and its (temporal) legend [8].

1.1. Solutions Proposed to Reduce Cognitive Load

Cartographers explored a number of ways to reduce cognitive load and make animations more accessible. While many ideas are related to the user interface design with the implicit or explicit goal to minimize split attention and/or maximize user control [9–14], others focus on the transition between time-steps [15,16], on techniques to highlight important changes [17], or on the relation between data complexity and animation speed. Multimäki and Ahonen-Rainio [18] equalize the temporal scale according to the temporal density of change, slowing down the animation in busy times, while increasing animation pace when the number of changes is low. Although this seems useful to limit cognitive load, it presumably fails to convey the temporal structure in the development of spatial processes. Moreover, stretching time to limit the amount of change between two subsequent frames requires adequate data. In case of low temporal sampling frequency, the amount of change between two adjacent time-steps/frames might already exceed cognitive limits, regardless of the slow animation pace. Thus, in addition to appropriate user interfaces, approaches are needed that reduce cognitive load by reducing the visual complexity of the data itself by means of generalization.

In order to remove "flicker" in choropleth map animations resulting from spurious value-changes over time, Monmonier [19] developed a classification algorithm trying to optimize class breaks in this respect. As it turned out that the problem could not be solved sufficiently by (crisp) data classification, he advocated for temporal averaging instead, and hypothesized on the usefulness of kernels that simultaneously smooth the data in the spatial and the temporal dimension [20]. Harrower [21] highlighted the benefits of spatial and temporal aggregation to facilitate change detection in a temporal animation of remotely sensed data. Giving general advice for effective animated map design, Harrower [10] advocated for highly generalized animated maps by using data filtering, data smoothing, or aggregating data into two or three classes. Even though he later withdrew his statement on the use of classification [22], he re-emphasized that filtering and smoothing is an important design principle of animated maps to provide a coherent "big picture" [9].

1.2. Research Gap

To the best of our knowledge, there was only a single empirical study addressing generalization in animated choropleth maps so far: McCabe [23] investigated the effects of temporal aggregation and temporal smoothing on two tasks (picking the map frame with the overall highest values, comparing the cumulative values between two regions over the whole animation) related to the interpretation of measles epidemics data in Niger. For both tasks, he could not find benefits from temporal averaging of data, but rather found a non-significant tendency that participants who used temporally smoothed data even performed worse in picking the "maximum" map frame, which is understandable due to the smoothing of instantaneous peaks. Given the notion that animated choropleth maps are most useful to gain an overview of the development of spatial process [22,24,25] and significant local outliers (polygons with values greatly differing from their neighbors in space and time) thereof [26], it is questionable whether the tasks in McCabe's experiment were the best choice to tackle potential benefits of animated choropleth maps generalization. To fill this gap, we conducted an experiment to evaluate the effect of local outlier preserving value generalization in space, in time, and in a combination of both dimensions on the detection of overall trends and local outliers in animated choropleth maps.

1.3. Document Organization

In the following section, we illustrate the design of our experiment and characterize the group of 440 test-persons who participated in our study. Then, we present the analysis of the obtained data, while adding insights from an extensive pilot study and thoroughly discuss our results in the light of perception and cognition of animated maps. We conclude with an outlook on future research opportunities.

2. Methods

To find out if and how different forms of value generalization of unclassed choropleth map animations affect the ability of users to detect general trends and local outliers thereof, we developed the following online-experiment: Participants saw short, synthetic map animations, each consisting of a general trend and two local outliers in space and time. Immediately after each animation ended, it was replaced by a set of six outlier candidates (first part of the experiment) or three trend candidates (second part of the experiment). From these sets participants had to select correct outliers and the correct trend, respectively. Having used differently generalized versions of the presented animations (non-generalized reference, temporal-, spatial-, and spatiotemporally generalized versions), we were able to examine the effect of generalization mode on the ability of users to correctly detect local outliers and overall trends.

2.1. Animated Map Stimuli

Each animation stimulus shows a map of 85 irregular but similar sized polygons that consists of 14 map frames. It was displayed at the size of 512 × 512 px and at the speed of five frames per second. Polygon values were simulated in GAMA (https://gama-platform.github.io) using a model that produces moving clusters embedded in a global trend and randomly adds local outliers to the otherwise highly autocorrelated data in space and time. From a large number of simulation results, we chose for the final map-stimuli those that contained exactly two local outlier polygons in space and time. Local outliers were determined by a heuristic that evaluates the value-difference of polygons to their first order space-time neighborhood, while considering the global autocorrelation of the data (see Traun and Mayrhofer [26] for further details). Spatiotemporal autocorrelation of all stimuli is rather similar with Moran's Is [27] between 0.84 and 0.93. From these reference stimuli, we derived three differently generalized versions by smoothing polygon values by their first order spatial, spatiotemporal, or temporal neighbors, respectively, while excluding the two local outlier-polygons from the generalization process. For data smoothing, we used the methods and software provided by Traun and Mayrhofer [26] and applied a CIE Lab-interpolation based, sequential yellow-to-brown color scheme to the unclassed data, using a linear min-max stretch (Figure 1).

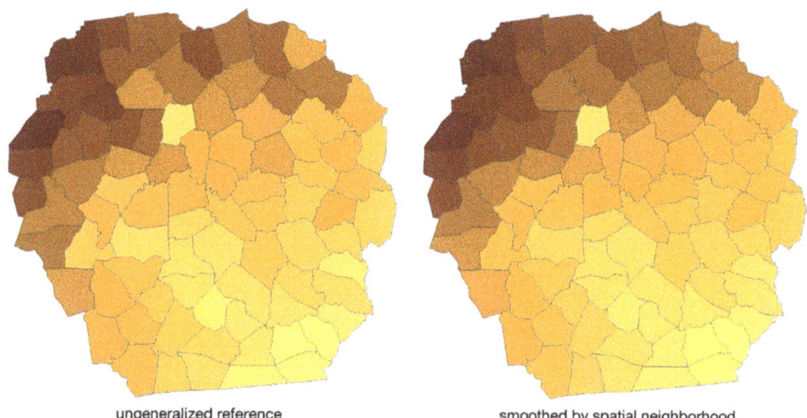

Figure 1. Slightly reduced local contrast in the spatially generalized version (**right**) of a frame from stimulus 1. The color of a local outlier polygon (bright polygon in the upper center) is not altered by generalization (**left**).

2.2. Response Items

After having seen a stimulus, participants had to choose the correct local outliers (first part of the experiment) and the appropriate global trend (second part) from a set of outlier and trend candidates, respectively.

2.2.1. Outlier Response Items

Six basemaps, highlighting one outlier candidate each, contain the two correct outliers from the stimulus and four wrong outlier candidates (Figure 2).

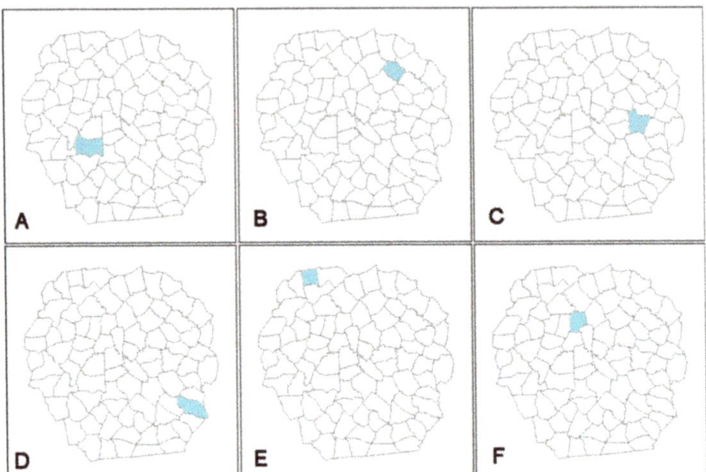

Figure 2. Local outlier response items for stimulus 1. Participants could choose between six outlier candidates (**A–F**). Option F shows the local outlier from Figure 1.

Contrary to the correct outlier candidates, wrong outlier candidates have a low value difference to their spatiotemporal neighborhood throughout the whole animation. Together with a dispersed distribution of outlier candidates over the basemap, this should prevent misinterpretation of non-outlier polygons as local outliers.

2.2.2. Trend Response Items

Global trend response items were produced by downscaling animated map stimuli to 160 × 160 px and applying a 15 px blur filter (Figure 3). To prevent participants from identifying trend response items not by the trend, but from the position of local outliers, they were replaced with the mean value/color of their neighbors before applying the filter. While correct global trend response items were produced from the respective stimulus animations, two alternative (wrong) items per response item set were derived from other stimuli or unused stimulus candidates. Each response item is started with a mouse click and could be replayed as often as desired. Response items are available together with other data from this research at https://tinyurl.com/mapstudyresults, as indicated in the data availability statement at the end of this document.

Figure 3. Global trend response item animations. As the six local outlier response items in the first part of the experiment, three trend response items immediately replace the seen stimulus in the second part of the study.

2.3. Study Design and Implementation

We decided for a mixed study design, which is based on four-group between subject-design, according to the modes of generalization (A-ungeneralized reference, B-spatial generalization, C-spatiotemporal generalization, and D-temporal generalization—refer to Stimulus 2, 3, and 4 in Figure 4).

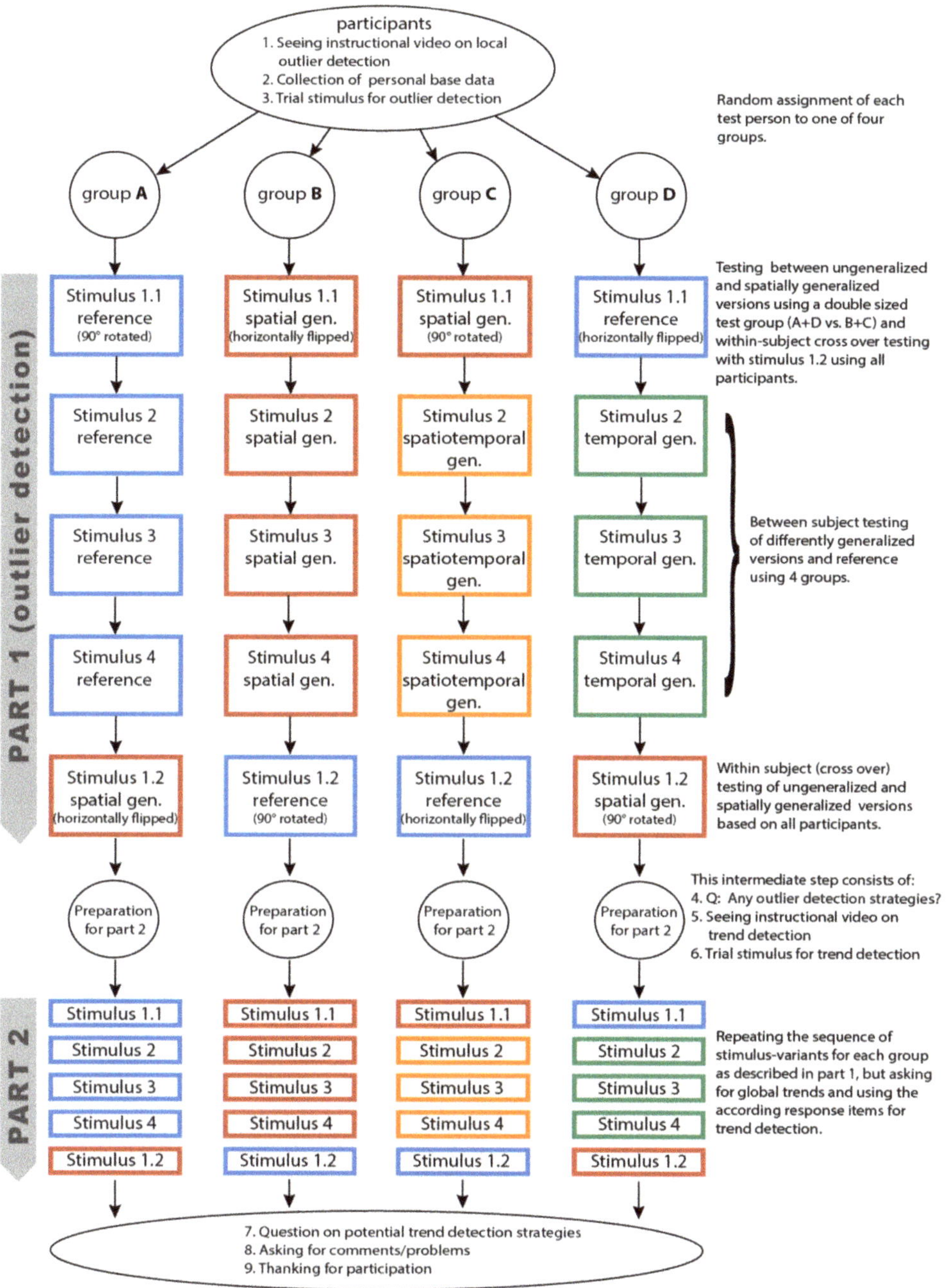

Figure 4. Study design.

From the results of a pilot study, we assumed that spatial generalization might have the highest impact on perception. Thus we developed a backup strategy for low participation numbers and small effect sizes and complemented the four-group design for Stimulus 2, 3, and 4 by a (double sized) two group between-subject plus a within-subject design for Stimulus 1. While Stimulus 1 was limited to two modes (ungeneralized and spatially generalized version), every participant saw the ungeneralized reference and the spatially generalized version of this stimulus in both parts of the experiment (refer to Stimulus 1.1 and 1.2 in Figure 4).

The experiment was set up as an online-study. Participation was restricted to desktop operating systems to prevent the use of small displays. Stimulus preloading ensured uninterrupted playback in case of low internet bandwidth.

Having accepted the invitation, participants saw a video, explaining the task of looking for two local outliers in an exemplary stimulus animation and demonstrating the selection of response items. Then, data on variables controlling for age, sex, visual impairments, map use experience, computer gaming frequency, and highest educational level were collected. Before participants went through the first set of stimulus sequences, they practiced outlier detection with a trial stimulus. Then, they were randomly assigned to one of four groups (Figure 4). Each stimulus sequence includes three steps:

1. 3-second countdown and automatic start of the animation.
2. Immediate replacement of the animation with response items for local outliers to choose from (Figure 2).
3. Rating of the difficulty of the task.

After finishing this first part, participants were asked to comment on potential strategies to identify and remember outliers. Then, they saw an instructional video on trend detection and practiced again with a trial stimulus, before entering the second part of the experiment. There, each group saw exactly the same stimuli-variants from the first part, but had to choose the correct trend response item out of three options (Figure 3). Finally, participants reported on trend detection strategies and were asked for feedback on the overall survey and any (technical) issues they encountered.

2.4. Participants and Data

The main study took place in December 2020. Invitations were distributed via social media and sent out by email to geography students and students/recent graduates from a distance learning program in GIS with the request for further dissemination. This effort resulted in 308 complete datasets with slightly different group sizes, different age distributions, and remarkably more male participants. To be more balanced in these aspects and considering positive experiences in the use of online crowdsourcing services in cognitive experiments [28] we complemented data with crowdworker-responses. By using stratified sampling in respect to age and sex, we acquired 169 crowdworkers at the platform clickworker.de. Each of them was compensated with 1.10 Euro for their time investment (median: 9 min). From the combined set of 477 responses, we randomly removed three male responses from group B and two male responses from group C to have equally sized groups. Referring to Crump et al. [28], we got rid of sloppy attempts by calculating a quality index (watching the videos to end, time used for provision of personal data and comments, overall outlier and trend detection performance) and removed the worst four male and female attempts from each group. The final dataset comprises four equally sized groups, each consisting of 55 male and 55 female responses. For age distributions and the share of crowdworkers per group see Figures 5 and 6. The distribution of cartographic competences among the participants is given in Table 1.

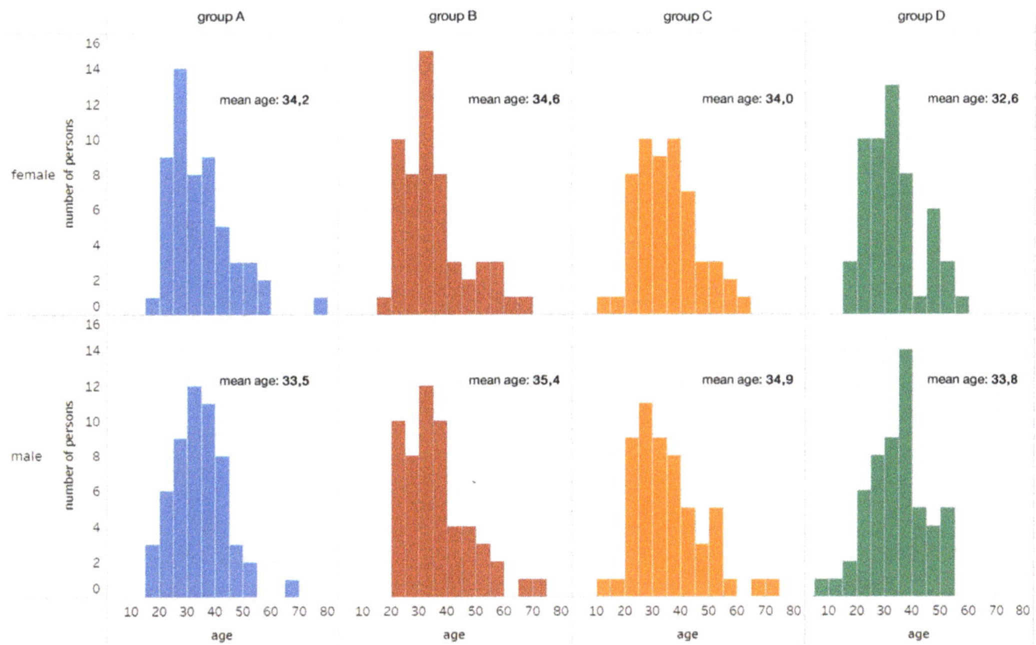

Figure 5. Age distribution by sex and group.

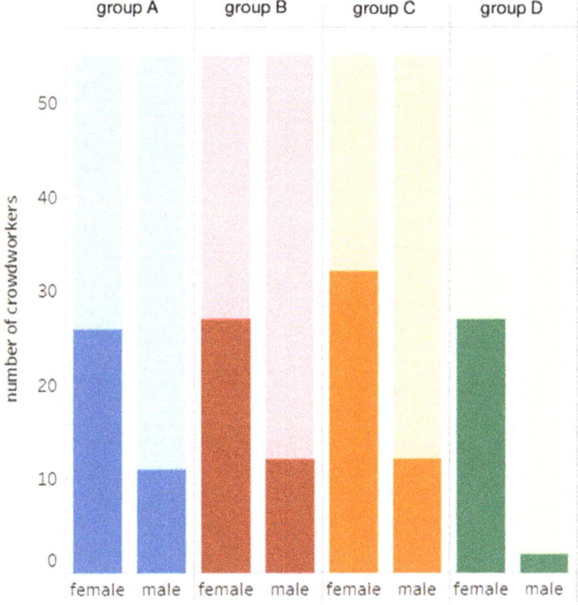

Figure 6. Saturated bars show the number of crowdworker responses (33.4% of total) by sex and group. Compare to the proportion of other participants, depicted by the light bars.

Table 1. Cartographic competence—absolute counts.

	Group A	Group B	Group C	Group D	Σ
None or little experience with maps	17	12	17	11	57
Active map use, no background in cartography	35	38	34	44	151
Basic knowledge in cartography	37	34	30	23	124
Advanced cart. knowledge, active mapping	17	23	23	26	89
Expert knowledge in cartography	4	3	6	6	19
Σ	110	110	110	110	440

To analyze whether or not people prefer generalized map animations, the main study was supplemented by data from our extensive pilot study, which was also conducted in an online-format. There, 334 (different) persons saw an infinitely looped, successive comparison (A = Reference, B = spatially generalized version) of Stimulus 4 and were asked to verbally describe differences and issue preferences.

3. Results

Statistical analysis was done in R [29], predominantly using the package *npmv* [30]. It compares the multivariate distributions for a single explanatory variable (like generalization mode) using nonparametric techniques and is even suitable for small samples. Using approximations for ANOVA Type, Wilks′ Lambda, Lawley Hotelling, and Bartlett Nanda Pillai Test statistics along with according permutation tests, the package allows us to compare the results of up to eight statistical testing approaches, whereas the actual number of applicable tests depends on the data structure [30]. With one exception, we received good agreement between different tests, which is an indication for the stability of the obtained p-values. As advised by Ellis, Burchett, Harrar, and Bathke [30], and for sake of readability, the p-values for Wilks′ Lambda are reported whenever this test was applicable. In all other cases, we provide the p-values from the ANOVA Type test. Reported p-values were Bonferroni-corrected for multiple testing.

3.1. *Local Outlier Detection*

For each person and stimulus instance, the numbers of correctly detected local outliers (0–2) and wrongly indicated outliers (up to 4 "false positives" possible) were recorded.

3.1.1. Correctly Detected Local Outliers

To see the influence of the independent variable "generalization mode" on the ability to correctly detect local outliers, we summed up the absolute counts per group (A,B,C,D) for 0, 1, and 2 correct outliers for each of the stimuli 2, 3, and 4. For stimulus 1.1 and 1.2, we combined the groups A + D and B + C as shown in Figure 4 and calculated the frequencies of 0, 1, and 2 correct outliers accordingly. The resulting pattern (Figure 7) clearly shows that (outlier preserving) generalization using the direct neighbors in space facilitated the detection of both local outliers best, followed by spatiotemporal and temporal generalization variants.

People having seen generalized stimuli outperformed respondents from the reference group in all but one instance: For Stimulus 4, the temporally smoothed variant led to the poorest results. We, however, do not attribute any particular meaning to this, as we had tested exactly these two variants (reference and temporally generalized version) of stimulus 4 in our pilot study. Results showed quite similar outlier detection performance for both groups consisting of 84 and 82 (different) persons, respectively. Therefore, we consider this inconsistency in the main experiment to be a statistical outlier.

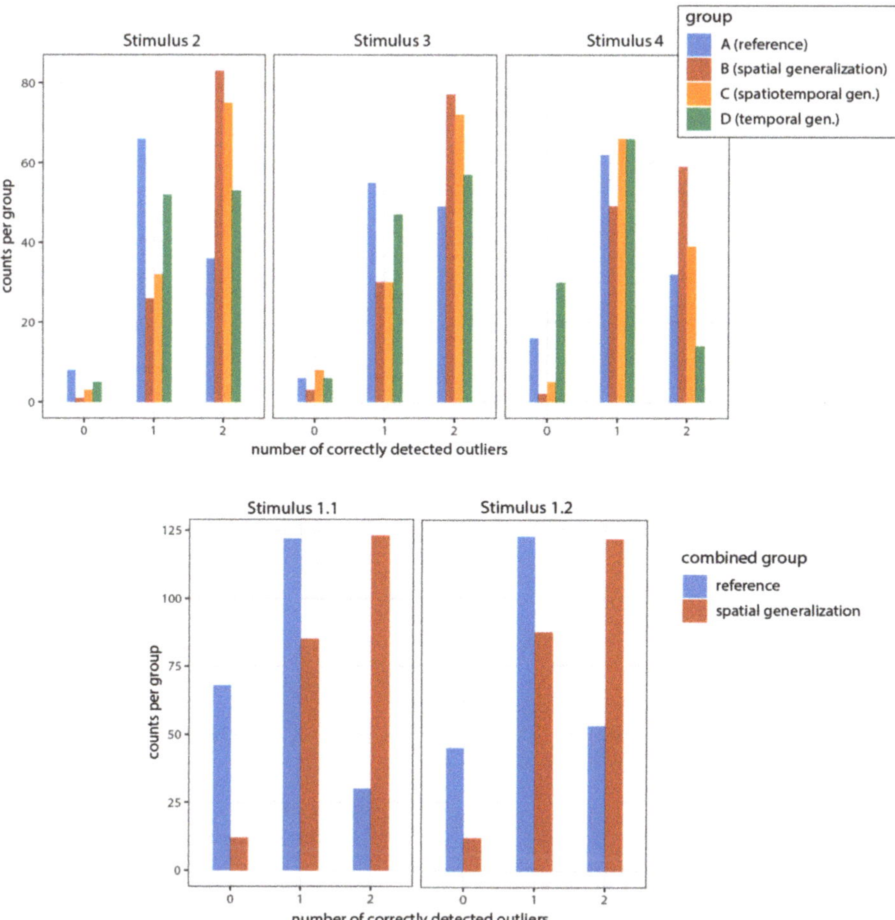

Figure 7. Number of persons per group/generalization type for correctly detecting none, one, or both local outliers in each stimulus.

Global statistical testing (combined for stimuli 2, 3, and 4) for each of the 3 generalization variants against the reference group rejected the null hypothesis (no difference) on the $\alpha = 0.01$ level ($p < 0.001$). When, however, testing the temporal variant against the reference just for Stimulus 2 and 3 while excluding the erratic result from Stimulus 4, the (Bonferroni corrected) result is not significant anymore ($p = 0.22$). Thus, an effect of outlier preserving, temporal generalization on outlier detection seems questionable.

Statistical analysis of the double sized group results from stimulus 1.1 and 1.2 confirms the highly significant effect of outlier preserving smoothing in space on local outlier detection ($p < 0.001$). The relatively improved performance of the reference group in Stimulus 1.2 probably results from evolving strategies for local outlier detection when being exposed to this (last) stimulus. An additional within-person crossover test based on stimulus 1 further confirms the outcome from the between group tests: For each person, the number of correctly detected outliers from the generalized version was subtracted from the according number from the ungeneralized reference stimulus. For the resulting distribution the 95%-bootstrap-confidence interval for the mean is given by [−0.640, −0.493] and the null hypothesis (mean is equal to 0) is rejected

3.1.2. False Positive Outliers

As test persons were informed of the presence of two local outliers in each animation, the theoretical possibility of three or four wrong picks did not happen. In most cases, zero or one wrong candidates were chosen (Figure 8). Although results are not reciprocal to correct outlier identification, they follow similar, yet inverted (less is "better" in this case) lines.

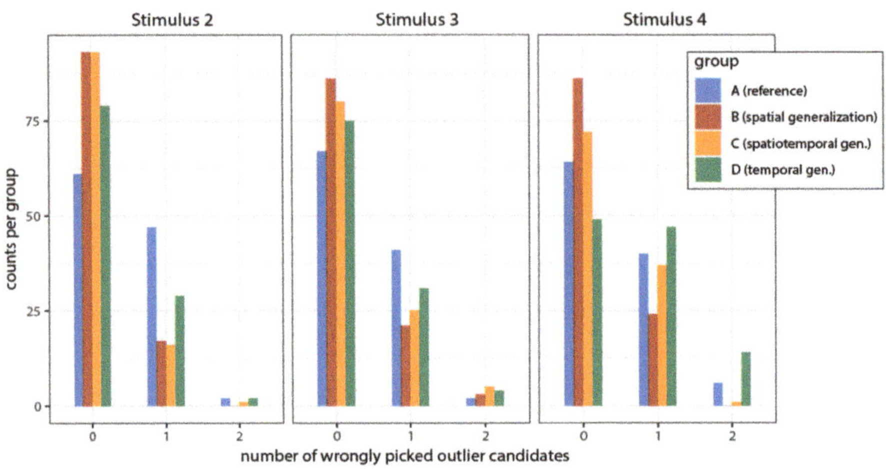

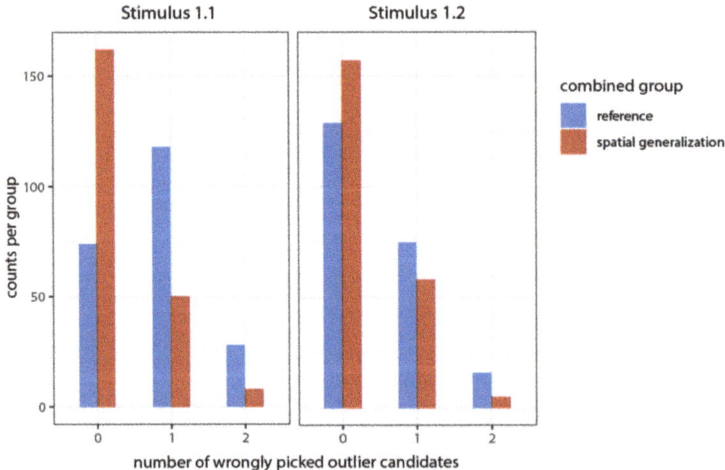

Figure 8. Number of persons per group/generalization type for erroneously picking none, one, or two wrong outlier candidates ("false positives").

Differences between the reference group and the spatial and spatiotemporal groups are highly significant ($p < 0.001$), with both groups constantly outperforming the reference group. Again, the temporal group performed worse than the reference group for Stimulus 4. Including this Stimulus in the statistical analysis leads to inhomogeneous p-values in the applied tests, ranging from $p = 0.008$ (Wilks' Lambda) to $p = 0.04$ (ANOVA Type permutation test). Limiting the test to Stimulus 2 and 3 results in an insignificant p-value of 0.08 (Wilks' Lambda).

A test of the double-sized groups in Stimulus 1.1 and 1.2 confirms the differences between reference and spatial generalized variants (both $p < 0.001$). Again, a general learning effect might be the reason for the improved results of the reference group in the second (flipped and rotated) instance of this stimulus.

3.2. Global Trend

We started our empirical studies with the assumption, that value generalization removing "visual noise" from choropleth map animations will facilitate the detection of the remaining overall trends. According to our quantitative results (Figure 9), this is not the case.

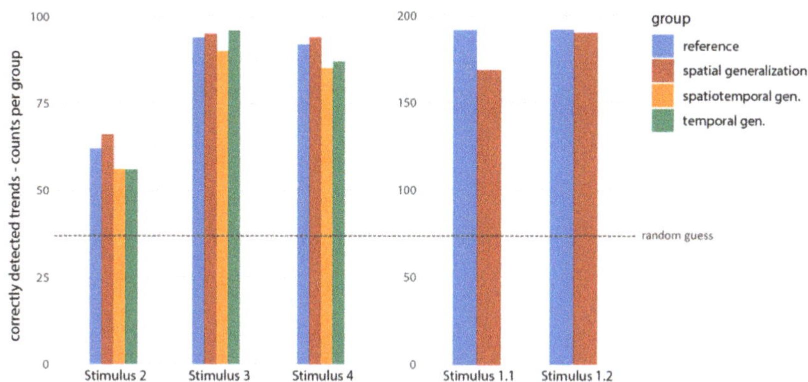

Figure 9. Number of persons per group/generalization type having chosen the correct trend response item.

Although there are seemingly small benefits for the spatial generalization variant for Stimulus 2, 3, and 4, group differences are not significant. Interestingly, there is a highly significant result ($p = 0.005$) pointing to the opposite direction for Stimulus 1.1. As this was the first stimulus participants were exposed to in the second part of the survey (after one trial stimulus for practicing), several persons from the spatial group probably were distracted from their trend detection task by the quite salient outliers in this instance and involuntarily turned their attention back to "outlier detection mode". The nearly equal performance of both groups for the (identical) last stimulus of the survey (Stimulus 1.2) and some self-observation when going through the survey support this interpretation.

3.3. Person-Related Covariates

Due to the large number of participants and their random assignment to groups, potentially confounding variables like cartographic competence or educational level are distributed quite evenly among groups. To test the influence of those variables on outlier detection capabilities, we derived a personal outlier score for each participant by adding up all correctly detected outliers for the five stimuli seen. In the same manner, a personal trend score (sum of correct trends) was derived.

To check for effects related to age, we derived five evenly spaced age groups accommodating the range between the youngest (8 years) and the oldest (76 years) participant. Using the *npmv* package for R again, those groups were tested on differences in scoring. When using all five age groups, there are highly significant differences for trend scores ($p = 0.006$), decreasing with age (Figure 10). As the highest (62.4–76) age group is rather small (7 participants), we dropped it due to potential statistical problems associated with highly different group sizes. Differences of the remaining four age groups are still highly significant ($p = 0.008$). Trend scores are negatively correlated to age with a spearman's rho (rs) of −0.14.

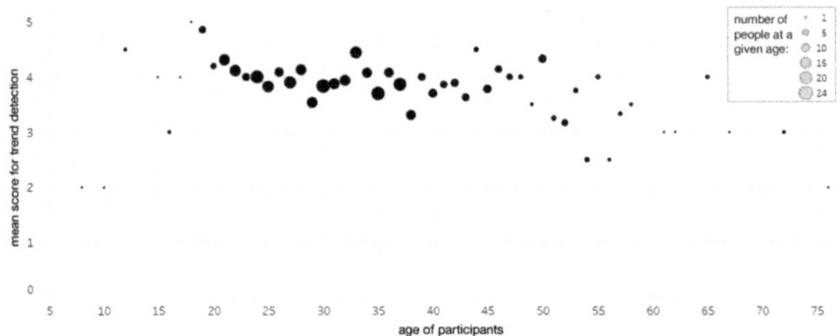

Figure 10. Trend scores decrease with age. Personal trend scores are based on all stimuli and bounded by 0 (all trends incorrect) and +5 (all trends correctly detected). Participants of the same age are represented by one dot.

There is also a similar tendency according to outlier scores, although not significant. We further did not find significant differences of outlier- and trend scores for sex, map use experience, educational level, computer gaming frequency, nor for paid (crowdworkers) versus voluntary participation.

3.4. Self-Confidence

Self-confidence ratings are positively correlated to the actual capability to correctly detect trends (rs = 0.30) and outliers (rs = 0.55). Under the condition of low confidence in the personal choice, a hit rate of 58% in trend-detection seems to be quite high (Figure 11). On the other hand, 8% of the according decisions were still wrong, although participants were highly confident to be right. In the case of outlier detection, the rate of misses (for having both outliers correct) increases up to 19% under the highest confidence rating (Figure 12). Thus, people do seemingly overestimate their visual detection abilities as described by Levin et al. [31].

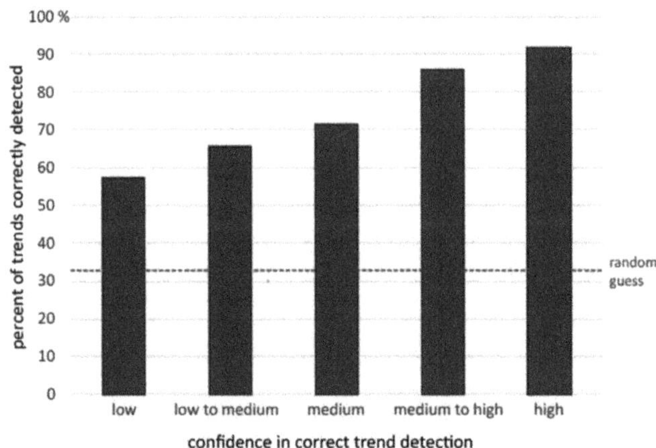

Figure 11. Hit rates for different levels of confidence in correct trend detection.

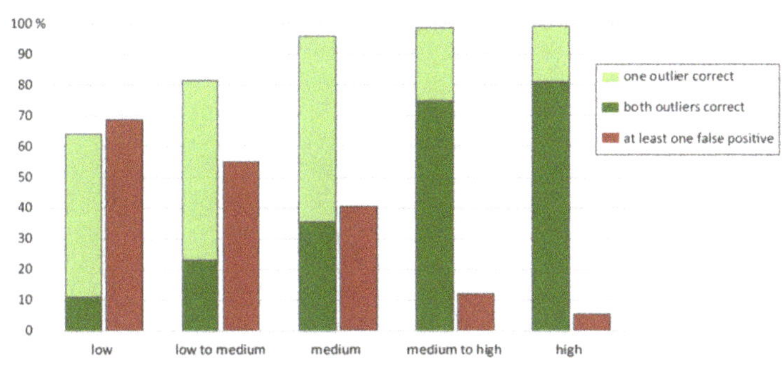

Figure 12. Hit rates and false alarms for different levels of confidence in correct outlier detection.

3.5. Strategies for Perception and Memorization

Qualitative answers on strategies for outlier and trend detection were split into perception and memorization strategies and clustered into categories (Figure 13).

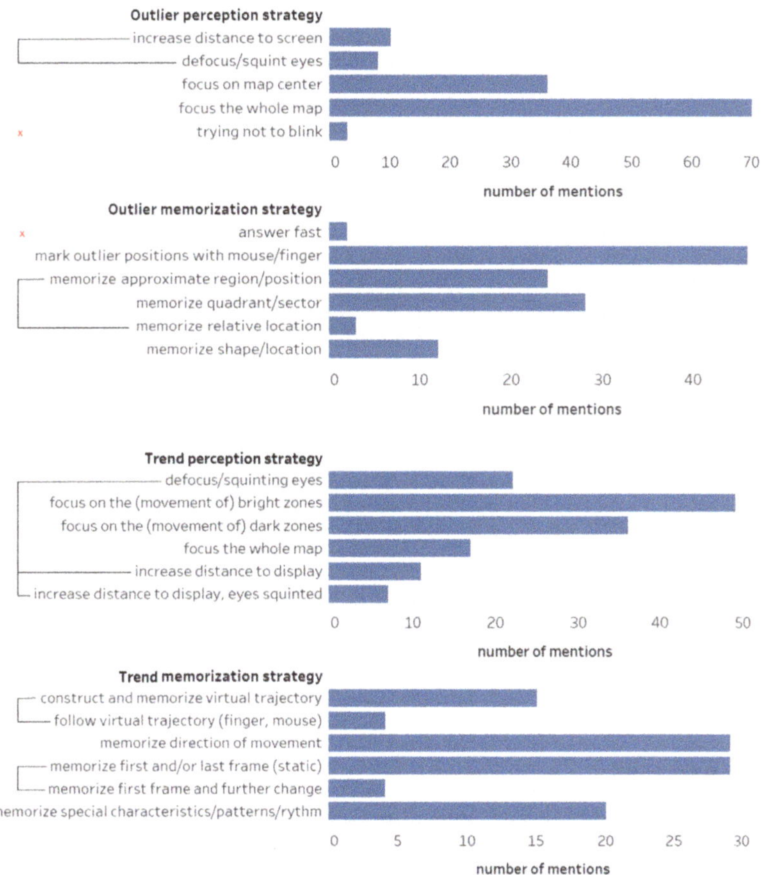

Figure 13. Strategies for outlier and trend detection and memorization.

As we wanted to test the effectiveness of the strategies mentioned and thus strived for group sizes with higher statistical reliability, we further aggregated related categories like "construct and memorize virtual trajectory" and "follow virtual trajectory (finger, mouse)" or excluded them ("try not to blink", "answer fast") from analysis as indicated in Figure 13. Outlier detection and memorization strategies were tested for differences in personal outlier scores and trend related strategies for differences in personal trend scores, respectively. While we did not find significant impacts of outlier perception/memorization and trend memorization strategies on user performance, trend perception strategies resulted in significantly different scores ($p = 0.016$). Among those strategies, squinting the eyes, defocusing the map, and/or leaning back while watching the animation worked best (Figure 14).

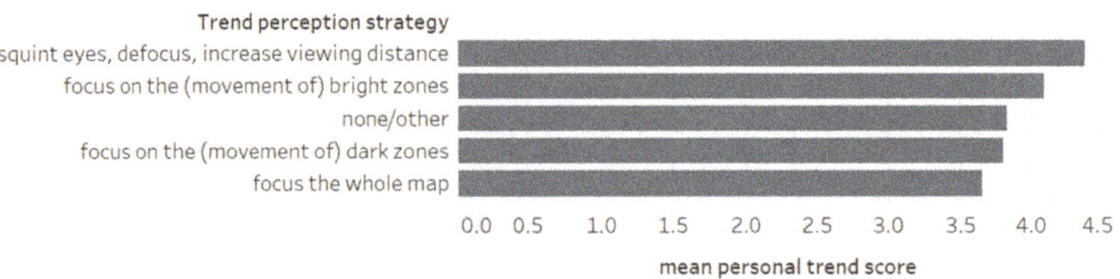

Figure 14. Performance of strategies for trend perception.

3.6. Described Perception and Preference

In our pilot study, participants were asked to verbally describe the difference between a reference animation ("Version A") and its spatially generalized counterpart ("Version B") using Stimulus 4. While 30% indicated to see no or only little differences, and 14% did not answer this open question, a content analysis of the remaining descriptions shows that the spatially generalized version was described to be more homogenous, smoother, less hectic, and less noisy. Furthermore, several persons emphasized that the generalized version facilitates the detection of a clear trend and/or local outliers thereof (Figure 15).

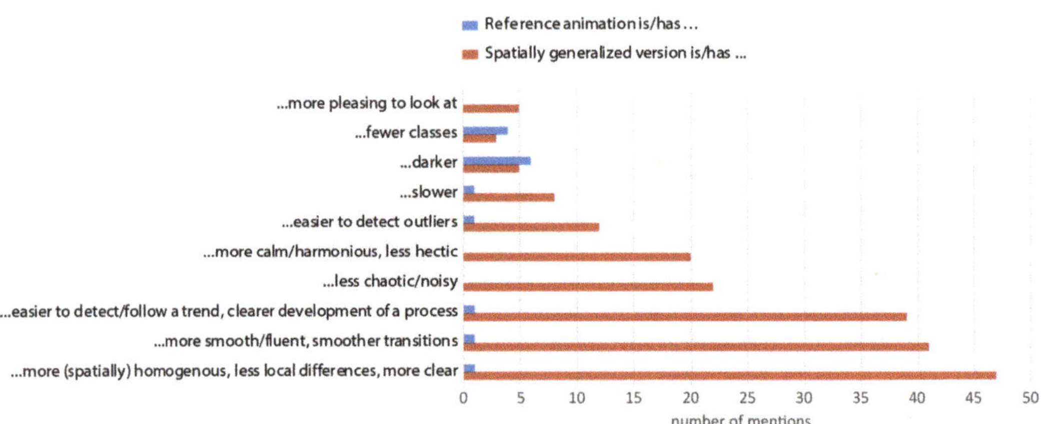

Figure 15. Number of comparative mentions in verbal descriptions of differences between the reference animation and its spatially generalized counterpart of Stimulus 4. Classification was done semantically, e.g., "Version A is more chaotic than version B" is classified as "Spatially generalized version (B) is less chaotic". In addition to these 217 comparative statements, 25 statements like "In version B the bright yellow seems to move farther up north" could not be attributed in a comparative way and were omitted.

We noticed considerable differences in the individual ability to consciously see and describe differences between both versions: Some participants proposed the surprisingly correct assumption that version B might be a spatially generalized version from raw data shown in version A, and one of them even mentioned that outliers seem to be excluded from spatial smoothing. In turn, others reported to have repeatedly watched the animation loop for minutes without seeing any difference. After survey completion, a member of the latter group approached us, asking whether there was any difference. After explaining and showing the animation loop again, he was astonished to have missed the yet clearly seen differences beforehand. Arguing that he has been looking for changes on the level of individual polygons instead of the overall picture, different foci of visual attention, and thus inattentional blindness [32], might be the reason for the 30% share of participants who did not see any difference.

Regarding the question "Which version is easier to understand?", participants from the pilot study clearly preferred the spatial generalized version (Figure 16). To foster involvement, we decided on an exclusive choice in this question, but provided the possibility to refuse but provide additional comments. Based on those comments, we created the categories "no preference" (people that saw the differences, but did not prefer either option), "no or minimal differences" (participants saw no or hardly any difference and thus cannot make a decision), and "no answer" (no comment). While typical comments of the group in favor of the spatially generalized version rephrase the verbal descriptions from above, the "no preference" group often answered in an "it depends" fashion, emphasizing different goals (general overview versus local detail) of an animation.

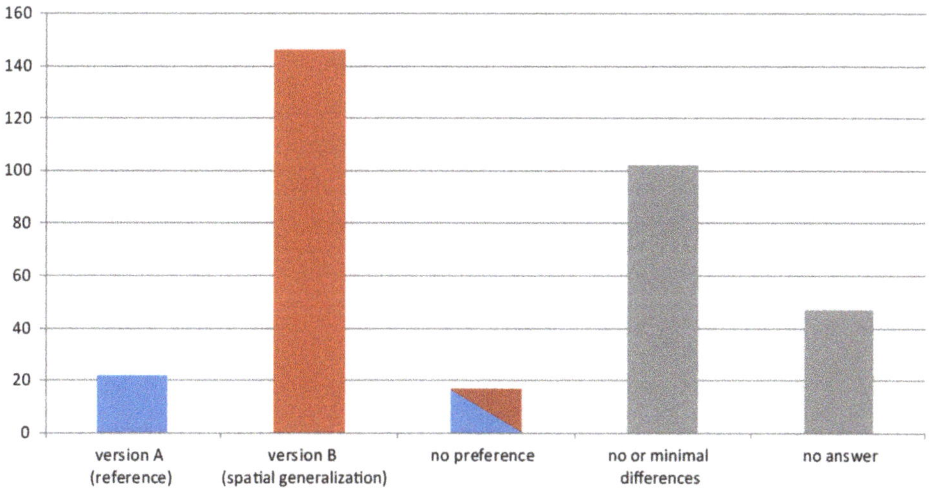

Figure 16. Answers for "Which version is easier to understand?" Absolute counts, $n = 334$.

4. Discussion

Our experiments document that local outlier excluding generalization of highly autocorrelated animated choropleth maps is (a) effective for emphasizing local outliers, and (b) works best when performed in the spatial dimension. It also suggests that map users prefer noise reduced animations. Nevertheless, assumed improvements in trend detection are not confirmed by the data, although roughly 10 percent of the participants from the pilot study explicitly mention a perceived facilitation in this respect. While the results suggest that the personal ability to identify local outliers and global trends in the used map animations does not depend on sex, cartographic experience, gaming affinity, or educational level, it seems that age and trend detection strategies play a significant role for trend detection. In the following, those results are further discussed.

4.1. Local Outlier Detection

Searching for and attending to objects from a visual scene is thought to be an interplay of bottom-up and top-down processes of vision [33]. Which group of processes dominates largely depends on the actual task. Tasks heavily based on prior knowledge about the world, like searching the desk in need of a pen (instead of searching the picture above, even if it is much more salient than the desk), top-down-processes will dominate where we attend to and direct our gaze at.

Conversely, in uninformed discovery of completely new visual information, bottom-up processes driven by the heterogeneity of the visual field itself play the leading role. Areas that "pop-out" by being different in intensity, color, or orientation are more likely to "catch the eye" than areas with little local contrast. The local intensity of perceptual conspicuousness or, in other words, the "physical attraction" of each region within the field of view can be described by visual saliency models that are based on neuroanatomical and physiological knowledge about our visual system [34,35]. Implementations of such models [36–38] produce saliency maps of visual scenes, estimating the perceived visual salience of objects within photos, videos, or even maps in a geographic sense [39,40]. When using saliency maps to predict eye fixation points and thus overt visual attention of test persons, best predictions were achieved when subjects explicitly had to search for high salience in the scene, followed by free viewing tasks and a search for predefined but salient objects [41]. The task of uninformed looking for two local outliers as features of highest local contrast in color/lightness might lie somewhere between direct salience search and salient object search. Thus, we consider the visual saliency of local outliers being decisive for their detection in choropleth map animations.

Visual salience does not only increase with the individual difference of a target object (in our case a local outlier) to other objects (neighboring polygons), but also with the homogeneity of those potentially distracting objects [33,42]. Changes in distractor homogeneity without affecting distractor–target difference might happen in cases where a different visual variable separates the target from distractors. As (our) local outliers differ from non-outlier polygons only in the amount of local color/lightness contrast, both modes of outlier saliency increase (increased target-distractor difference, increased distractor homogeneity) will operate simultaneously, if local contrast between distractors is reduced.

On the physiological level of color perception, simultaneous contrast also needs to be considered. Simultaneous contrast is commonly referred to as a shift of the brightness and the hue of an object towards the complementary brightness and hue of the surrounding area [43], perceptually enhancing color contrast and thus facilitating object delineation. Often, this is illustrated by pointing to different percepts of the same physical color in differently colored, uniform surroundings. Using this simple model of simultaneous contrast, it points to a perceptual gain of overall color contrast in the non-generalized stimuli, except for already rather uniform regions. Perceptual amplification of local contrast is therefore reduced between spatially smoothed neighbors. The perceived color of a local outlier would, however, hardly be affected by slight color adaptions among surrounding neighbors as the mean hue and brightness stays relatively constant. Thanks to the advice of a reviewer, we came across the work of Brown and MacLeod [44], who show that color appearance is also a function of the variance of surrounding colors. Colors appear more vivid against a low contrast surrounding than against heterogeneous, high contrast surroundings, even if the space-averaged, surrounding means are identical. Thus, the salience of local outliers is further enhanced by spatial-generalization of neighbors, due to mechanisms of this special type of simultaneous contrast.

Studies in visual search show an inverse relation of target salience and search time needed [45]. Thus, the probability to detect a local outlier in a given time (namely the duration of its existence within the animation) will increase with its salience. This explains the higher performance in outlier detection based on the spatially generalized choropleth map animations, which provide elevated outlier salience by reduced local contrast of distractor-polygons.

However, why does this mechanism not work similarly well when reducing local noise in time while preserving the large lightness changes of local outliers? Traun and Mayrhofer [26] calculated metrics on spatial and temporal map complexity for classed choropleth time series maps and their spatial, temporal, and spatiotemporally generalized variants, respectively. While visual complexity unsurprisingly decreased in the dimension (space/time) used for data smoothing, complexity increased in the unused dimension. Only for the spatiotemporally generalized variant, they reported a (comparably slight) decrease in both spatial and temporal complexity. Implicitly assuming that spatial and temporal complexity similarly affect perception, they favored the spatiotemporal variant. Contrary to this, our results suggest that spatial complexity reduction has a much higher impact on local outlier detection ability than a decrease of complexity in time. Temporal smoothing of individual polygons seemingly reduces the spatial coherence of animations, which explains the increase in spatial complexity measures reported by Traun and Mayrhofer [26]. Generalization in individual lightness transitions seems to be rather irrelevant, as we most probably do not "see" choropleth map animations as spatially unrelated sets of "locations" separately changing over time, but as an inherently spatial yet dynamically changing visual field, from which we perceptually delineate dynamic objects of higher order (regions) based on local contrast differences (especially contrast edges) in space. Such a view does not only conform to the theory of perceptual object formation as proposed by Gestalt psychology [46]. It also makes sense from an evolutional–functional perspective on vision in facilitating interaction with a world of three-dimensional objects. Although some of these objects might show an appearance or movement, their primary organizational structure is spatial, not temporal. Interestingly (or consequently?), the neural architecture of our visual system reflects the spatial layout of our visual field [47]. Along the stages of visual processing from retinal receptive fields through the lateral geniculate nucleus, the primary visual cortex and several subsequent cortical areas dedicated to vision, the spatial arrangement of neurons preserves the spatial arrangement of ganglion cells in the retina. Nearby locations in our visual field are thus processed by nearby neurons in a number of visual field maps within our brain [48].

In the main study, participants were informed that each stimulus contains exactly two outliers. Disclosing this information, we reacted to several comments from the pilot study where participants were left in uncertainty about this fact. While many of those complained about the overall difficulty of the task, others implicitly revealed that they assumed to find two outliers after having worked through the first stimuli. Therefore, we decided to remove this additional level of complexity in the main study. Despite this and other changes in the final study design, the results of the pilot study closely resemble the patterns observed in the main study for correct outliers and false-positives (best performance of the spatial generalization variant, no significant difference between reference and temporal variant and the spatiotemporal version somewhere in-between).

4.2. Global Trend Detection

Change in the global trend of spatiotemporally highly autocorrelated choropleth map animations is perceived as motion. The impression of a "moving front", the "expansion of high value areas" or the "movement of a cluster of low values" is rooted in the conveyed connectivity by the contiguous borders [23] and the overall gradually changing lightness in space and time. As we assimilate regions of similar lightness to perceptual figures with borders defined by steeper color gradients, the underlying spatial process—like the outbreak of an infectious disease expressed by locally expanding incidence rates—leads to the impression of border movement (e.g., movement of the epidemic front), although only the lightness of individual polygons changes in a coordinated way. Depending on the perceived motion directions of all of its borders, the bounded figure virtually grows, shrinks, or moves. In comparison, the spatially separated symbols of an animated proportional symbol map do not provide such strong perception of "underlying motion" [23], nor would it result from choropleth map animations of data with little autocorrelation in space or time. In the latter case, the underlying process or the chosen scale of aggregation simply provides too little coherence within individual lightness changes, leading to increasing numbers of appearing and disappearing figures, quickly exceeding our cognitive capacities.

Another important factor for apparent motion is animation speed. In their comparative study on the detection of space-time clusters in small multiples and animated maps, Griffin, et al. [49] found that the apparent movement and thus the detection rate of clusters was closely related to animation pace. They plausibly hypothesize, that the gestalt grouping principle of common fate [46] cannot be established by our visual system if the animation runs too slow, while cognitive processing cannot keep up, if it is too fast. While we did not vary animation speed in our experiment, overall global trend detection rates were rather high (Figure 9). Thus, we assume that animation speed was within an appropriate range for detection of apparent motion.

In our experiment, generalization did not significantly improve the ability of participants to assign the correct blurred trend candidate to the respective stimulus, although 11% of participants emphasized a subjectively easier to follow trend in the spatially generalized animation in the pilot study.

A possible explanation for the negative quantitative results regarding our hypothesis that improved local homogeneity improves trend perception is the human ability to attend to distinct spatial frequencies and efficiently filter visual noise at spatial frequency levels that are not attended [50,51]. According to Snowden, Thompson, and Troscianko [47], keeping eyes squinted or increasing viewing distance facilitates the perception of low spatial frequencies. Therefore, it is not surprising that the intuitive adoption of exactly these strategies significantly facilitated trend perception. The superior performance of the group having focused on bright zones over the followers of the dark zones is in line with literature too. Nothdurft [52] found that bright targets are more salient among dark targets than vice versa.

It turned out that trend detection abilities in choropleth map animations are decreasing with age. Although the decline is moderate, it is highly significant. In view of general perceptual and cognitive changes across the human lifespan, this result is not surprising. From a psychophysiological perspective, general trend detection primarily seems related to the perception of motion coherence and translational motion (in opposite to the radial flow we experience when moving through the environment or biological motion, elicited by body movements of human figures). Both modes of motion perception decline with age [53,54], thus our findings are in line with the respective literature.

5. Conclusions and Outlook

We conclude that outlier preserving value generalization in space facilitates the identification of local outliers in choropleth map animations, while we could not find such an effect for generalization in time.

Our overall negative results on the benefit of temporal smoothing of choropleth map animations complement similar findings by McCabe [23] for different map use tasks. Like us, he used unclassed choropleth maps for his experiment. If the transfer of these findings to classed choropleth map animation holds, the according complexity metrics as introduced by Goldsberry and Battersby [2] needed to be rethought, as they solely depend on (class) changes of individual polygons in time. As a reduction in temporal complexity does not seem to improve perception, the question arises, whether purely temporal complexity metrics are a useful proxy for the perceived complexity of animated choropleth maps. Indices based on composite "perception objects" that have been segmented directly from spatiotemporal data or from the output of dynamic saliency models [55] might be a first step towards a more adequate description of animated map complexity. The development of according algorithms and tools to estimate the perceptibility of animated choropleth maps offers ample room for future research.

In their study mentioned above, Griffin, MacEachren, Hardisty, Steiner, and Li [49] noticed, that the optimal animation pace for apparent motion detection was different for various cluster intensities. This clearly prompts for interactive control of the animation speed by the user, but also shows how sensitive perceptual grouping by common fate is regarding animation speed. The complex relations between animation speed and different degrees of spatial and temporal autocorrelation on the perception of apparent motion in animated choropleth maps still waits to be fully uncovered. While we did not find evidence for benefits of value generalization for global trend detection, it cannot be excluded that this also holds true in situations where apparent motion is harder established due to noisier conditions, less "coordinated" color changes or due to suboptimal animation speed for detecting a certain spatial process.

Author Contributions: Conceptualization, investigation, writing of the original and revised draft, visualization, Christoph Traun; statistical data analysis, Manuela Larissa Schreyer and Christoph Traun; software for stimulus generation, Gudrun Wallentin. All authors have read and agreed to the published version of the manuscript.

Funding: Except for publication support by the Open Access Publication Fund of the University of Salzburg, this research received no external funding.

Institutional Review Board Statement: Not applicable.

Informed Consent Statement: Not applicable.

Data Availability Statement: Map stimuli, response items, and the analyzed, yet anonymized, dataset from the main study are available at https://tinyurl.com/mapstudyresults.

Acknowledgments: The main author likes to thank Wolfgang Trutschnig for his valuable discussion on the study design, the anonymous reviewers for their helpful inputs, and the voluntary participants for their kind engagement in the experiment. All authors acknowledge financial support by the Open Access Publication Fund of the University of Salzburg.

Conflicts of Interest: The authors declare no conflict of interest.

References

1. Harrower, M. The Cognitive Limits of Animated Maps. *Cartographica* **2007**, *42*, 349–357. [CrossRef]
2. Goldsberry, K.; Battersby, S. Issues of Change Detection in Animated Choropleth Maps. *Cartogr. Int. J. Geogr. Inf. Geovis.* **2009**, *44*, 201–215. [CrossRef]
3. DuBois, M.; Battersby, S.E. A Raster-Based Neighborhood Model for Evaluating Complexity in Dynamic Maps. In Proceedings of the AutoCarto 2012, Columbus, OH, USA, 16–18 September 2012.
4. Rosenholtz, R. Capabilities and limitations of peripheral vision. *Annu. Rev. Vis. Sci.* **2016**, *2*, 437–457. [CrossRef]
5. Sweller, J. Cognitive load during problem solving: Effects on learning. *Cogn. Sci.* **1988**, *12*, 257–285. [CrossRef]
6. Sweller, J.; van Merriënboer, J.J.; Paas, F. Cognitive architecture and instructional design: 20 years later. *Educ. Psychol. Rev.* **2019**, *31*, 1–32. [CrossRef]
7. Luck, S.J.; Vogel, E.K. The capacity of visual working memory for features and conjunctions. *Nature* **1997**, *390*, 279–281. [CrossRef]
8. Metaferia, M.T. Visual Enhancement of Animated Time Series to Reduce Change Blindness. Master's Thesis, University of Twente, Enschede, The Netherlands, 2011.
9. Harrower, M.; Fabrikant, S.I. The role of map animation for geographic visualization. In *Geographic Visualization. Concepts, Tools and Applications*; Dodge, M., Derby, M.M., Turner, M., Eds.; John Wiley & Sons: Chichester, UK, 2008; pp. 49–65.
10. Harrower, M. Tips for designing effective animated maps. *Cartogr. Perspect.* **2003**, 63–65. [CrossRef]
11. Kraak, M.-J.; Edsall, R.; MacEachren, A.M. Cartographic animation and legends for temporal maps: Exploration and or interaction. In Proceedings of the 18th International Cartographic Conference, Stockholm, Sweden, 23–27 June 1997; pp. 253–261.
12. Midtbø, T. Advanced legends for interactive dynamic maps. In Proceedings of the 23th International Cartographic Conference "Cartography for everyone and for you", Moscow, Russia, 4–10 August 2007; pp. 4–10.
13. Muehlenhaus, I. *Web Cartography: Map Design for Interactive and Mobile Devices*; CRC Press: Boca Raton, FL, USA, 2013.
14. Opach, T.; Gołębiowska, I.; Fabrikant, S.I. How Do People View Multi-Component Animated Maps? *Cartogr. J.* **2014**, *51*, 330–342. [CrossRef]
15. Fish, C.; Goldsberry, K.P.; Battersby, S. Change blindness in animated choropleth maps: An empirical study. *Cartogr. Geogr. Inf. Sci.* **2011**, *38*, 350–362. [CrossRef]

16. Battersby, S.E.; Goldsberry, K.P. Considerations in Design of Transition Behaviors for Dynamic Thematic Maps. *Cartogr. Perspect.* **2010**, 16–32. [CrossRef]
17. Cybulski, P.; Medyńska-Gulij, B. Cartographic redundancy in reducing change blindness in detecting extreme values in spatio-temporal maps. *ISPRS Int. J. Geo-Inf.* **2018**, *7*, 8. [CrossRef]
18. Multimäki, S.; Ahonen-Rainio, P. Temporally Transformed Map Animation for Visual Data Analysis. In Proceedings of the GEO Processing 2015: The 7th International Conference on Advanced Geographic Information Systems, Applications, and Services, Lisbon Portugal, 22–27 February 2015; pp. 25–31.
19. Monmonier, M. Minimum-change categories for dynamic temporal choropleth maps. *J. Pa. Acad. Sci.* **1994**, *68*, 42–47.
20. Monmonier, M. Temporal generalization for dynamic maps. *Cartogr. Geogr. Inf. Syst.* **1996**, *23*, 96–98. [CrossRef]
21. Harrower, M. Visualizing change: Using cartographic animation to explore remotely-sensed data. *Cartogr. Perspect.* **2001**, 30–42. [CrossRef]
22. Harrower, M. Unclassed animated choropleth maps. *Cartogr. J.* **2007**, *44*, 313–320. [CrossRef]
23. McCabe, C.A. Effects of Data Complexity and Map Abstraction on the Perception of Patterns in Infectious Disease Animations. Master's Thesis, The Pennsylvania State University, University Park, PA, USA, 2009.
24. Ogao, P.J.; Kraak, M.J. Defining visualization operations for temporal cartographic animation design. *Int. J. Appl. Earth Obs. Geoinf.* **2002**, *4*, 23–31. [CrossRef]
25. Slocum, T.A.; Slutter, R.S., Jr.; Kessler, F.C.; Yoder, S.C. A Qualitative Evaluation of MapTime, A Program For Exploring Spatiotemporal Point Data. *Cartographica* **2004**, *39*, 43–68. [CrossRef]
26. Traun, C.; Mayrhofer, C. Complexity reduction in choropleth map animations by autocorrelation weighted generalization of time-series data. *Cartogr. Geogr. Inf. Sci.* **2018**, *45*, 221–237. [CrossRef]
27. Moran, P.A.P. Note on Continuous Stochastic Phenomena. *Biometrika* **1950**, *37*, 17–23. [CrossRef]
28. Crump, M.J.C.; McDonnell, J.V.; Gureckis, T.M. Evaluating Amazon's Mechanical Turk as a Tool for Experimental Behavioral Research. *PLoS ONE* **2013**, *8*, e57410. [CrossRef]
29. Team, R.C. *R: A Language and Environment for Statistical Computing;* R Foundation for Statistical Computing: Vienna, Austria, 2021.
30. Ellis, A.R.; Burchett, W.W.; Harrar, S.W.; Bathke, A.C. Nonparametric inference for multivariate data: The R package npmv. *J. Stat. Softw.* **2017**, *76*, 1–18.
31. Levin, D.T.; Momen, N.; Drivdahl, S.B.; Simons, D.J. Change Blindness Blindness: The Metacognitive Error of Overestimating Change-detection Ability. *Vis. Cogn.* **2000**, *7*, 397–412. [CrossRef]
32. Simons, D.J. Attentional capture and inattentional blindness. *Trends Cogn. Sci.* **2000**, *4*, 147–155. [CrossRef]
33. Wolfe, J.M.; Horowitz, T.S. Five factors that guide attention in visual search. *Nat. Hum. Behav.* **2017**, *1*, 0058. [CrossRef]
34. Koch, C.; Ullman, S. Shifts in selective visual attention: Towards the underlying neural circuitry. *Hum. Neurobiol.* **1985**, *4*, 219–227. [PubMed]
35. Veale, R.; Hafed, Z.M.; Yoshida, M. How is visual salience computed in the brain? Insights from behaviour, neurobiology and modelling. *Philos. Trans. R. Soc. B Biol. Sci.* **2017**, *372*, 20160113. [CrossRef]
36. Bruce, N.D.; Tsotsos, J.K. Saliency, attention, and visual search: An information theoretic approach. *J. Vis.* **2009**, *9*, 5. [CrossRef]
37. Itti, L.; Koch, C.; Niebur, E. A model of saliency-based visual attention for rapid scene analysis. *IEEE Trans. Pattern Anal. Mach. Intell.* **1998**, *20*, 1254–1259. [CrossRef]
38. Zhang, L.; Tong, M.H.; Marks, T.K.; Shan, H.; Cottrell, G.W. SUN: A Bayesian framework for saliency using natural statistics. *J. Vis.* **2008**, *8*, 32. [CrossRef]
39. Davies, C.; Fabrikant, S.I.; Hegarty, M. Toward Empirically Verified Cartographic Displays. In *The Cambridge Handbook of Applied Perception Research;* Szalma, J.L., Scerbo, M.W., Hancock, P.A., Parasuraman, R., Hoffman, R.R., Eds.; Cambridge University Press: Cambridge, UK, 2015; pp. 711–730.
40. Fabrikant, S.I.; Goldsberry, K. Thematic relevance and perceptual salience of dynamic geovisualization displays. In Proceedings of the 22th ICA/ACI International Cartographic Conference, A Coruña, Spain, 9–16 July 2005.
41. Koehler, K.; Guo, F.; Zhang, S.; Eckstein, M.P. What do saliency models predict? *J. Vis.* **2014**, *14*, 14. [CrossRef]
42. Yantis, S. How visual salience wins the battle for awareness. *Nat. Neurosci.* **2005**, *8*, 975–977. [CrossRef]
43. Slocum, T.A.; McMaster, R.B.; Kessler, F.C.; Howard, H.H. *Thematic Cartography and Geovisualization*, 3rd ed.; Pearson Prentice Hall: Upper Saddle River, NJ, USA, 2009.
44. Brown, R.O.; MacLeod, D.I. Color appearance depends on the variance of surround colors. *Curr. Biol.* **1997**, *7*, 844–849. [CrossRef]
45. Liesefeld, H.R.; Moran, R.; Usher, M.; Müller, H.J.; Zehetleitner, M. Search efficiency as a function of target saliency: The transition from inefficient to efficient search and beyond. *J. Exp. Psychol. Human Percept. Perform.* **2016**, *42*, 821. [CrossRef]
46. Wagemans, J.; Elder, J.H.; Kubovy, M.; Palmer, S.E.; Peterson, M.A.; Singh, M.; von der Heydt, R. A century of Gestalt psychology in visual perception: I. Perceptual grouping and figure–ground organization. *Psychol. Bull.* **2012**, *138*, 1172–1217. [CrossRef]
47. Snowden, R.; Thompson, P.; Troscianko, T. *Basic Vision: An Introduction to Visual Perception;* Oxford University Press: Oxford, UK, 2012.
48. Wandell, B.A.; Dumoulin, S.O.; Brewer, A.A. Visual field maps in human cortex. *Neuron* **2007**, *56*, 366–383. [CrossRef]
49. Griffin, A.L.; MacEachren, A.M.; Hardisty, F.; Steiner, E.; Li, B. A comparison of animated maps with static small-multiple maps for visually identifying space-time clusters. *Ann. Assoc. Am. Geogr.* **2006**, *96*, 740–753. [CrossRef]

50. Shulman, G.L.; Sullivan, M.A.; Gish, K.; Sakoda, W.J. The Role of Spatial-Frequency Channels in the Perception of Local and Global Structure. *Perception* **1986**, *15*, 259–273. [CrossRef]
51. Sowden, P.T.; Schyns, P.G. Channel surfing in the visual brain. *Trends Cogn. Sci.* **2006**, *10*, 538–545. [CrossRef]
52. Nothdurft, H.-C. Attention shifts to salient targets. *Vis. Res.* **2002**, *42*, 1287–1306. [CrossRef]
53. Billino, J.; Bremmer, F.; Gegenfurtner, K.R. Differential aging of motion processing mechanisms: Evidence against general perceptual decline. *Vis. Res.* **2008**, *48*, 1254–1261. [CrossRef]
54. Snowden, R.J.; Kavanagh, E. Motion Perception in the Ageing Visual System: Minimum Motion, Motion Coherence, and Speed Discrimination Thresholds. *Perception* **2006**, *35*, 9–24. [CrossRef]
55. Bak, C.; Kocak, A.; Erdem, E.; Erdem, A. Spatio-Temporal Saliency Networks for Dynamic Saliency Prediction. *IEEE Trans. Multimedia* **2018**, *20*, 1688–1698. [CrossRef]

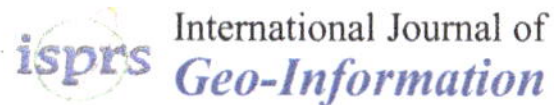
International Journal of
Geo-Information

Article

Complexity Level of People Gathering Presentation on an Animated Map—Objective Effectiveness Versus Expert Opinion

Beata Medyńska-Gulij *, Łukasz Wielebski, Łukasz Halik and Maciej Smaczyński

Research Division Cartography and Geomatics, Adam Mickiewicz University, 61-712 Poznań, Poland; lukwiel@amu.edu.pl (Ł.W.); lhalik@amu.edu.pl (Ł.H.); maciej.smaczynski@amu.edu.pl (M.S.)

* Correspondence: bmg@amu.edu.pl

Received: 2 January 2020; Accepted: 19 February 2020; Published: 20 February 2020

Abstract: The aim of the following study was to present three alternative methods of visualization on animated maps illustrating the movement of people gathered at an open-air event recorded on photographs taken by a drone. The effectiveness of an orthorectified low-level aerial image (a so-called orthophoto), a dot distribution map, and a buffer map was tested in an experiment featuring experts, and key significance was attached to the juxtaposition of objective responses with subjective opinions. The results of the study enabled its authors to draw conclusions regarding the importance of visualizing topographic references (stable objects) and people (mobile objects) and the usefulness of the particular elements of animated maps for their analysis and interpretation.

Keywords: cartographic design; animated map; people gathering presentation; orthophoto; dot distribution map; buffer map; drone image; expert opinion; objective effectiveness

---

1. Introduction

Monitoring the dynamics of people gatherings with the aid of unmanned aerial vehicle (UAV) technology opens up new possibilities for visualizing space [1]. The analysis of spatial behaviors of participants on the basis of previously prepared visualizations is of crucial importance to organizers of open-air events. That is why the authors of the present study decided to focus on the scientific bases of creating images illustrating an event's participants' spatial behaviors [2]. The issue is topical given the current increase in the use of UAVs, widely known as drones, for the purpose of recording the course of events and gatherings. The final product in such cases takes the form of a video or camera images. One of the uses of such images for analytical purposes is crowd monitoring [3]. The popularity of UAV use to gather information has even led to the launch of an Internet project called Dronestagram, which allows its users to share their georeferenced drone-taken images [4]. Raster images taken from a bird's-eye-view perspective have become the basis for cartographic visualization, which calls for a special methodological approach to the principles of map design and mapping techniques [5,6]. The optimal method for presenting changes in spatial phenomena over time is by means of animated maps [7], whose most important elements include a frame of cartographic content, title, scale, legend or temporal legend, and scale bar, but also additional interactive tools, multimedia, special effects, and interactions, which all help to enrich the map [8].

The present study mostly touches upon the issue of how to properly present stable and mobile objects at open-air events, which are, according to the researchers, of key significance in perceiving spatio-temporal relations [9]. A stable spatial object is understood as a reference object of great functional importance to an event's participants; a mobile object on a map, on the other hand, is simply a single participant of the event in question. The line or frame delimiting the cartographic content is thought to stabilize the whole cartographic image [10]. In the case of a mass event's area being delimited with a line, it might additionally affect the perception of changes on the animated map to a significant degree. The key role in multimedia cartography is the preparation of a map that is as simple as possible for interpretation [11]. In the context of the studies undertaken here, emphasis was placed on the conception and construction of maps in order to communicate spatial results efficiently mostly according to cartographic design principles [12].

Another important issue is the choice of a research method that will allow experts in cartography and geomatics to express their subjective opinions on the usefulness of the designed visualizations [13]. However, the effectiveness of a visualization for drawing conclusions can also be measured objectively by means of questions answered by the participants [14,15]. A study using questionnaires with three alternative map variants was previously conducted to test the influence of the proper design of a tourist map on the changing preferences of its users [16]. The tourist preferences of the students were juxtaposed with three variants of graphical design, all representing the same topographic content. The results obtained clearly illustrated the great potential of map design to influence a tourist map user in terms of their choice of tourist objects and routes.

Yet another problem to be dealt with in the present study is the choice of a proper way to represent people. The most natural option would probably be the dot variant, as, logically, it uses one dot to represent a single person [17]. However, what might be problematic in this case is the very small size of dots and their merging/overlapping, which makes the analysis more complicated [7]. When one creates visualizations based on photographs, there is a need for a special design of cartographic signs to be superimposed onto the image from the camera, which is associated with the use of graphical variables [18]. The authors of the present article also wanted to draw the reader's attention to the issue of the cartographic representation of the distances between a mass event's participants, which in this case meant distinguishing the various possible types of interpersonal distance, e.g., intimate distance, personal distance, social distance, or public distance [19].

The main aim of the study in hand was to assess the possible variants for visualizing topographic references (stable objects) and UAV-recorded people gatherings (mobile objects) in animations based on three types of map. The authors wanted to find the answer to the following questions:

- How do experts in cartography and geomatics subjectively assess the usefulness of the following three types of animated presentation: an orthophoto, a dot distribution map, and a buffer map?
- What is the objective effectiveness of those three methods of visualization for analyzing spatio-temporal relations during an open-air event?
- How important is it in animated map design to graphically highlight the objects serving as points of interest to the participants and, therefore, treated as orientation points?
- Which elements of an map are, according to experts, unnecessary, and which are needed?

2. Methodology

In order to accomplish the set goals, the authors conducted the study in the following stages:

- Deciding on how to graphically visualize a single person;
- Elaborating the map contents for the three animations and delimiting the borders of the event's area with a line on each of them;
- Designing the layout of the animated map;
- Preparing three online questionnaires;
- Conducting the questionnaires among the experts;
- Analyzing the responses obtained, and comparing the subjective expert opinions with the objective results of the completed tasks; and
- Assessing the effectiveness of the three visualizations.

2.1. Graphically Visualize a Single Person

The animated maps used in the study were created based on eight selected photographs taken with a drone between 18:00 and 21:00. Those raster images were obtained at the "Geographer's Picnic", an event held next to the building of *Collegium Geographicum* at the Adam Mickiewicz University in Poznan, Poland. The event was hosted in an area of approximately 3500 m^2 and attended by 900 students. A key requirement for creating the visualizations was to define three different ways to present the basic mobile object, i.e., a single person. Figure 1 shows three possible graphical variants for representing a single person and a gathering of people on an orthophoto, a dot distribution map, and a buffer map. The orthophoto presents a low-level aerial image of the event's area as seen from above, in a parallel projection. In this case, a person is visualized as a set of pixels (Figure 1a). Distinguishing people from other objects is made easier thanks to the shadows cast by them. The shadows also help the reader to correctly identify the number of people standing in close proximity. On a dot distribution map, one person is represented by one dot (Figure 1b). In this case, one needs to interpret the orthophoto and the oblique photos first before one can determine the exact location of a person. The determinant of their location is the position of their head/the axis of their body.

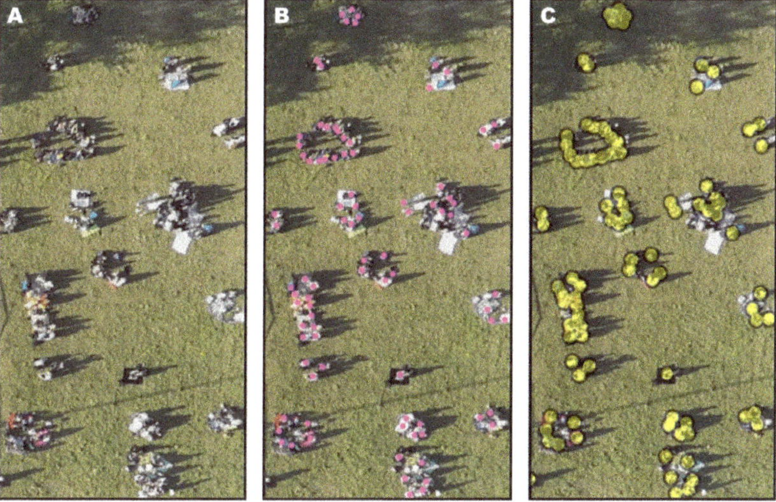

Figure 1. Three variants for the graphical presentation of a single person and a buffer of people: (**a**) pixels, (**b**) dots, and (**c**) point buffers.

The buffer map, in its turn, reflects the distances between the event's participants, which might correspond to the relations between them. This refers to the notion of so-called social distances. According to Hall [19], the personal distance zone can be further broken down into a close phase (45–75 cm) and a far phase (75–120 cm). Taking into consideration the likelihood of witnessing buffers of people forming at the event, the authors decided to surround every dot with a 60-cm "buffer" (counting from the center of the dot), i.e., a single person was represented by a circle with a diameter of 120 cm (Figure 1c).

2.2. The Map Contents and the Line Delimiting the Event's Venue

Figure 2 shows four ways of visualizing the participants in and the venue for an open-air event. They vary in how much the original information was processed and how a single person was represented. An oblique low-level aerial image was the basis for creating the orthophoto, which then served as the basis for the dot distribution map, the latter finally being the basis for the buffer map.

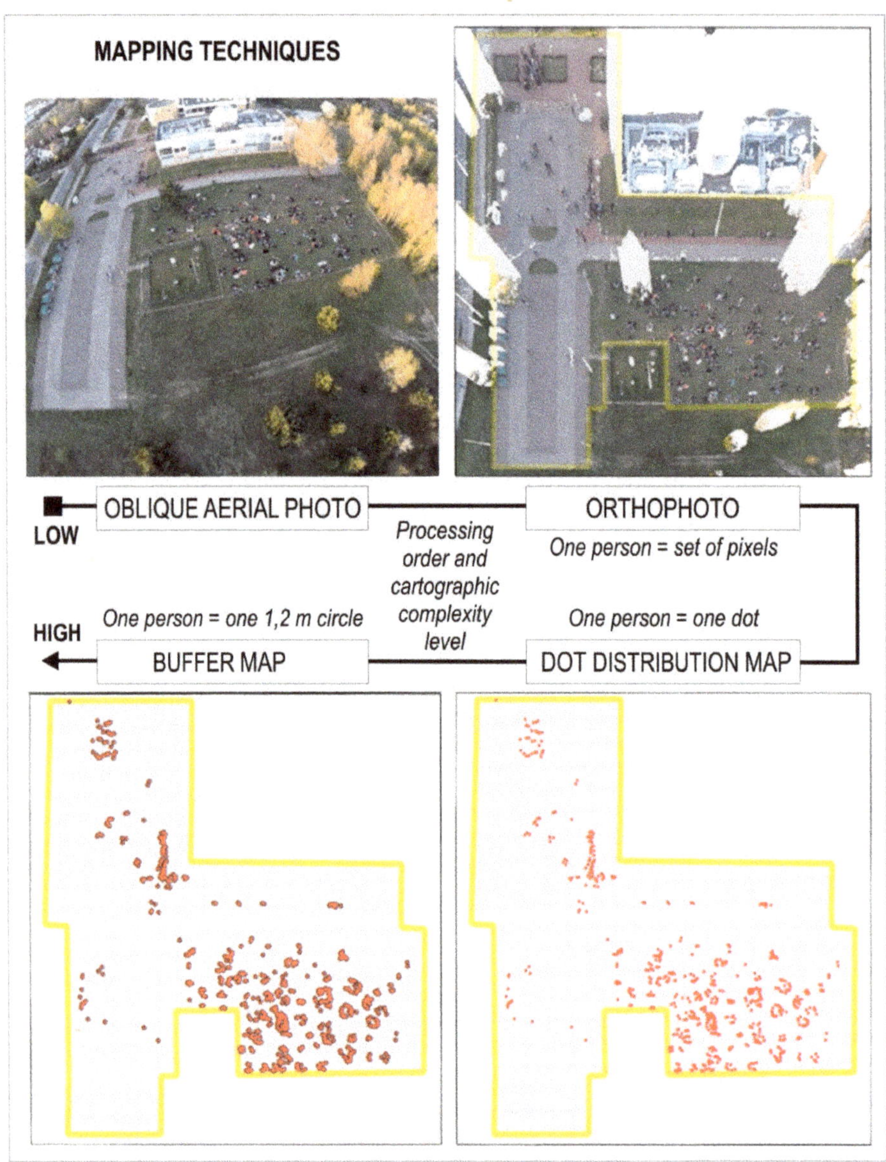

Figure 2. Four ways of visualizing the participants and the venue of the event.

Eight othophotos were created in Agisoft PhotoScan on the basis of oblique photos taken from the air, from various perspectives, with an unmanned aerial vehicle (the four-motor drone). The photos used in the study were taken every 20 min on average (subsequent states registered at the following time intervals: 23, 19, 21, 18, 14, 26, and 21 min). The drone was a platform for a professional sports camera, with a focal length of 3 mm, taking photos with a resolution of 12 megapixels. The wide-angle lens the camera was equipped with caused a distinctive 'spherical' distortion of the image in the oblique photos (Figure 2), resulting in the so-called fish-eye effect. The photographs were registered on an EPSG 2177 coordinate system on the basis of ground control reference points, using professional photogrammetric software.

In order to represent the location of people in the dot method, it was necessary to set the right size and color for the dots. Red was chosen due to its capacity to attract the reader's attention and the great contrast relative to the background. The size was set to be as large as possible without overlapping, corresponding approximately to the size of participants' heads on the orthophoto (Figure 2).

The buffer map was produced on the basis of the points in the vector layer in the QGIS application, where buffer zones with a radius of 60 cm were generated. The aggregation tool was used, which made it possible to merge the buffers and, thus, obtain a cluster for groups of people located in closest proximity.

An important step for the latter analyses of the course of the event and the behavior of participants was enriching the map with stable objects of spatial reference. The yellow lines on Figure 2 (in the case of the orthophoto, with additional transparency applied) delimit the area that served as the venue of the event; they constitute an important reference element and also improve the map's layout graphically, making it more balanced.

2.3. Designing the Layout of the Animated Map

Figure 3 shows the general layout of the animated map, including its basic elements and a template designed on the basis of this layout, listing all the graphical and descriptive elements used for each of the mapping methods. The template reflected the shape of the event's venue, while the choice and distribution of the individual elements resulted from the nature of the experimental questions. The template included the following elements: a timeline; the time (the hour and the minutes); the total number of participants; the number of participants on the picnic field; the number of participants outside the picnic field; a legend showing the graphical representation of a single person and a group; a scale bar; and a north arrow. What was crucial for the study was highlighting the stable objects in yellow: grill, DJ (Disc Jockey), bus stop, toilet, benches, information points, and picnic field, as well as highlighting the line delimiting the event venue, which became the main linear stable object. The stable objects were accompanied by text labels.

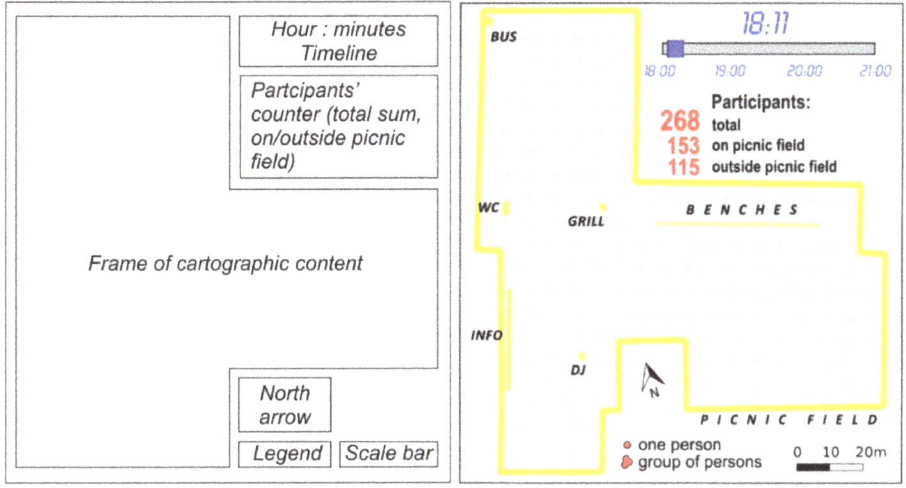

Figure 3. The layout and the template for the three animated maps.

Stable objects for topographical reference were marked on the visualizations in such a way as to facilitate their perception, and their geometry was adjusted to fit the spatial perception and memorization abilities of the viewer. In this way, four point objects were placed on the map: grill, bus stop, DJ, and toilet, and two linear objects: information points (an array of stalls provided by the event organizers) and benches. Those six objects were all represented in a similar way—as rectangles with a yellow outline, corresponding to the outline of the represented object, and accompanied by labels in the form of black text. The label for the picnic field was placed below the mapped area so that it would not obscure the cartographic content or inhibit the perception of changes in the location or number of mobile objects, i.e., participants of the event.

2.4. Preparing Three Online Questionnaires

In accordance with the authors' intention to compare the subjective opinions of experts with their objective answers to the experimental questions, the questionnaire was split into two parts. The objective questions were phrased in such a way as to enable an assessment (on the basis of the number of correct responses) of the three methods of visualization in terms of their effectiveness for showing spatial relations between the distribution of the event's participants and the distinctive objects located in the event's area. They also made it possible to determine the dynamics of the growth in the number of participants over time. The animation produced in Photoscape consisted of eight frames showing the course of the event over the successive hours and it was saved in the .gif format, in a variant ensuring constant repetition of the sequence of images (Figure 4).

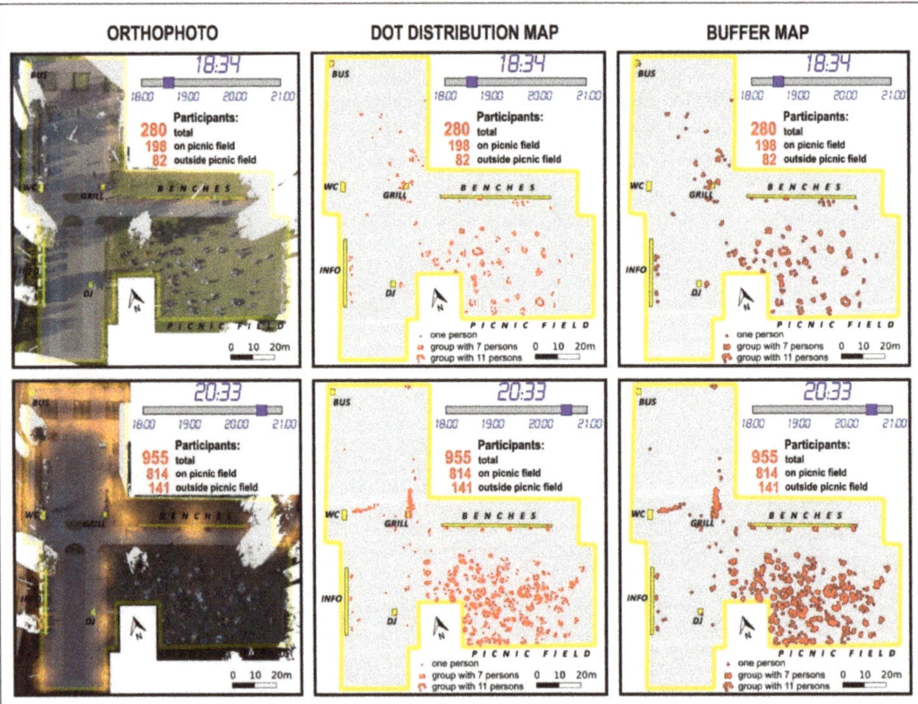

Figure 4. The comparison of the three animations on the basis of a single frame of the animated map presenting the distribution of the event participants at 18:34 and 20:33.

The five objective questions, together with the possible variants of answers to be chosen by the subjects, were as follows:

QUESTION A: When was the event attended by the highest number of people?

Possible answers: 18:11, 18:34, 18:53, 19:14, 19:32, 19:46, 20:12, 20:33
QUESTION B: When was the toilet queue the longest?
Possible answers: 18:11, 18:34, 18:53, 19:14, 19:32, 19:46, 20:12, 20:33
QUESTION C: Which part of the picnic field was the most crowded?
Possible answers: Middle and left; middle and right; middle and lower; middle and upper.
QUESTION D: Up till 19:00, the most frequent size of a group of participants on the picnic field was: (possible variants) 3–10 Persons, 8–12 Persons, 10–18 Persons, 16–22 Persons, 20–32 Persons
QUESTION E: After 19.30, the most frequent size of a group of participants on the picnic field was: (possible variants) 3–10 Persons, 8–12 Persons, 10–18 Persons, 16–22 Persons, 20–32 Persons

In the second part, the participant made a subjective decision, picking one of several possible answers on the basis of unanimated views of three juxtaposed maps illustrating the situation at 18:34 (Figure 4). The three tasks, and their possible answer variants, were as follows:

TASK I: Please assess those three visualizations in terms of their usefulness for analyzing spatial relations. Possible answers: insufficient, weak, good, very good.
TASK II: Please estimate how effective for analyzing spatial relations it is for the visualization to include distinctive objects (e.g., toilet, grill, DJ, etc.) and a line delimiting the event's venue. Possible answers: distracting, not important, of marginal importance, important, necessary.
TASK III: Please point out the elements which are unnecessary in the visualization. Multiple answers allowed. Possible answers: (1) timeline, (2) time: hour: minutes, (3) total number of participants, (4) participants outside the picnic field, 5) participants on the picnic field, (6) distinctive objects (toilet, grill, DJ, etc.), (7) labels of the distinctive objects, (8) scale bar, (9) north arrow, (10) the line delimiting the event's venue, (11) legend showing the signs representing a single person and a group of people.

2.5. *Conducting a Questionnaire among Experts*

The online questionnaires were prepared and conducted. Each of the three Internet surveys included one animated map, on the basis of which the participants had to answer the experimental questions (Figure 4). The questionnaires on a website were completed by 42 foreign experts on cartography and geomatics from West European universities (academic teachers conducting classes and scientific research in the field of cartography, geomatics, and geoinformation) and specialists from professional offices producing cartographic visualizations (academic cartographers, cartographers–geomatics experts [20]). As people unfamiliar with the area of research, they formed a group of objective respondents. The participants completed the survey anonymously, with no time limit imposed, and did not receive any payment. In addition, each of the questionnaires in the subjective part of the questions contained all three visualization methods, so that respondents could make their subjective assessment.

2.6. *Objective Tasks and Subjective Opinions*

The juxtaposition of the answers to the objective questions as shown in Figure 5 mostly reflects differences in interpreting the dynamics behind the changing numbers and clustering of people during the event. The very good accuracy obtained for the question regarding the total number of participants might partly have resulted from the duplicate information—the number of people was not only shown graphically on the map, but also as a number next to the map. When one takes into account the analysis of the map content alone, better results were obtained for the buffer map. It is immediately visible that all three visualizations were effective enough to enable the accurate perception of the queue to the toilet—this object was located quite far from the others, which made it stand out on each of the visualizations. Each visualization also clearly shows the significant rise in the number or people standing one behind the other when the event was drawing to an end.

Clearly, the effectiveness was lower in the case of interpreting the information about the group size and groups' distribution on the picnic field (questions C, D, and E). In those cases, the buffer map yielded better results than the dot distribution map, and much better results than the orthophoto.

The animation involving the buffer map clearly turned out to be the most effective one for communicating the information about the course of the event (60% of correct responses). It was followed by the dot distribution map animation, which yielded close to 50% accuracy. Most difficulties in terms of the right interpretation were experienced by participants while dealing with the orthophoto (less than 40% of correct answers), and it was especially problematic in the tasks regarding the participants' distribution on the picnic field and the estimation of the group size.

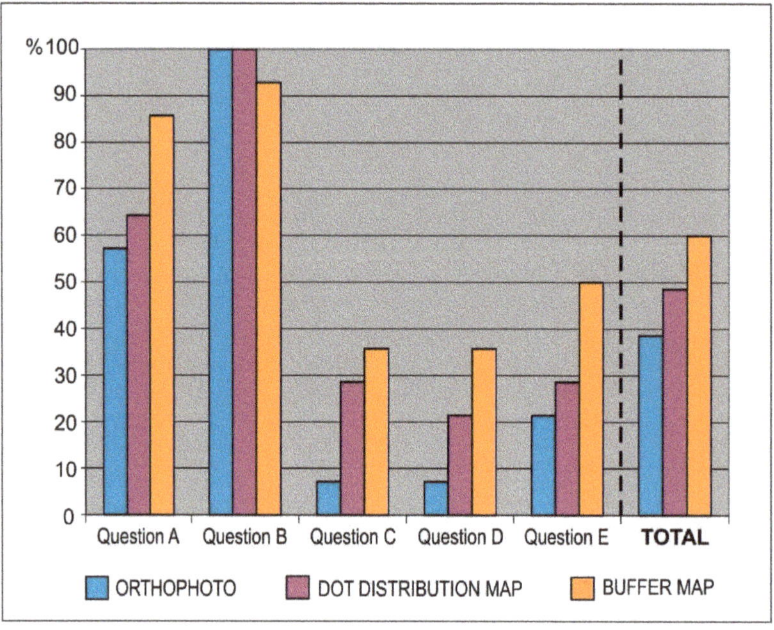

Figure 5. The juxtaposition of the correct answers given by respondents to the objective questions A–E.

The request for subjective opinions followed in the second part of the survey so that the experts, having already provided their answers to the objective questions, could draw on their familiarity with the mapping method. Figure 6 showcases two charts. The three columns in the upper chart illustrate the opinions of all 42 experimental subjects regarding each of the visualizations. In this case, it is worth pointing out the difference between the subjective opinions of experts, who ranked the dot distribution map the highest, and the answers to the objective part of the questionnaire, which shows the buffer map to be the most effective.

The results obtained for the same question, but classified according to the three variants of the questionnaire, are shown in the lower chart in Figure 6. Regardless of the variant of the tested visualization method, the highest number of respondents (50% and greater) assessed the orthophoto as the least useful. This result was also reflected in the results of the objective part of the survey as shown in Figure 5. None of the respondents using the dot distribution map or the buffer map animations considered those methods to be ineffective. The users of the dot distribution map ranked this method higher than the buffer map, just as the buffer map users preferred their method to the dot distribution map. Experts answering the objective questions on the basis of the orthophoto assessed it as the least effective, but still ranked it higher than the respondents working with the dot distribution map and buffer map animations.

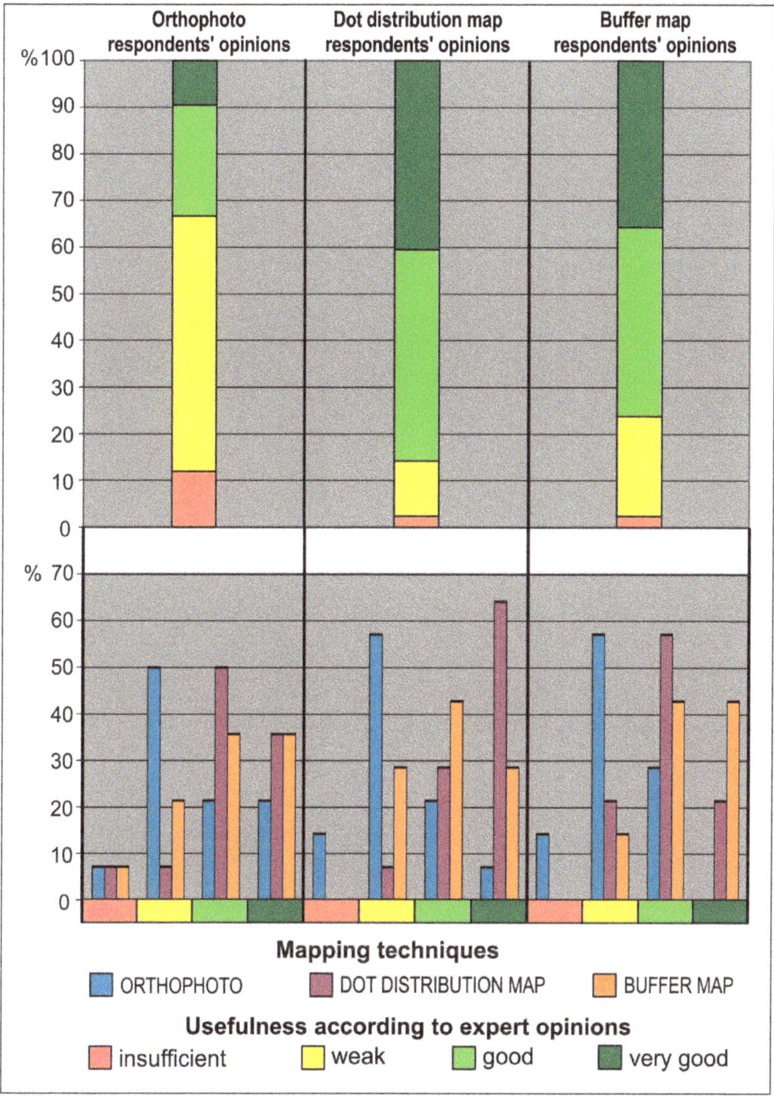

Figure 6. The juxtaposition of the subjective views on the usefulness of the three mapping techniques for observing and analyzing spatio-temporal relations.

As far as the expert opinions are concerned, placing stable objects and the venue borders on all three visualizations was generally considered to be important or even necessary (Figure 7). While for the dot distribution map and the buffer map none of the respondents chose the answers 'distracting' or 'not important', more than 20% of the experts assessed stable objects in those ways. Such subjective opinions might result from the fact that the orthophoto already features a lot of colors and adding even more graphical elements might seem unnecessary or even chaotic, as they partly obscure the raster image. However, placing several stable objects on a specially chosen gray background on a dot distribution map or a buffer map did not inhibit its readability or obscure any other graphical elements.

A detailed analysis of the respondents' answers to the second question broken down into questionnaire variants yields the conclusion that the respondents' opinions were the most varied in the case of the orthophoto (bar charts in the lower part of Figure 7). For the buffer map and the dot distribution map, respondents agreed to a higher extent, consistently choosing only the three highest levels of usefulness. Interestingly, the participants working with the dot distribution map animation were the most consistent in attaching a great significance to placing stable objects and the venue borders on the dot distribution map and the buffer map, although they differed in their opinions regarding the importance of those objects on the raster map.

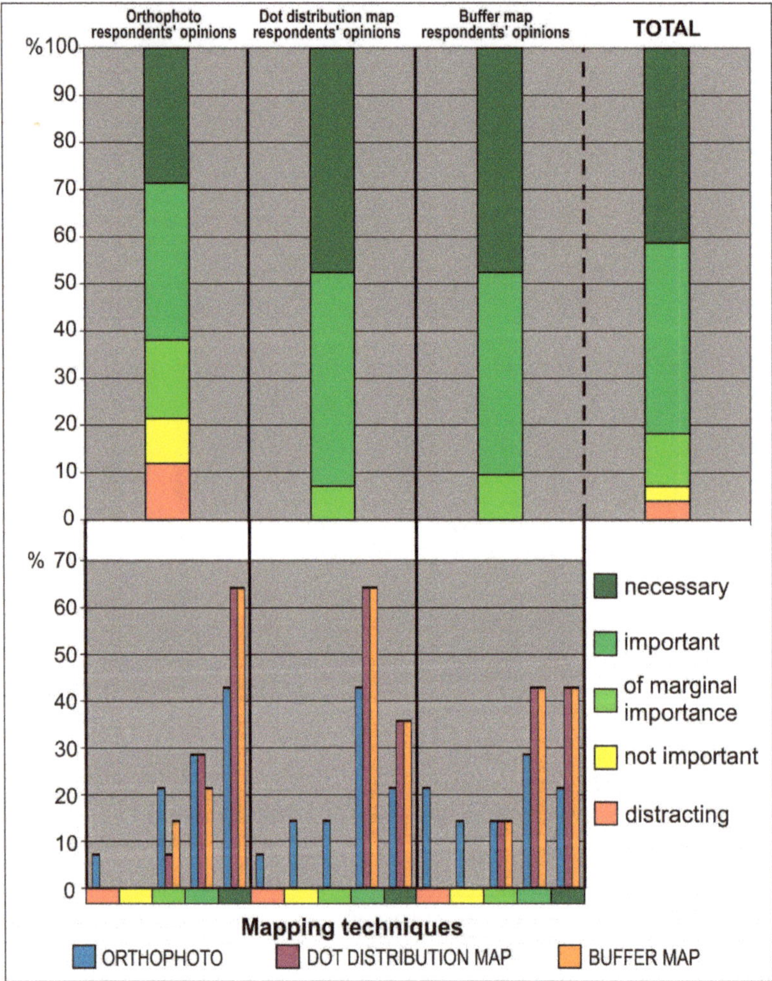

Figure 7. The juxtaposition of the subjective views on the role of stable objects and venue borders for analyzing spatio-temporal relations.

The respondents had varying opinions on the effectiveness of the particular elements of visualizations for the analyses (Figure 8). It must be noted, however, that the experts generally approved of the overall design of the animated map, as only one of its elements was deemed to be unnecessary by them; i.e., the north arrow (80%). The second least useful element was the scale bar (40%). All of the other elements, in the respondents' opinion, should be included on an animated map. Only slightly above 2% (1 person out of all 42 respondents) thought it unnecessary to include the distinctive objects; the same result was recorded for the labels of those objects. This confirms the results already obtained in Question B. Interestingly, though, the border of the venue was not that consistently ranked as important—about 20% of the respondents thought it unnecessary. When it comes to the elements not constituting parts of the map as such, the respondents attached less importance to the information about the number of participants outside the picnic field than to the number of participants on the field. This was probably caused by the question, in which the subjects were asked to estimate the ratio, and for that they needed to estimate the number of people on the picnic field.

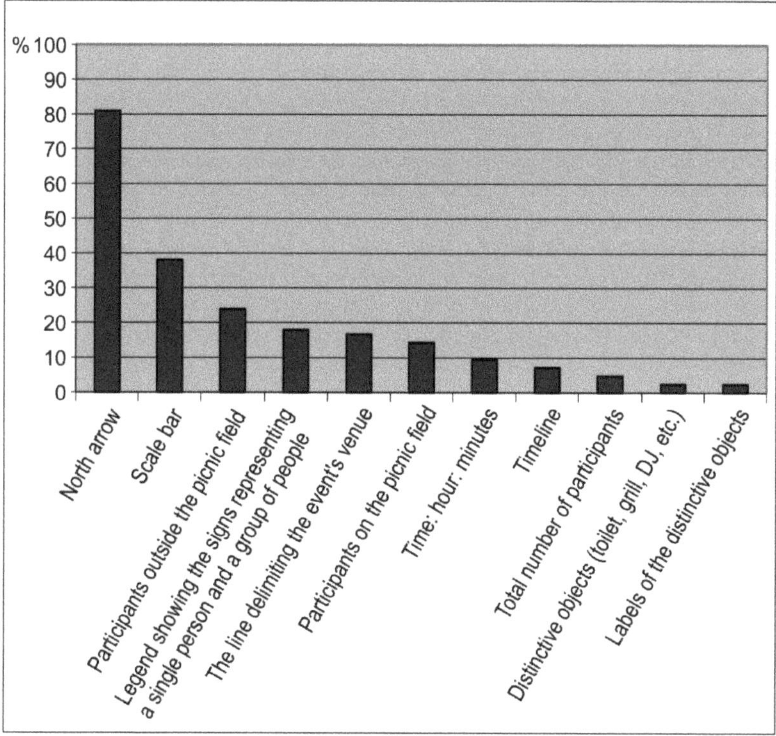

Figure 8. The elements that the experts considered to be unnecessary in animated maps.

3. Conclusions

The results obtained here are part of wider considerations about user-centered design for interactive maps in the context of the search for simple and effective design principles [21]. The results also point out changes in the application of traditional cartographic design principles by experts in multimedia cartography going beyond interactive multimedia products on the Internet [22]. The authors of this study drew attention to designing the map layout and functionality for animated visualizations in three ways to transform more and more drone images. A handful of tips obtained here may be relevant for users of dynamic developed web applications with the raw spatial data that most often require cartographic presentation [23,24].

The primary aim of the study was accomplished by proposing three alternative methods of visualization, whose effectiveness for analyzing the course of an open-air event was investigated with the participation of experts in cartography and geomatics. What was of key significance to the obtained results was the juxtaposition of objective answers with subjective opinions [25], which allowed for a more informed formulation of conclusions regarding the effectiveness of the three different visualizations created on the basis of drone-taken photographs.

The animation employing the buffer map, i.e., the one in which the original information was processed the most and the personal distances were represented, objectively turned out to be the most effective one in terms of analyzing the dynamics of the spatial relations among people gathering at an open-air event attended by approximately 1000 people in an area of about 0.5 ha. The dot distribution map, which was subjectively assessed by experts as better, turned out to be slightly less effective than the buffer map. The results of both the objective and the subjective experiment, however, confirmed that the orthophoto had the lowest effectiveness.

All three methods were found to be highly effective for analyzing the successive (linear) growth in the number of participants near an object placed far away from big groups (the queue to the toilet). Questions regarding the dynamics of spatial relations were the most difficult ones for the respondents to answer. Those were the questions about the crowdedness of the particular parts of the picnic field and the number of participants forming buffers in particular time intervals. This data could be obtained only by analyzing the map, and the results show such relations to be relatively difficult to interpret. It is thus necessary to search for methods to facilitate this process and make it more effective. It is worth mentioning that the user of a particular animation tended to express a more favorable opinion of the method that he/she had come to know on a frame-by-frame basis than about a method he/she had seen only on one stationary frame.

The tested animations do not exhaust all possible types of land use or lighting conditions, but provide a good starting point for further research and testing methods in other environments. In the case of lighting conditions, variable light intensity was taken into account because the experiment covered hours from afternoon to evening, with artificial lighting visible in the later images. The color of clothes undoubtedly affects the degree of difficulty in noticing the participants of the picnic on the orthophoto map and we have added such information in the text. However, it seems that, even assuming that everyone would wear red jackets, the effectiveness of the orthophoto map and other methods would not be the same. This is because this method still requires users to interpret which set of pixels is a person, whereas in the case of dot and buffer maps, this burden has been completed by the creator of the visualization.

The idea behind the experiment was to examine the effectiveness of methods for subsequent transformations—from the simplest orthophoto method requiring the user to interpret the image more thoroughly, through the dot distribution map showing the distribution of picnic participants, to the buffer map suggesting relationships between participants. Stable objects appeared on all visualizations, while the buffer and dot distribution map method removed the orthophoto background (foundation), which could distract the user unnecessarily. Probably, in this case, the effect would be better than if the orthophotomap was used alone, but worse in relation to, e.g., the dot distribution map method. The orthophotomap was only an intermediate stage characterized by a large number of details and a very low degree of information processing, and recognizing such elements as a grill, toilet, or DJ spot without indicating them with the help of symbolized stable objects would be much more difficult for a user who did not participate in the event.

Because we invited experts in cartography and geomatics from outside Poland to participate in the survey, it was done online, and we were unable to use the eyetracking method [26]. We also assumed that one animation iteration may be insufficient due to the need to focus on a moving image—a different level of concentration at the initial frames of the animation and another when replaying the same animation, but with a different question. The ability to concentrate may be different for different people and also depends on external factors that we were not able to control by conducting a distance study, which is why there was no time restriction on completing the task. Expert opinion was important to us, so for this reason the amount of time was unlimited, in order that respondents could learn about the method and form an opinion about it. More important to us was how much information from a given type of method could be read correctly and whether the subjective indications of the respondents in the second part of the survey follow their correct answers to objective questions.

Experts were in favor of minimizing the number of map elements, even at the cost of breaching traditional cartographic principles, as in the case of not including on the animated map a scale bar or a north arrow for non-north-oriented maps [27]. The study results pointed to the great importance of the graphic enhancement of the objects serving as points of interest for the event's participants and being thus treated as reference objects while analyzing the course of the event. There are many graphical ways to delimit an area in a raster file without distracting but instead focusing on the area needed; an example is the boundary of Greece (https://tinyurl.com/y2z65qvp), which is presented in such a way that the area seems to be lifted, whereas it is only a properly symbolized line [28]. This conclusion was also backed by the expert opinions. Those elements and their labels were assessed as being among the most needed components of the animation. In view of the above, the authors concluded that the search for alternative means of presentation is of key significance in improving the effectiveness of visualizations, which can be methodically confirmed by expert opinions. However, it is advisable to turn to further, objective, user-focused research in order to verify this effectiveness.

Author Contributions: Conceptualization, Beata Medyńska-Gulij; Data curation, Łukasz Halik; Investigation, Beata Medyńska-Gulij, Łukasz Wielebski, and Maciej Smaczyński; Methodology, Beata Medyńska-Gulij; Project administration, Beata Medyńska-Gulij; Resources, Maciej Smaczyński; Software, Łukasz Wielebski; Validation, Łukasz Halik and Maciej Smaczyński; Visualization, Łukasz Wielebski; Writing—original draft, Beata Medyńska-Gulij and Łukasz Wielebski. All authors have read and agreed to the published version of the manuscript.

Acknowledgments: This paper is the result of research on visualization methods carried out within statutory research in the Research Division Cartography and Geomatics, Faculty of Geographical and Geological Sciences, Adam Mickiewicz University in Poznań, in Poland.

Conflicts of Interest: The authors declare no conflict of interest.

References

1. Alshearya, A.; Almagbile, A.; Alqrashi, M.; Wang, J.; Khalil, H.; Al-Zahrani, M. Assessment of the applicability of unmanned aerial vehicle (UAV) for mapping levels of crowd density. In Proceedings of the 9th International Symposium on Mobile Mapping Technology (MMT2015), Sydney, Australia, 9–11 December 2015.
2. Medyńska-Gulij, B.; Dickmann, F.; Halik, Ł.; Wielebski, Ł.W. Mehrperspektivische Visualisierung von Informationen zum räumlichen Freizeitverhalten. Ein Smartphone-gestützter Ansatz zur Kartographie von Tourismusrouten. Multiperspective visualisation of spatial spare time activities. A smartphone-based approach to mapping tourist routes. *Kartogr. Nachr.* **2015**, *67*, 323–329.
3. Colomina, I.; Molina, P. Unmanned aerial systems for photogrammetry and remote sensing: A review. *Isprs J. Photogramm. Remote Sens.* **2014**, *92*, 79–97. [CrossRef]
4. Hochmair, H.H.; Zielstra, D. Analysing user contribution patterns of drone pictures to the dronestagram photo sharing portal. *J. Spat. Sci.* **2015**, *60*, 79–98. [CrossRef]
5. Dent, B.; Torguson, J.; Hodler, T.W. *Cartography: Thematic Map Design*, 6th ed.; McGraw-Hill Publishing: London, UK, 2009; ISBN 0077391756.
6. Forrest, D.; Pearson, A.; Collier, P. The Representation of Topographic Information on Maps – The Coastal Environment. *Carto. J.* **1997**, *34*, 77–85. [CrossRef]
7. Slocum, T.A. *Thematic Cartography and Geovisualization*, 3rd ed.; Pearson Prentice Hall: Upper Saddle River, NJ, USA, 2009; ISBN 9780132298346.
8. Cybulski, P. Design rules and practices for animated maps online. *J. Spat. Sci.* **2016**, *61*, 461–471. [CrossRef]
9. Smaczyński, M.; Medyńska-Gulij, B. Low aerial imagery—An assessment of georeferencing errors and the potential for use in environmental inventory. *Geod. Cartogr.* **2017**, *66*, 89–104. [CrossRef]
10. Dent, B.D. *Cartography. Thematic Map Design*, 5th ed.; WCB/McGraw-Hill: Boston, MA, USA, 1999; ISBN 007231902X.
11. Robinson, A.H. *Elements of Cartography*, 5th ed.; Wiley: Chichester, UK, 1987; ISBN 9780471098775.
12. Kriz, K. Maps and Design-Influence of Depiction, Space and Aesthetics on Geocommunication. In *Understanding Different Geographies*; Kriz, K., Cartwright, W., Kinberger, M., Eds.; Springer: Berlin, Germany, 2013; pp. 9–23, ISBN 978-3-642-29770-0.

13. Medyńska-Gulij, B.; Lorek, D.; Hannemann, N.; Cybulski, P.; Wielebski, Ł.; Horbiński, T.; Dickmann, F. Die kartographische Rekonstruktion der Landschaftsentwicklung des Oberschlesischen Industriegebiets (Polen) und des Ruhrgebiets (Deutschland). Cartographic reconstruction of the landscape development of the Upper Silesian industrial area (Poland) and the Ruhr area (Germany). *KN J. Cartogr. Geogr. Inf.* **2019**, *69*, 131–142. [CrossRef]
14. van Elzakker, C.P.J.M.; Ooms, K. Understanding map uses and users. In *The Routledge Handbook of Mapping and Cartography*; Kent, A.J., Vujakovic, P., Eds.; Routledge: London, UK, 2018; pp. 55–67, ISBN 978-1-13-883102-5.
15. Wielebski, Ł. Mapping techniques of spatio-temporal relationships for a centric road network model. *Kartogr. Nachr.* **2014**, *64*, 269–276.
16. Medyńska-Gulij, B. The effect of cartographic content on tourist map users. *Cartography* **2003**, 49–54. [CrossRef]
17. Medyńska-Gulij, B. Point Symbols: Investigating Principles and Originality in Cartographic Design. *Cartogr. J.* **2013**, *45*, 62–67. [CrossRef]
18. Halik, Ł.; Medyńska-Gulij, B. The Differentiation of Point Symbols using Selected Visual Variables in the Mobile Augmented Reality System. *Cartogr. J.* **2016**, *32*, 1–10. [CrossRef]
19. Hall, E.T. *The Hidden Dimension;* Anchor Books: New York, NY, USA, 1990; ISBN 9780385084765.
20. Medyńska-Gulij, B. Educating Tomorrow's Cartographers. In *The Routledge Handbook of Mapping and Cartography*; Kent, A., Vujacovic, P., Eds.; Routledge: London, UK, 2018; pp. 561–568.
21. Roth, R.; Ross, K.; MacEachren, A. User-Centered Design for Interactive Maps: A Case Study in Crime Analysis. *Int. J. Geo-Inf.* **2015**, *4*, 262–301. [CrossRef]
22. Cartwright, W. *Multimedia Cartography. With 4 Tables, and a Complementary CD-ROM*; Springer: Berlin, Germany, 2007; ISBN 3-540-65818-1.
23. Schomacker, R.; Gert, E.; Schweikart, J. Public art map—Art in public spaces. Acquisitions of data in the context of volunteered geographical information. *Kartographische Nachrichten* **2015**, *5*, 281–288.
24. Cybulski, P.; Wielebski, Ł. Effectiveness of Dynamic Point Symbols in Quantitative Mapping. *Cartogr. J.* **2019**, *56*, 146–160. [CrossRef]
25. Wielebski, Ł.; Medyńska-Gulij, B. Graphically supported evaluation of mapping techniques used in presenting spatial accessibility. *Cartogr. Geogr. Inf. Sci.* **2018**, *64*, 1–23. [CrossRef]
26. Horbiński, T.; Cybulski, P.; Medyńska-Gulij, B. Graphic design and placement of buttons in mobile map application. *Cartogr. J.* **2020**. [CrossRef]
27. Edler, D.; Keil, J.; Tuller, M.-C.; Bestgen, A.-K.; Dickmann, F. Searching for the 'Right' Legend: The Impact of Legend Position on Legend Decoding in a Cartographic Memory Task. *Cartogr. J* **2018**. [CrossRef]
28. Antoniou, V.; Nomikou, P.; Papaspyropoulos, K.; Vlasopoulos, O.; Zafeirakopoulou, E.; Bardouli, P.; Chrysopoulou, E. Geo-biodiversity and Cultural Environment of the regions surrounding the Corinth Gulf (Greece). In *Regional Conference on Geomorphology. Focal Theme: "Geomorphology of Climatically and Tectonically Sensitive Areas", UNESCO Global Geoparks: Geoheritage Assessment and Management—Geo-Tourism Development*; Abstract Book: Athens, Greece, 2019; p. 256.

International Journal of
Geo-Information

MDPI

Article

Effectiveness of Memorizing an Animated Route—Comparing Satellite and Road Map Differences in the Eye-Tracking Study

Paweł Cybulski

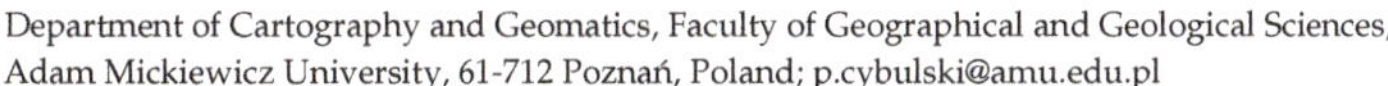

Department of Cartography and Geomatics, Faculty of Geographical and Geological Sciences, Adam Mickiewicz University, 61-712 Poznań, Poland; p.cybulski@amu.edu.pl

Abstract: There is no consensus on the importance of satellite images in the process of memorizing a route from a map image, especially if the route is displayed on the Internet using dynamic (animated) cartographic visualization. In modern dynamic maps built with JavaScript APIs, background layers can be easily altered by map users. The animation attracts people's attention better than static images, but it causes some perceptual problems. This study examined the influence of the number of turns on the effectiveness (correctness) and efficiency of memorizing the animated route on different cartographic backgrounds. The routes of three difficulty levels, based on satellite and road background, were compared. The results show that the satellite background was not a significant factor influencing the efficiency and effectiveness of route memorizing. Recordings of the eye movement confirmed this. The study reveals that there were intergroup differences in participants' visual behavior. Participants who described their spatial abilities as "very good" performed better (in terms of effectiveness and efficiency) in route memorizing tasks. For future research, there is a need to study route variability and its impact on participants' performance. Moreover, future studies should involve differences in route visualization (e.g., without and with ephemeral or permanent trail).

Keywords: animated map; eye tracking; road map; route memorizing; cartography; effectiveness; multimedia cartography

Citation: Cybulski, P. Effectiveness of Memorizing an Animated Route—Comparing Satellite and Road Map Differences in the Eye-Tracking Study. *ISPRS Int. J. Geo-Inf.* **2021**, *10*, 159. https://doi.org/10.3390/ijgi10030159

Academic Editors: Wolfgang Kainz and Vit Vozenilek

Received: 21 January 2021
Accepted: 11 March 2021
Published: 12 March 2021

1. Introduction

Cartographic animations often depict spatio-temporal phenomena. They use dynamic variables defined by DiBiase et al. [1] to make the impression of dynamic change of visual variables such as the movement of point symbols [2,3]. Nowadays, popular information (e.g., The Guardian, Business Insider) and educational services (e.g., openculture.com, battlefields.org) use animated maps. Animation has become ubiquitous, not only in cartography but also as a visualization tool in other areas such as medicine [4], gaming [5], virtual reality [6], or architecture [7]. The animation method is very successful in non-conventional cartographic products such as anaglyphs, along with the use of timeline [8]. Despite its popularity, animation faces several perceptual problems [9,10]. Most of them are related to the fact that working memory is not able to store a large amount of information, which appears in subsequent scenes of the animated map [11]. This problem translates into the ability to memorize information, such as the route [12].

Current research in cartography has confirmed the fact that animated maps are more demanding than their static counterparts in some map-reading tasks [13–15]. This is mainly since, as Kraak [16] notes, changes on the map may be different (e.g., appearance/disappearance, increase/decrease, or shrink/expand). However, some animated maps may still be more effective than others. More abstract maps support route remembering much more than those having many details [17]. Satellite photographs are more detailed than abstract images and might lead to cognitive overload [18]. However, abstract maps require symbol interpretation, and therefore, a priori knowledge [19]. Several factors could influence the perceptive processing of each cartographic background. Some are related to individual features of map users [20]. Others could be related to spatial or cartographic factors such as screen resolution [21] or location of areas of interest [22]. The growing number of details on the map is associated with technological development but also with the fact that people trust images that are more realistic or graphically attractive [23,24]. Effectiveness in reading animated maps is also dependent on the difficulty of the map-reading task. Regarding memorizing the route, it can be the number of turns or the length of the entire route. Keil et al. [25] noticed that adding landmark pictograms after a route learning task leads to a decrease in effectiveness and an increase in higher false alarm rates. Recent research shows that the effectiveness and efficiency of a given map can be explained by how the user observes it. In this case, eye-tracking studies are helpful [26,27].

Eye-tracking is an objective method for assessing the user's visual strategy while memorizing and then recalling the route from memory. It can assess how people visually process maps and the visual processing may affect how efficiently and effectively people memorize a route. The user's eye movement analysis provides many useful indicators that complement standard methods, such as the questionnaire [28]. The most basic metrics are fixations—relatively stationary visual gaze on a single location, and saccades—a rapid movement of gaze to another location [29]. However, during observation of a cartographic animation, smooth pursuit occurs. This is a slow eye-tracking of a moving stimulus [30]. However, both saccadic and pursuit eye movement are independent and, in many circumstances, these two types of eye movement act conjointly, e.g., saccades supplement smooth pursuit [31]. Differences, e.g., in time to the first fixation or the length of the saccades between individual persons in the sample group, may explain different visual strategies depending on the characteristics such as gender or education. Some studies showed that age differences in route memorization are a crucial factor, and young adults are more effective in this type of task [32]. However, extending the research to an older group of people could support other interesting results. Knowing how people recall the route from memory and what is the actual effectiveness and efficiency concerning the degree of difficulty and the cartographic background is helpful for dynamic map designers [33]. Intergroup differences sometimes explain the reasons for the different effectiveness and efficiency of map-based route memorizing tasks [34,35].

Concerning issues raised, several questions arise: does the complexity (number of turns, dot speed) of the animated route have the same impact on the effectiveness and efficiency of maps with road and satellite background? Does the user's visual behavior (eye movement metrics) differ depending on the cartographic background? Are there intergroup differences in the study sample? To answer the question posed, conducted research involves users route memorizing tasks and eye-tracking technology. It is also significant to distinguish between map-based route learning, which is an application in this study, from navigational route learning and wayfinding. Map-based route learning is a situation in which a user studies a map alone, which might precede further spatial navigation [36]. The main goal of this research is to determine the effectiveness and efficiency in route memorizing tasks on the basis of animated maps with a different cartographic background. To reach this goal, this study developed a set of animated maps with road and satellite backgrounds based on Google Maps. Additionally, to examine how the effectiveness and efficiency change concerning the complexity of the route, this study took into account the variable number of turns between individual maps. In order to define individual and intergroup differences between users' visual behavior, this study used eye movement registration.

2. Related Research

Klippel et al. [13] presented a study based on animated route maps. They found that spatial chunking of route segments is a helpful strategy. The experimental procedure of that study included two groups (static and dynamic conditions). In the static presentation, the route was visible on a map. In the dynamic conditions, the researchers used an animated route. The participants' task was to give verbal instructions on the basis of the route they saw. Klippel et al. [13] concluded that static presentation allows more spatial chunking than an animated map.

Assessing the performance of route recalling was conducted in the experiment by Tom and Denis [37]. They found that participants memorized routes better when they processed landmarks rather than street names. In the route comparison task, landmark information was also responsible for higher efficiency (less time needed for response). In their conclusions, Tom and Denis [37] confirmed the significant impact of landmarks on route perception. Tom and Tversky [38] presented another route-remembering study based on street names and landmarks. They referred to previous studies and claimed that names of landmarks were more vivid and distinctive than street names. They concluded that vividness/distinctiveness is a major factor of route memorization. This corresponds to Cybulski and Medyńska-Gulij's [39] findings that cartographic redundancy is influential in the change detection of animated stimuli.

Designing an animated route map is possible, thanks to a vast number of available tools (on-line applications, commerce software, etc.). However, there is less research on the effectiveness of these maps. Çöltekin et al. [17] found that more abstract road maps are more effective in the process of memorizing the route by men than those using a satellite background. They also determined the impact of spatial abilities on memorizing the route and found that people who described their spatial abilities as high (in mental rotation task) performed slower with satellite images.

Previous studies in visualization by Brucker et al. [40] confirmed these conclusions. However, the research did not determine what impact the route's difficulty has on the effectiveness and efficiency of these maps. The study of the degree of realism, which is related to the complexity of the image, was the subject of studies on memorizing the route in virtual reality [41]. They confirmed that mixed realism (high realism with abstract elements) is the most optimal solution in route-learning tasks in virtual reality. Lokka and Çöltekin [42] point out that the realistic view was the most attractive to users even though it was not the most effective. General studies on the effectiveness of memorizing information based on animation are mostly unanimous and state that animation is more problematic than a static image [8,42]. However, a certain novelty in this type of study reveals how to estimate user correct response. In cartographical effectiveness studies, the user's response could be accurate or inaccurate [8,43–45]. On several occasions, researchers have highlighted the need to develop a more detailed effectiveness measurement methodology [8,46–48]. This article relates to previously applied research approaches that allow for estimating the effectiveness of memorizing the route in map-based tasks.

In addition to effectiveness and efficiency tests, cartographic research increasingly includes methods that allow for the registration of the user's eye movement [49–52]. In an animated route, a crucial element is the movement of an object that goes from the origin to the destination point. Krassanakis et al. [50] researched this issue with the use of eye-tracking equipment. They noted that object movement is the main factor that attracts the user's attention, while several static details could distract user attention. This knowledge is based on eye movement recording and suggests that a more realistic background could cause distraction and may result in less effective route memorization. In Dong et al.'s [52] research, they found that various visual variables of animated maps differ in the level of effectiveness, and the size proved to be the most effective. However, in memorizing the route on the basis of animation, the most crucial variable is location because the object moves from the starting point to the destination point. The eye-tracking used in cartographic studies often uses devices with 60-120 Hz sampling frequency and 0.5° of spatial accuracy and were validated as suitable for scientific research [53,54].

Unfortunately, there are not many studies dedicated to animated routes in particular in the cartographic context. For this reason, the presented research is a voice in the discussion on the impact of the number of turns on the effectiveness and efficiency of animated routes and their visual perception.

3. Methodology

3.1. Animated Route Maps

The study used a total of 60 maps (30 maps with a road background and 30 maps with satellite background). The maps had 3 levels of route difficulty: the first level had 3 turns and 0.9–1.1 km route length (dot speed 2.8 cm/s), the second level had 5 turns and 1.2–1.4 km (dot speed 3.7 cm/s) route length, and the third level had 7 turns and 1.5–1.7 km (dot speed 4.7 cm/s) route length. There were 10 road and 10 satellite maps in each level of route difficulty (60 maps in total). In this study, the turns were understood as junction-dependent, a change of direction. The sharp turn could also function as a turn. However, all cases were road junction-based. Figure 1 presents examples of designed maps. Each route was unique and was not presented on both backgrounds. The increase in the number of turns in urban areas with a similar density of buildings naturally caused the route elongation. Maintaining a similar density of the road network requires cartographic background with the more urbanized area. This would entail a lack of uniformity in the cartographic background. Therefore, designing the maps required reducing the route uniformization at the expense of more uniform backgrounds. Therefore, the design assumed route length ranges. This resulted in a variable dot speed but within constant values.

Figure 1. Six examples of maps used in the study. On the left, there are maps (**A–C**) with road background, and on the right are maps (**D–F**) with satellite background. The actual track is highlighted with black and white lines (not visible by participants), with a pink dot directing the starting position of the animated route.

Study participants viewed a total of 60 maps. Presenting identical routes on both cartographic backgrounds could result in a participants' learning effect, despite the randomization. For this reason, the design required the development of unique routes on the maps.

All maps were developed in HTML5 standard on Google Maps background with labels removed. They all had a similar level of visual complexity, road density, and included urbanized areas. The method of assessing these factors is similar to Çöltekin et al.'s [17] research. Complete unification of maps was not possible; however, the author believes that the content, level of route difficulty (in terms of turns and dot speed), and the constant scale were controlled across all stimuli. All maps had a button that started and repeated the animation, and participants could observe the route twice. This is considered as a basic interaction on an animated map, and it was optional [55]. Each route had a starting position marked by a dot. After pressing the "start" button, the dot moved to the destination place and stopped (there were no route marking). Each route, regardless of the difficulty level, lasted 8 s. As a result, each difficulty level had a slightly different dot speed, however, the route difficulty was obtained by certain criteria. This value was selected after a series of trials. This value corresponds to the values indicated in the research by Cybulski and Wielebski [45] regarding the effective duration of cartographic animations, which was defined between 8 and 11 s for 1 dynamic stimulus. Moreover, in similar research on memorizing the animated route, Çöltekin et al. [17], in one of their tasks, used 8.4 s of animation. For displaying stimuli, the study used a 15.6-inch laptop with an LCD screen and pixel resolution 1920 × 1080 in controlled lab conditions.

3.2. Experimental Procedure

The study begins with calibration with an eye tracking device. If the calibration results were unsatisfactory, the participant repeated calibration. Recording of eye movements starts from this moment. After that, participants proceeded to deliver demographic information (age, gender). Additionally, study participants determined their spatial abilities. Unlike Çöltekin et al. [17], participants did not use MRT (mental rotation task), but subjective declarations instead. The question was, "rate your spatial abilities on a scale of 1 to 5". There were 5 possible answers arranged on a Likert-type rating scale: 5—very good; 4—good; 3—average; 2—bad; 1—very bad. The study consisted of 3 main parts. The first part consisted of completing the familiarization task. After making sure that participants understood what the task would be like, participants moved to the second part of the study. It consisted of watching the 8-second route animation twice and trying to memorize it. The animated route lasted 8 seconds and could be watched twice, however, participants had unlimited time to respond. The task was to "watch the animation and try to remember the route". When they were ready to recall the route after watching the animation, they pressed the F10 button. The third part of the task consisted of drawing the route on the previously displayed background. Each participant had the "clear" button at his disposal so that in case of a mistake, they could draw the route again. If they finished drawing the route, they pressed F10 to proceed to the next animation. Each participant observed maps in random order (with no repetitive maps). Therefore, each participant viewed all of the 60 maps. The drawing algorithm used the D3 library (a JavaScript library).

3.3. Participants

In the study, 132 people participated (78 of them were male and 54 were female). The age range of participants was from 22 to 36 years. All of them were extramural students of Adam Mickiewicz University Poznan. Participant selection was based on other related studies [56–58]. The participants subjectively assessed their spatial abilities. They were told that this is the subjective measure of their space-related abilities. Most of them rated them as "good" (68 answers). A relatively large group assessed their abilities as "average" (40 answers), and the lowest amount of people evaluated themselves as "very good" (24 answers). Interestingly, only males rated their abilities as "very good". Most females assessed spatial abilities as "good" (29 answers).

3.4. Effectiveness and Efficiency Evaluation

Effectiveness in this study is understood differently from the recent effectiveness studies of Çöltekin et al. [17]. In their work, effectiveness was a ratio between the number of correct answers and all answers expressed as a percentage. However, when it comes to remembering the route, one can recall it in some parts. Then the effectiveness of the animated route map will be the ratio of the correctly remembered section to the entire route (length and location). If the participant memorized the entire route (i.e., correctly drew shape, location, and length), then the effectiveness is 100%. However, if the participant drew a longer route than the one observed in the animation, the effectiveness decreased by a percentage of the extra route length (in a ratio to the entire length). The participants did not have to precisely draw the road along its geometrical axis. The effectiveness assessment was topologic. Participants needed to approximately follow the route and make all the correct junction turns.

Efficiency is a time in which a participant proceeded to the route drawing stage. In the study, this time was counted in seconds from the beginning of the animation to its end, which was followed by the drawing stage (pressing the F10 button). The shorter the time necessary to proceed further, the higher the efficiency. It is worth noting that efficiency is not the time needed to recall the route from memory and draw it. Mainly this is because, when drawing sketches, participants are focused on correct drawing. They are not attempting to draw the route fast [59,60]. Moreover, the ability to draw with a computer mouse could affect efficiency.

3.5. *Eye movement Metrics*

The analyzed eye movement metrics in this study are saccadic amplitude in degrees (which corresponds with saccade length), an overall number of saccades and fixations, fixation duration, and the number of fixations on the route. Tobii Eye-Tracker X2-60 recorded and collected eye movement metrics with a spatial accuracy of 0.5 degrees of visual angle, and Tobii Studio software processed the data for identification of fixations and saccades. The equipment used in the study captured gaze position every 16.67 ms. The Tobii I-VT Fixation Filter (velocity-based) divided eye movement into fixations and saccades [61].

The use of these metrics is essential. Fixations allow 1 exact determination of the user's focused attention, and together with saccades are elementary components of eye movement [62,63]. Fixations are responsible for the processing of visual information. On the other hand, the saccadic movement does not process visual information [64,65].

4. Results

4.1. *Effectiveness and Efficiency Analysis*

Table 1 presents the animated route map effectiveness results. The analysis used the median due to the non-normal distribution of the data. It turns out that the effectiveness decreased as the number of turns increased, but the most significant effectiveness decrease concerned maps with seven turns. This observation applies to both cartographic backgrounds used in the experiment. In the case of maps with a road background, the Mann–Whitney test [66] showed statistical significance in the decrease in the median effectiveness between three and five turns ($p < 0.01$), between five and seven turns ($p < 0.0001$), and between three and seven turns ($p < 0.0001$). In the case of satellite background, the Mann–Whitney test proved statistically significant differences between maps with three and five turns ($p < 0.0001$), five and seven turns ($p < 0.0001$), and three and seven turns ($p < 0.0001$). The differences in effectiveness between the road and satellite background with the same number of turns are minimal and are not statistically significant.

Table 1. The median effectiveness results for both cartographic backgrounds (road and satellite) with different number of turns.

Animated Route Map Effectiveness—Median		
Road map three turns	Road map five turns	Road map seven turns
97.0%	94.0%	84.5%
Satellite map three turns	Satellite map five turns	Satellite map seven turns
96.0%	94.0%	83.0%

Effectiveness determination consisted of the various levels of route complexity concerning gender differences with Kruskal–Wallis ANOVA [67]. The statistical test did not report significant variation in effectiveness between genders. There were also no differences in the influence of cartographic background on effectiveness among males and females. However, there were statistically significant differences ($p<0.05$) in the route memorization effectiveness in terms of the satellite map with three turns among spatial abilities groups. Participants who assessed their spatial abilities as very good achieved 98%, good achieved 96%, and average achieved 94.5% effectiveness. Figure 2 presents the detected variation with maximum and minimum values among spatial abilities and effectiveness on satellite background map with three turns.

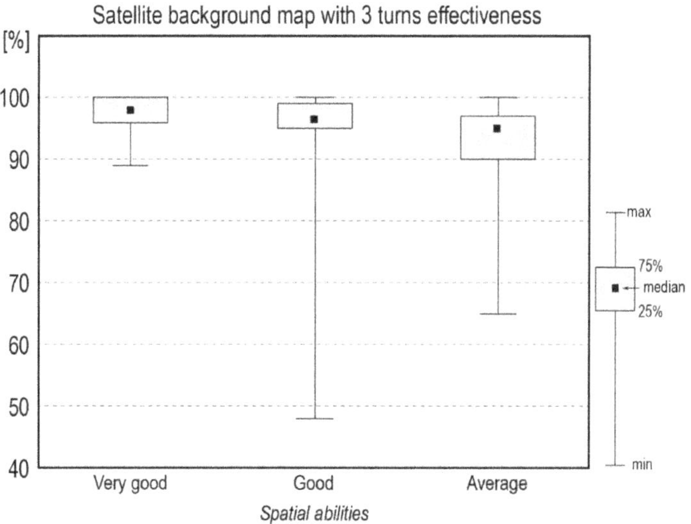

Figure 2. Statistically significant differences between effectiveness of route memorization in terms of satellite background with 3 turns in spatial abilities groups.

Table 2 presents the efficiency results. As in the case of effectiveness, the analysis used the median due to the distribution of data. The efficiency remained at a similar level regardless of the number of turns. In the case of a satellite background, the time needed to solve the task was slightly longer. Interestingly, seven turns had a standard deviation greater than the rest. The Mann–Whitney test showed no significant median differences between maps with road background. The same was the case for the maps with satellite background. There were also no significant differences between the road and satellite backgrounds (comparing the same number of turns).

Table 2. The median efficiency results for both cartographic backgrounds (road and satellite) with different number of turns.

Animated Route Map Efficiency—Median ± SD		
Road map three turns	Road map five turns	Road map seven turns
23.6 s ± 5.6 s	23.6 s ± 5.4 s	24.0 s ± 7.3 s
Satellite map three turns	Satellite map five turns	Satellite map seven turns
24.9 s ± 5.4 s	25.0 s ± 6.1 s	25.5 s ± 7.1 s

Kruskal–Wallis ANOVA in gender groups did not reveal any significant differences in the efficiency of route memorization. However, spatial abilities groups reported significant difference ($p < 0.005$) among efficiency in road map background with three turns. Figure 3 presents differences among efficiency and spatial abilities on road background map with three turns. In terms of the spatial abilities groups, very good participants achieved 20.6 s, good achieved 24.0 s, and average achieved 24.0 s.

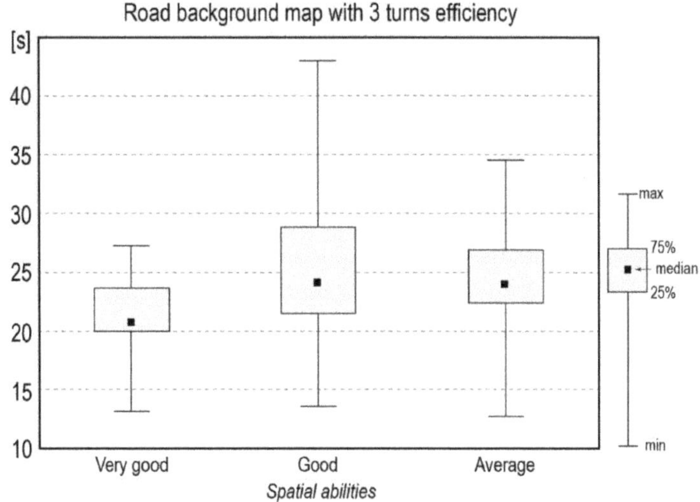

Figure 3. Statistically significant differences between efficiency of route memorization in terms of road background with three turns in spatial abilities groups.

4.2. Eye Movement Analysis

Table 3 presents the median saccadic amplitude, which corresponds to the saccade length. Saccadic amplitudes were similar, despite the growing number of turns. The Mann–Whitney test indicated statistically significant differences in saccadic amplitude between road and satellite background with five turns ($p < 0.05$).

Table 3. The median saccadic amplitude results for both cartographic backgrounds (road and satellite) with different number of turns.

Saccadic Amplitude—Median ± SD		
Road map three turns	Road map five turns	Road map seven turns
3.5° ± 1.3°	3.5° ± 1.2°	3.3° ± 1.1°
Satellite map three turns	Satellite map five turns	Satellite map seven turns
3.4° ± 1.1°	3.2° ± 0.8°	3.4° ± 0.9°

However, there were statistically significant differences in the saccadic amplitude between females and males using Kruskal–Wallis ANOVA, as shown in Figure 4. In the case of the road background in the conditions of three and five turns, females had longer saccadic amplitudes (3.3° for male vs. 3.7° for female in the case of three turns—$p < 0.05$, and 3.2° for male vs. 4.0° for female in the case of five turns—$p < 0.0001$). Only in the case of road background with seven turns were the differences not statistically significant. The situation was similar in the case of the satellite background. Females had longer saccadic amplitude than males in all number of turns (3.1° for males vs. 3.5° for females in the case of three turns—$p < 0.05$, 2.9° for males vs. 3.6° for females in the case of five turns—$p < 0.001$, and 3.1° for males vs. 3.6° for females in the case of seven turns—$p < 0.01$).

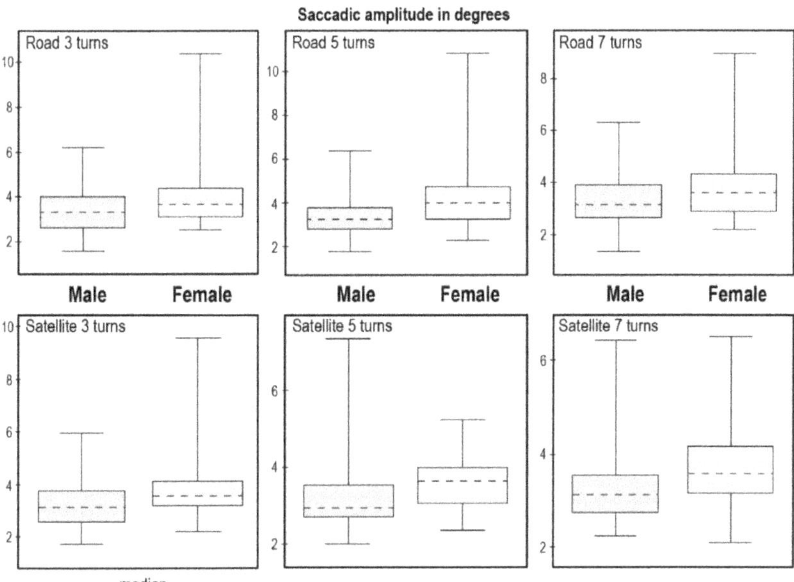

Figure 4. Differences in saccadic amplitude between males and females in different background conditions and various numbers of turns.

There were also statistically significant differences in the saccadic amplitude between groups with different spatial abilities. Overall, participants whose spatial abilities were very good had shorter saccadic amplitudes. In particular, on the road background, participants with average spatial abilities had a median saccade of 4.0° (three turns) and 3.8° (five turns), while participants with very good spatial abilities had 2.9° (three and five turns). The p-value for both maps was <0.001. Moreover, on the satellite background, participants with very good spatial abilities had shorter saccadic amplitudes than those who had average abilities. In particular, participants with average spatial abilities had a median saccade of 3.6° (five turns) and 3.5° (seven turns), while participants with very good spatial abilities had 3.0° (five and seven turns). The p-value for both maps was <0.05.

Since smooth pursuit was observed on the animated route map, the number of saccades was found to be greater than the number of fixations. On the road background with three turns, 91 saccades occurred; with five turns, 88 saccades occurred; and seven turns had 96 saccades (all medians). On the satellite background, there were 100 saccades (three turns), 90 saccades (five turns), and 102 saccades (seven turns). However, these differences between each background and the number of turns (in terms of the number of saccades) were not statistically significant.

Table 4 presents the median number of overall fixations. The number of fixations increased when comparing maps with three and seven turns. However, on a satellite background, the differences were smaller. Additionally, the number of fixations on the satellite background was only slightly higher than on the road background. The Mann–Whitney test partially confirmed this. There were statistically significant differences between road background with three and seven turns ($p < 0.05$). However, there was no difference between five and seven turns. The same was the case with the satellite background.

The median number of fixations positively correlated with efficiency measures (seconds of viewing the map) for both backgrounds and all turns. This was an expected effect because the longer the user watched the animation, the more fixations there would be in general. For the road background, the strength of the correlation was $r = 0.71$ for three turns, $r = 0.70$ for five turns, and $r = 0.72$ for seven turns in the Spearman correlation test.

For the satellite background, the strength of the correlation was $r = 0.72$ for three turns, $r = 0.78$ for five turns, and $r = 0.77$ for seven turns (p-value for all correlations in both backgrounds was <0.0001).

Table 4. The median number of fixation results for both cartographic backgrounds (road and satellite) with different number of turns.

Number of Fixations—Median ± SD		
Road map three turns	Road map five turns	Road map seven turns
60.5 ± 22.9	65 ± 21.4	67 ± 28.1
Satellite map three turns	Satellite map five turns	Satellite map seven turns
68 ± 22.3	67 ± 23.7	71 ± 29.7

Figure 5 presents the spatial distribution of fixations while watching different route animations. As the selected examples show, and confirmed by other cases, most fixations occurred at the beginning of each route. Despite this, the most errors in drawing the route concerned the correct determination of the beginning of the route. Despite different backgrounds and the number of turns, the median number of fixations on the route were similar. The median varied between 33 and 35 fixations. In this case, this meant that about half of the fixations were on the route. The rest included the direct surroundings of the route, the animation start button, and the background. Dispersed fixations around the route (in its direct neighborhood) are related to a smooth pursuit, which causes oscillation of the gaze [68].

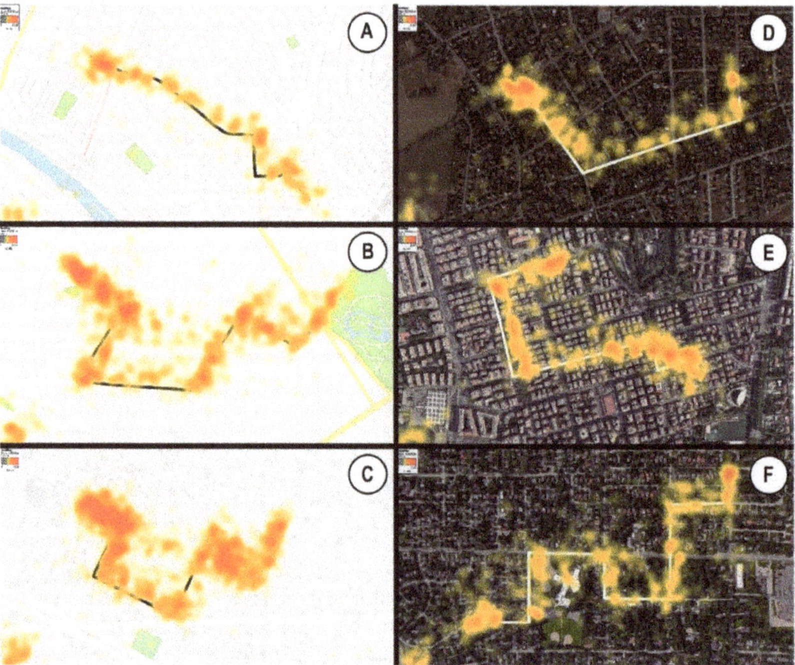

Figure 5. Spatial distribution of random 15 participants' fixations on selected route animations. The more red the color, the more fixations occurred during the watching of the animation. Subfigures on the left (**A–C**) presents example maps with road background. Subfigures on the right (**D–F**) presents example maps with satellite background.

5. Discussion

The Çöltekin et al. [17] research has shown that road maps (being more abstract) make it easier to memorize the route concerning satellite maps. Brucker et al. [40] came to similar conclusions except for users with high spatial abilities. However, the studies presented here do not support these conclusions. Çöltekin et al.'s [17] research determined the correctness of memorizing the route on the principle of true/false. Here, the effectiveness of the animated route map was the ratio of the correctly drawn section to the entire length. Like Çöltekin et al. [17], participants' spatial abilities were part of the analysis. However, the participants did not take part in the study under competitive conditions but had the time limit for observing the animation. According to the study presented, this is the main reason why there are no statistically significant differences between a satellite background and an abstract road map in memorizing route effectiveness. Already, Wilkening and Fabrikant [69] noticed in their research that time pressure negatively affects map-based related tasks. The difficulty is an aspect of the animation that is crucial for effectiveness, as has been proven in many recent studies [8,25,39]. Moreover, this study proves that the number of turns of an animated route reduces the effectiveness of memorizing by several percent. This applies to both the road and satellite background.

On the basis of this experiment design, this study found that three different variables (number of turns, route length, and dot speed) impacted task difficulty. It is impossible to say which of these factors was the most influential on effectiveness or efficiency. Despite the isolation of the number of turns, there was still the conjunction of both factors (dot speed and line length), which lacked isolation as individual variables. Therefore, this is the main limitation of this research. The results follow from a group of variables. This might be a bias in efficiency and effectiveness, which resulted in the uniformization of participants' performance. In the examination of reasons behind the effectiveness and efficiency, there is a need for isolation of each of these factors.

Efficiency was related to the time needed to proceed to the drawing stage. Although the study used a fixed time of route presentation, participants sometimes decided not to repeat the animation (12 participants proceeded to the drawing phase watching animation only once). On the other hand, participants had unlimited time for proceeding further after viewing the animation twice. Some of them needed additional time before they were ready to draw the route. It was noticed that some of the participants helped themselves with the mouse cursor and followed the dot during the animation and right after it ended. The presented study determined efficiency based on participants' performance in time.

As in the research by Çöltekin et al. [17], there were no statistically significant differences in efficiency while memorizing a route on a road or satellite background. This seems justified because, in this type of task, where the background context is not of primary importance, the users' attention ignores the background [70]. There was also no relationship between efficiency and effectiveness. Lokka and Çöltekin [41] confirmed that in virtual reality, the most supporting solution for route memorizing is a mixed reality that consists of realistic and abstract objects. The conducted study did not take into account such a variant of maps that would have mixed background.

However, when it comes to the visual behavior of the participants, some similarities, as well as intergroup differences, can be noticed. The noticeable increase in median number fixation was visible only on the map with a road background between the route with three and seven turns. This is also confirmed by the saccadic amplitude, which was relatively the same in all studied cases. It is the basis for the claim that a more realistic cartographic background does not affect the visual behavior of the user in a specific task that is memorizing the animated route. However, there were statistically significant differences in the saccadic amplitude between females and males. It turns out that, in terms of this study, females had slightly longer saccades than males, regardless of the number of turns or the cartographic background. This may indicate female's lower spatial abilities, typical for non-expert users [36,71], although the conducted study does not confirm this. Participants with declared average spatial abilities did indeed have longer saccadic amplitudes, but this was not related to efficiency or effectiveness.

6. Conclusions

Cartographic animation methods are widespread on the Internet. The animation not only concerns the car route, but there are numerous examples of people's migration routes; culture or language migration routes; and money, marathons, and capital flow routes. Many of these cartographic messages reach wide viewers through news portals or social media. In the context of the presented study, animated route visualizations determine specific visual behavior. This behavior is not dependent on the cartographic background nor complexity of the route. The principal part of the observed area is the animated route. The differences in visual behavior are gender-specific. It turns out that females have longer distances between fixations while watching the route. However, this does not translate in any way into the effectiveness or efficiency of memorizing the route. Moreover, participants declaring "very good" spatial abilities also had shorter saccadic amplitudes than those declaring "average". An interesting observation was that the starting position of the animated route was the most observed point, even though most errors were related to estimating the starting position in the drawing phase.

The study also showed that the effectiveness of memorizing the route depends mainly on the difficulty (expressed in the number of turns) and not on the cartographic background, as was suggested by a previous study by Çöltekin et al. [17]. However, the contrast between these conclusions may result from the various environmental approaches. The previous study on this topic validated effectiveness in competition conditions that are responsible for causing negative effects on learning [69,72]. Using more visually realistic images as a background does not translate into higher or lower effectiveness of route memorizing in time-limited conditions. However, it turned out that the easiest tasks, with only three turns, were best solved by people who described their spatial abilities as "good" and "very good". This applies to both the road and satellite background.

In the context of the methodology, it is necessary to note that the presented method is more detailed in determining the effectiveness of memorizing a route than the true/false paradigm [17,73]. Thanks to it, the scoring system includes an implicit way to contain a degree of correctness. However, the problem of automating the route comparison process remains. The user drew a route, and the geometrical comparison of both drawn and actual routes were possible. However, it would be desirable in the future to check the convergence of the drawn route in real time. In addition to the complexity of tasks (based on the number of turns), measures such as average darkness, image compression, or edge detection determine the map complexity [74]. Therefore, differences in the graphic load of road maps with the satellite and vector background may affect the results.

The knowledge gained from this research allows for a better look at the perception of maps. When memorizing the animated route in time-limited and non-competitive conditions, one finds that the cartographic background does not affect visual behavior. It also does not affect the efficiency or effectiveness presented in the context of this study. It turns out that the subjective spatial abilities do not affect effectiveness or efficiency, however, they are related to the visual behavior (in the group of participants with average, good, and very good spatial abilities). However, a particular drawback of the study is the unrepresentativeness of the participants whose spatial abilities are bad or very bad. Moreover, the data supports males' overconfidence position, which is not reflected in performance. Thanks to this knowledge, map designers can choose the cartographic background more consciously, knowing the advantages and disadvantages of their choice. They can also expect a particular reaction from the end-user. The research confirms that the importance of the cartographic background is of secondary importance during the activity related to memorizing the animated route. The recorded eye movement confirms that the route itself is of the greatest importance.

For the future, it is worth continuing research on the subjective opinions of map users. As shown by a few studies (e.g., [75,76]), subjective feelings guide users in choosing mapping applications, which does not always translate into the most effective. Thus, by studying the choices of users, one can improve the design of effective solutions so that they meet the expectations of users, and at the same time maintain their effectiveness. Future studies should also involve characterization of route complexity using ht-index and fractal dimension as an individual factor [77,78]. The form of the route may be a significant variable to alter. There is a difference with the form of the animated route that leaves a trail and a moving point object that leaves no trail. The view of the entire route would be a valuable support for users. On the other hand, the dot-only mode is a more basic option. It occurs, for example, on mini-maps in computer games [79]. Future research studies should incorporate this issue in experimental design. It turns out that finding the starting point of the animation is problematic. Therefore, future research should consider methods of emphasizing the initial position and appearance of the cartographic symbol. Future research should extend the research sample to include people without academic experience, more diverse age structure, or different study backgrounds. Along with various sample groups, further research might consider a selection of preferred backgrounds. This could reveal participants' preferences of cartographic background correlated with spatial abilities.

Funding: This research received no external funding.

Data Availability Statement: The data that support the findings of this study are openly available in "Harvard Dataverse" at https://doi.org/10.7910/DVN/40VX6X (accessed on 2 February 2021).

Acknowledgments: The paper is the result of research on visualization methods carried out within statutory research in the Department of Cartography and Geomatics, Faculty of Geographical and Geological Sciences, Adam Mickiewicz University Poznań, Poland.

Conflicts of Interest: The author declares no conflict of interest.

References

1. DiBiase, D.; MacEachern, A.M.; Krygier, J.B.; Reeves, C. Animation and the role of map design in scientific visualization. *Cartogr. Geogr. Inf. Syst.* **1992**, *19*, 201–214. [CrossRef]
2. Andrienko, G.; Andrienko, N.; Demsar, U.; Dransch, D.; Dykes, J.; Fabrikant, S.I.; Jern, M.; Kraak, M.-J.; Schumann, H.; Tominski, C. Space, time and visual analytics. *Int. J. Geogr. Inf. Sci.* **2010**, *24*, 1577–1600. [CrossRef]
3. Maggi, S.; Fabrikant, S.I.; Imbert, J.-P.; Hurter, C. How do display design and user characteristics matter in animations? An empirical study with air traffic control displays. *Cartogr. Int. J. Geogr. Inf. Geovis.* **2016**, *51*, 25–37. [CrossRef]
4. See, L.-C.; Chang, Y.-H.; Chuang, K.-L.; Lai, H.-R.; Peng, P.-I.; Jean, W.-C.; Wang, C.-H. Animation program used to encourage patients or family members to take an active role for eliminating wrong-site, wrong-person, wrong-procedure surgeries: Preliminary evaluation. *Int. J. Surg.* **2011**, *9*, 241–247. [CrossRef] [PubMed]
5. Edler, D.; Dickmann, F. The Impact of 1980s and 1990s Video Games on Multimedia Cartography. *Cartogr. Int. J. Geogr. Inf. Geovis.* **2017**, *52*, 168–177. [CrossRef]
6. Medyńska-Gulij, B.; Zagata, K. Expert and Gamers on Immersion into Reconstructed Strongholds. *ISPRS Int. J. Geoinf.* **2020**, *9*, 655. [CrossRef]
7. Spallone, R. In the Space and in the Time. Representing Architectural Ideas by Digital Animation. In Proceedings of the International and Interdisciplinary Conference IMMAGINI, Brixen, Italy, 27–28 November 2017; Volume 1, pp. 27–28, 962. [CrossRef]
8. Harrower, M. The Cognitive Limits of Animated Maps. *Cartogr. Int. J. Geogr. Inf. Geovis.* **2007**, *42*, 349–357. [CrossRef]
9. Vozenilek, V.; Kralik, T. Anaglyph videoanimations from oblique stereoimages. In Proceedings of the Sixth International Conference on Graphic and Image Processing 2015, China, Beijing, 24–26 October 2014. [CrossRef]
10. Garlandini, S.; Fabrikant, S.I. Evaluating the effectiveness and efficiency of visual variables for geographic information visualization. In *Spatial Information Theory. COSIT 2009, Aber Wrac'h, France, 21–25 September 2009*; Horsby, K.S., Clarmunt, C., Denis, M., Ligozat, G., Eds.; Springer: Berlin/Heidelberg, Germany, 2009; pp. 195–211.
11. Miller, G The Magical Number Seven, Plus or Minus Two: Some Limits on Our Capacity for Processing Information. *Psychol. Rev.* **1956**, *63*, 81–97. [CrossRef]
12. Farrell, M.J.; Arnold, P.; Pettifer, S.; Adams, J.; Graham, T.; MacManamon, M. Transfer of route learning from virtual to real environments. *J. Exp. Psychol. Appl.* **2003**, *9*, 219–227. [CrossRef]
13. Klippel, A.; Tappe, H.; Habel, C. Pictorial representations of routes: Chunking route segments during comprehension. In *Spatial Cognition*; Freksa, C., Brauer, W., Habel, C., Wender, K., Eds.; Springer: Berlin/Heidelberg, Germany, 2002; pp. 11–33. [CrossRef]

14. Midtbø, T.; Larsen, E. Map animation versus static maps—When is one of them better. In Proceedings of the Joint ICA Commissions Seminar on Internet-Based Cartographic Teaching and Learning, Madrid, Spain, 6–8 July 2005; pp. 155–160.
15. Nossum, A.S. Semistatic Animation—Integrating Past, Present and Future in Map Animations. *Cartogr. J.* **2012**, *49*, 43–54. [CrossRef]
16. Kraak, M.-J. *Mapping Time: Illustrated by Minard's Map of Napoleon's Russian Campaign of 1812*; Esri Press: Redlands, CA, USA, 2014.
17. Çöltekin, A.; Francelet, R.; Richter, K.-F.; Thoresen, J.; Fabrikant, S.I. The effects of visual realism, spatial abilities, and competition on performance in map-based route learning in men. *Cartogr. Geogr. Inf. Sci.* **2018**, *45*, 339–353. [CrossRef]
18. Mitchell, C.M.; Miller, R.A. Design strategies for computer-based information displays in real-time control systems. *Hum. Factors J. Hum. Factors Ergon. Soc.* **1983**, *25*, 359–369. [CrossRef]
19. Medyńska-Gulij, B. The effect of cartographic content on tourist map users. *Cartography* **2003**, *32*, 49–54. [CrossRef]
20. Hátlová, K.; Hanus, M. A Systematic Review into Factors Influencing Sketch Map Quality. *ISPRS Int. J. Geoinf.* **2020**, *9*, 217. [CrossRef]
21. Del Fatto, V.; Paolino, L.; Sebillo, M.; Vitiello, G. Spatial factors affecting user's perception in map simplification: An empirical analysis. In *Web and Wireless Geographical Information Systems*; Bertolotto, M., Ray, C., Li, X., Eds.; Springer: Berlin/Heidelberg, Germany, 2008; pp. 152–163. [CrossRef]
22. Cybulski, P.; Horbiński, T. User Experience in Using Graphical User Interfaces of Web Maps. *ISPRS Int. J. Geoinf.* **2020**, *9*, 412. [CrossRef]
23. Zanola, S.; Fabrikant, S.I.; Çöltekin, A. The effect of realism on the confidence in spatial data quality. In Proceedings of the International Cartographic Conference ICC, Chile, Santiago, 15–21 November 2009.
24. Wielebski, Ł.; Medyńska-Gulij, B. Graphically supported evaluation of mapping techniques used in presenting spatial accessibility. *Cartogr. Geogr. Inf. Sci.* **2019**, *46*, 311–333. [CrossRef]
25. Keil, J.; Mocnik, F.-B.; Edler, D.; Dickmann, F.; Kuchinke, L. Reduction of Map Information Regulates Visual Attention without Affecting Route Recognition Performance. *ISPRS Int. J. Geoinf.* **2018**, *7*, 469. [CrossRef]
26. Çöltekin, A.; Fabrikant, S.I.; Lacayo, M. Exploring the efficiency of users' visual analytics strategies based on sequence analysis of eye movement recordings. *Int. J. Geogr. Inf. Sci.* **2010**, *24*, 1559–1575. [CrossRef]
27. Korycka-Skorupa, J.; Gołębiowska, I. Numbers on Thematic Maps: Helpful Simplicity or Too Raw to Be Useful for Map Reading? *ISPRS Int. J. Geoinf.* **2020**, *9*, 415. [CrossRef]
28. Kiefer, P.; Giannopoulos, I.; Raubal, M.; Duchowski, A. Eye tracking for spatial research: Cognition, computation, challenges. *Spat. Cogn. Comput.* **2017**, *17*, 1–19. [CrossRef]
29. Fisher, B.; Ramsperger, E. Human express saccades: Extremely short reaction times of goal directed eye movements. *Exp. Brain Res.* **1984**, *51*, 191–195. [CrossRef]
30. Lisberger, S.G. Visual guidance of smooth pursuit eye movements: Sensation, action, and what happens in between. *Neuron* **2010**, *66*, 477–491. [CrossRef]
31. Pola, J. Models of the saccadic and smooth pursuit systems. In *Models of the Visual System*; Topics in Biomedical Engineering International Book Series; Hung, G.K., Ciuffreda, K.J., Eds.; Springer: Boston, MA, USA, 2002; pp. 385–429. [CrossRef]
32. Wilkniss, S.M.; Jones, M.G.; Korol, D.L.; Gold, P.E.; Manning, C.A. Age-related differences in an ecologically based study of route learning. *Psychol. Aging* **1997**, *12*, 372–375. [CrossRef]
33. Fish, C. Cartographic challenges in animated mapping. In Proceedings of the International Cartographic Conference Pre-Conference Workshop on Envisioning the Future of Cartographic Research, Brazil, Curitiba, 21 August 2015.
34. Opach, T.; Gołębiowska, I.; Fabrikant, S.I. How Do People View Multi-Component Animated Maps? *Cartogr. J.* **2014**, *51*, 330–342. [CrossRef]
35. Burian, J.; Popelka, S.; Beitlova, M. Evaluation of the Cartographic Quality of Urban Plans by Eye-Tracking. *ISPRS Int. J. Geoinf.* **2018**, *7*, 192. [CrossRef]
36. Thorndyke, P.W.; Hayes-Roth, B. Differences in spatial knowledge acquired form maps and navigation. *Cognit. Psychol.* **1982**, *14*, 560–589. [CrossRef]
37. Tom, A.; Denis, M. Language and spatial cognition: Comparing the roles of landmarks and street names in route instructions. *Appl. Cognit. Psychol.* **2004**, *18*, 1213–1230. [CrossRef]
38. Tom, A.; Tversky, B. Remembering Routes: Streets and Landmarks. *Appl. Cognit. Psychol.* **2012**, *26*, 182–193. [CrossRef]
39. Cybulski, P.; Medyńska-Gulij, B. Cartographic Redundancy in Reducing Change Blindness in Detecting Extreme Values in Spatio-Temporal Maps. *ISPRS Int. J. Geoinf.* **2018**, *7*, 8. [CrossRef]
40. Brucker, B.; Scheiter, K.; Gerjets, P. Learning with dynamic and static visualizations: Realistic details only benefit learners with high visuospatial abilities. *Comput. Hum. Behav.* **2014**, *36*, 330–339. [CrossRef]
41. Lokka, I.E.; Çöltekin, A. Toward optimizing the design of virtual environments for route learning: Empirically assessing the effects of changing levels of realism on memory. *Int. J. Digit. Earth* **2019**, *12*, 137–155. [CrossRef]
42. Höffler, T.N.; Leutner, D. Instructional animation versus static pictures: A meta-analysis. *Learn. Instr.* **2007**, *17*, 722–738. [CrossRef]
43. Lai, P.; Yeh, A. Assessing the Effectiveness of Dynamic Symbols in Cartographic Communication. *Cartogr. J.* **2004**, *41*, 229–244. [CrossRef]
44. Dong, W.; Ran, J.; Wang, J. Effectiveness and Efficiency of Map Symbols for Dynamic Geographic Information Visualization. *Cartogr. Geogr. Inf. Sci.* **2012**, *39*, 98–106. [CrossRef]

45. Cybulski, P.; Wielebski, Ł. Effectiveness of dynamic point symbols in quantitative mapping. *Cartogr. J.* **2019**, *56*, 146–160. [CrossRef]
46. Campbell, C.S.; Egbert, S.L. Animated Cartography: Thirty Years of Scratching the Surface. *Cartogr. Int. J. Geogr. Inf. Geovis.* **1990**, *27*, 24–46. [CrossRef]
47. MacEachren, A.; Kraak, M.-J. Research Challenges in Geovisualization. *Cartogr. Geogr. Inf. Sci.* **2001**, *28*, 3–12. [CrossRef]
48. Brychtova, A.; Popelka, S.; Dobesova, Z. Eye-Tracking methods for investigation of cartographic principles. In Proceedings of the 12th International Multidisciplinary Scientific GeoConference and EXPO, Varna, Bulgaria, 17–23 June 2012; pp. 1041–1048. [CrossRef]
49. Dong, W.; Liao, H.; Xu, F.; Liu, Z.; Zhang, S. Using eye tracking to evaluate the usability of animated maps. *Sci. Chin. Earth Sci.* **2014**, *57*, 512–522. [CrossRef]
50. Krassanakis, V.; Filippakopoulou, V.; Nakos, B. Detection of moving point symbols on cartographic backgrounds. *J. Eye Mov. Res.* **2016**, *9*, 1–16. [CrossRef]
51. Çöltekin, A.; Brychtová, A.; Griffin, A.L.; Robinson, A.C.; Imhof, M.; Pettit, C. Perceptual complexity of soil-landscape maps: A user evaluation of color organization in legend designs using eye tracking. *Int. J. Digit. Earth* **2016**, *10*, 560–581. [CrossRef]
52. Dong, W.; Jian, Y.; Zheng, L.; Liu, B.; Meng, L. Assessing Map-Reading Skills Using Eye Tracking and Bayesian Structural Equation Modelling. *Sustainability* **2018**, *10*, 3050. [CrossRef]
53. Ooms, K.; Krassanakis, V. Measuring the spatial noise of a low cost eye tracker to enhance fixation detection. *J. Imaging* **2018**, *4*, 96. [CrossRef]
54. Krassanakis, V.; Cybulski, P. A review on eye movement analysis in map reading process: The status of the last decade. *Geod. Cartogr.* **2019**, *68*, 191–209. [CrossRef]
55. Peterson, M.P. Active legends for interactive cartographic animation. *Int. J. Geogr. Inf. Sci.* **2010**, *13*, 375–383. [CrossRef]
56. Fish, C.; Goldsberry, K.P.; Battersby, S. Change Blindness in Animated Choropleth Maps: An Empirical Study. *Cartogr. Geogr. Inf. Sci.* **2011**, *38*, 350–362. [CrossRef]
57. Wang, J.; Li, Z. Effectiveness of visual, screen and dynamic variables in animated mapping. In Proceedings of the 25th International Cartographic Conference, Paris, France, 3–8 July 2011.
58. Stachoň, Z.; Šašinka, Č.; Čeněk, J.; Štěrba, Z.; Angsuesser, S.; Fabrikant, S.I.; Štampach, R.; Morong, K. Cross-cultural differences in figure–ground perception of cartographic stimuli. *Cartogr. Geogr. Inf. Sci.* **2018**, *46*, 82–94. [CrossRef]
59. Bestgen, A.-K.; Edler, D.; Müller, C.; Schulze, P.; Dickmann, F.; Kuchinke, L. Where Is It (in the Map)? Recall and Recognition of Spatial Information. *Cartogr. Int. J. Geogr. Inf. Geovis.* **2017**, *52*, 80–97. [CrossRef]
60. Edler, D.; Keil, J.; Bestgen, A.-K.; Kuchinke, L.; Dickmann, F. Hexagonal map grids—An experimental study on the performance in memory of object location. *Cartogr. Geogr. Inf. Sci.* **2019**, *46*, 401–411. [CrossRef]
61. Olsen, A. Identifying parameter values for an I-VT fixation filter suitable for handling data sampled with various sampling frequencies. In Proceedings of the ETRA '12: Proceedings of the Symposium on Eye Tracking Research and Applications, Santa Barbara, CA, USA, 28–30 March 2012; pp. 317–320. [CrossRef]
62. Kiefer, P.; Giannopoulos, I. Gaze map matching: Mapping eye tracking data to geographic vector features. In Proceedings of the 20th International Conference on Advances in Geographic Information Systems, New York, NY, USA, 6–9 November 2012; pp. 359–368. [CrossRef]
63. Dong, W.; Wang, S.; Chen, Y.; Meng, L. Using Eye Tracking to Evaluate the Usability of Flow Maps. *ISPRS Int. J. Geoinf.* **2018**, *7*, 281. [CrossRef]
64. Just, M.A.; Carpenter, P. Eye fixations and cognitive processes. *Cognit. Psychol.* **1976**, *8*, 441–480. [CrossRef]
65. Duchowski, A.T. *Eye Tracking Methodology: Theory and Practice*, 3rd ed.; Springer: Berlin/Heidelberg, Germany, 2007.
66. Mann, H.B.; Whitney, D.R. On a Test of Whether on of Two Random Variables is Stochastically Larger than the Other. *Ann. Math. Stat.* **1947**, *18*, 50–60. [CrossRef]
67. Kruskal, W.H.; Wallis, A. Use of Ranks in One-Criterion Variance Analysis. *J. Am. Stat. Assoc.* **1952**, *47*, 583–621. [CrossRef]
68. Erkelens, J.C. Coordination of smooth pursuit and saccades. *Vis. Res.* **2006**, *46*, 163–170. [CrossRef] [PubMed]
69. Wilkening, J.; Fabrikant, S.I. How users interact with a 3D geo-browser under time pressure. *Cartogr. Geogr. Inf. Sci.* **2013**, *40*, 40–52. [CrossRef]
70. Rensink, R.; O'Regan, J.K.; Clark, J.J. To see or not to see: The need for attention to perceive changes in scenes. *Psychol. Sci.* **1997**, *8*, 368–373. [CrossRef]
71. Ooms, K.; De Maeyer, F.; Fack, V.; Van Assche, E.; Witlox, F. Interpreting maps through the eyes of expert and novice users. *Int. J. Geogr. Inf. Sci.* **2012**, *26*, 1773–1788. [CrossRef]
72. Murayama, K.; Elliot, A.J. The competition–performance relation: A meta-analytic review and test of the opposing processes model of competition and performance. *Psychol. Bull.* **2012**, *138*, 1035–1070. [CrossRef] [PubMed]
73. Thoresen, J.C.; Francelet, R.; Çöltekin, A.; Richterm, K.-F.; Fabrikant, S.I.; Sandi, C. Not all anxious individuals get lost: Trait anxiety and mental rotation ability interact to explain performance in map-based route learning in men. *Neurobiol. Learn. Mem.* **2016**, *132*, 1–8. [CrossRef] [PubMed]
74. Barvir, R.; Vozenilek, V. Developing Versatile Graphic Map Load Metrics. *ISPRS Int. J. Geoinf.* **2020**, *9*, 705. [CrossRef]
75. Gołębiowska, I.; Opach, T.; Rød, J.K. For your eyes only? Evaluating a coordinated and multiple views tool with a map, a parallel coordinated plot and a table using an eye-tracking approach. *Int. J. Geogr. Inf. Sci.* **2016**, *31*, 237–252. [CrossRef]

76. Medyńska-Gulij, B.; Wielebski, Ł.; Halik, Ł.; Smaczyński, M. Complexity Level of People Gathering Presentation on an Animated Map—Objective Effectiveness Versus Expert Opinion. *ISPRS Int. J. Geoinf.* **2020**, *9*, 117. [CrossRef]
77. Jiang, B. The Fractal Nature of Maps and Mapping. *Int. J. Geogr. Inf. Sci.* **2015**, *29*, 159–174. [CrossRef]
78. Ma, L.; Zhang, H.; Lu, M. Building's fractal dimension trend and its application in visual complexity map. *Build. Environ.* **2020**, *178*. [CrossRef]
79. Zagata, K.; Gulij, J.; Halik, Ł.; Medyńska-Gulij, B. Mini-Map for Gamers Who Walk and Teleport in a Virtual Stronghold. *ISPRS Int. J. Geoinf.* **2021**, *10*, 96. [CrossRef]

International Journal of
Geo-Information

Article

Interactive Web-Map of the European Freeway Junction A1/A4 Development with the Use of Archival Cartographic Sources

Dariusz Lorek * and **Tymoteusz Horbiński**

Department of Cartography and Geomatics, Faculty of Geographical and Geological Sciences, Adam Mickiewicz University, 61-712 Poznań, Poland; tymoteusz.horbinski@amu.edu.pl
* Correspondence: kubal@amu.edu.pl; Tel.: +48-61-829-6249

Received: 24 June 2020; Accepted: 9 July 2020; Published: 14 July 2020

Abstract: In the article, authors have analyzed cartographic materials presenting the spatial development of Gliwice with the use of multimedia tools. The materials prove that this area has played an important part in the road system of the region, country and even part of Europe since the 19th century. The six maps from the studied area were analyzed e.g., the Urmesstischblätter map, polish topographic maps, and the OpenStreetMap. Based on these maps and their legends, vectorization of the main roads of the analyzed area was carried out. The evolution of the main road corridors on the six maps was analyzed with respect to the location of the European freeway junction (A1/A4), constituting a basis for the web map. According to the authors, the use of the interactive web map is the most comprehensive method of all technologies used by modern cartography. Spatial data collected from different cartographic publications (from the first half of the 19th century till the present) consider the most significant aspects of changes in the road network of the analyzed area in a detailed and user-friendly way.

Keywords: freeway junction; topographic maps from the 19th century; interactive web map; Leaflet; GeoJSON

1. Introduction

Topographic maps include the general geographic information, which determines their huge informational potential as far as the comprehensive study of changes in the environment and spatial structures is concerned [1]. The symbols standing for major roads were included in the legends of such maps, the editions from both the 19th and 20th centuries, at the very beginning. Major roads constitute a constructional basis for the diversely used land, marking its structure. Moreover, the road system, with its density or differences in quality, reflect the stage of development of a given area. Successive map editions demonstrate the evolution of road networks and typical changes in land use forms. The information included in historic maps constitutes a valuable source of information helpful in reconstructing historic states of space both in terms of individual objects and phenomena and entire areas defined as integrated settlement structures, thoroughfares, and land use forms [2].

The analysis of cartographic materials presenting the spatial development of Gliwice proves that this area has played an important part in the road system of the region, country and even part of Europe since 19th century [3]. Obtaining and visualizing spatial data concerning major roads and how they crossed over a span of nearly two hundred years is significant in order to be able to demonstrate changes in space use. The roads, mainly the 'Gliwice Sośnica' freeway junction connecting international east–west and north–south corridors, significantly increased the importance of the area (A1 and A4 highway).

Modern cartographic forms, such as multimedia cartography, offer a wide range of opportunities to present and compile spatial data, i.e., AR (Augmented Reality) and VR (Virtual Reality) [4–6], data from the low flight level [7] or interactive web maps [8–10]. The aim of the research described in this article was to verify the hypothesis of whether or not it is possible to demonstrate changes occurring in space and to reconstruct historic states of space, using modern multimedia technologies and cartographic sources that present the network of the most relevant roads of the region from the mid-19th century to the present day.

2. Methodology

2.1. Study Area

The study concerns the county of Gliwice located in southern Poland (Figure 1) which has constituted an important industrial center based on coal and ore mining and processing for over 150 years. Spatial development of the region was strictly linked with the processes of industrialization and urbanization that were constantly progressing since the mid-19th century, manifesting themselves mainly through buildings (houses, housing estates, mines, plants and factories) and road networks (roads, railroads, river/inland transport). The fact that the old city was located upon a river that played a significant part in transport was an important advantage of the city, increasing its role in the region [11,12]. Currently, despite economic transformations and the decrease in the role of the industry in the late '80s, the region has retained its character and form of development shaped by previous decades. At the same time, one can observe some changes in land use resulting from current urbanization processes.

Figure 1. Freeway junction A1/A4 in Gliwice (Poland).

The development of road networks is one of the current symptoms of the interference in spatial structures of Gliwice. Two international corridors, the A1 and A4 highways, have become particularly relevant. The A4 has been visibly moved away from the center of Gliwice to the south, to the territory not so highly developed (including rural areas). The course of A1 is more complex, as the highway meanders between highly developed areas (Figure 1), which are connected with the specificity of the region, where, since the mid-19th century, the borders between neighboring towns and formerly independent settlements have been blurred as a result of the significant development of industry and settlements [12,13].

Apart from the city area, an extra buffer zone of 1 km exists, resulting from the location of the freeway junction near the border with Gierałtowice, meaning that certain freeway exits are situated fragmentarily out of the borders of Gliwice. Moreover, on the stretch of several kilometers, the A4 highway runs in the axis of the area border. The buffer adopted allowed one to present and analyze the studied phenomenon in a complex way.

2.2. Source Materials

When collecting source material for the research, two main criteria were taken into consideration: the scale of the map and the period that the map presented. In terms of accuracy, the scale of 1:25,000 was adopted, as it allowed one to distinguish between several categories of roads, at the same time presenting the state of space. In 19th century Gliwice was under Prussian partition, hence, it was necessary to run a search query of the cartographic resources from Polish and German archives and libraries. Then, it was possible to collect maps of the selected area in the adopted scale.

The period represented on the map constituted the second criteria adopted in the process of collecting materials. Particular states are separated by time intervals of 30–50 years. It occurred for a few collections that some specific sections, which were parts of the current city area, differed by several years, according to the dates on sheets. The maps from late '20s constituted the greatest problem, as not all sheets depicting the current area of Gliwice were prepared in accordance with land borders of that time. In such cases the Prussian topographic map of 1:25,000, published at that time, was used.

Furthermore, all the maps collected were included in the publishing series that covered the entire country (Figure 2). Having made the aforementioned assumption, one could select 6 map collections, starting from the second decade of 19th century, as the maps from that period constituted a significant step in making cartometric maps (based on constantly developed geodetic surveys and improved surveying instruments).

2.2.1. Urmesstischblätter Map

The *Urmesstischblätter* map, prepared at the scale of 1:25,000, depicts the state of space since the 1820s. It was created as a result of the demand for a unified and accurate map of the entire Prussia after the decisions of the Congress of Vienna that introduced a new division of the European countries. To meet the demand, the network of triangulation stations was created. One could connect with the stations during field work by means of plane table surveying. Individual sheets were prepared according to instructions by units of soldiers and commanding officers, whose names were included at the bottom of each sheet [1]. In the 19th century, the area of Gliwice was under Prussian rule and such a situation continued till 1918. At that time, the Prussian topographic maps *Urmesstischblätter* and *Messtischblätter* were created. As Germany was united in the second half of the 19th century, the aforementioned maps presently constitute the heritage of the Federal Republic of Germany.

The map was prepared in Müffling polyhedric projection (*Preussische Polyederprojektion).* For each sheet geographic coordinates in relation to Ferro Meridian were determined, adopting the sheet size of 10′ (longitude) and 6′ (latitude) [14,15]. The popular 19th century method of hachuring that allowed one to provide slope angles was applied to present the landform.

It is a manuscript map whose entire collection is a part of the Berlin State Library collection [16]. The sections for the area analyzed come from 1827 and 1828. The area in question consisted of 4 sheets with the following numbers: 3307, 3308, 3351 and 3352. There was no legend on the sheets, however, the legend was published separately in 1818 with detailed clarifications of multicolored symbols indicating a high informational potential of the map [17].

2.2.2. Messtischblätter Map

The 19th century progress in surveying methods and instruments was used by Prussian authorities in 1876 to prepare another topographic map. *Messtischblätter* maps, as a continuation of the previous map series, were also made at the scale of 1:25,000. A new series of surveys and adopted technical solutions allowed one to work out cartometric maps [10,18]. According to the name of the collection translated from German (messen—to measure/survey, Tisch—table), plane table surveying was still the basic method in use, however, more accurate instruments, such as a drawing board mounted on a tripod with a stadia, were also applied.

Maps were published by the method of printing (lithography), but in comparison with *Urmesstischblätter* the use of color was limited at that time. Monochromatic or dichromatic sheets were produced, with the blue color for marking surface water [19]. Maps were published until the '30 s and were updated a few times at that time. A reference to the Prime Meridian in Greenwich occurred in the description of coordinates on the sheet.

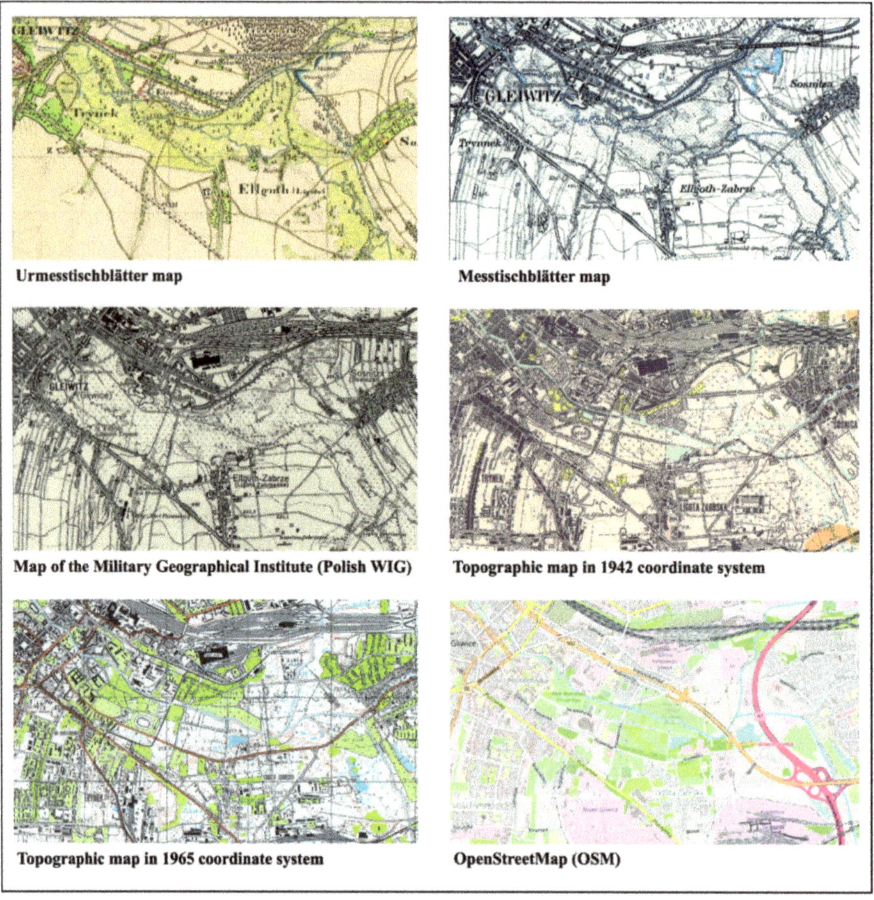

Figure 2. Source materials for the six periods.

Messtischblätter maps retained the same sheet size and division into sections as the first series of maps from the early 19th century. Hence, the area of Gliwice occupied the same four sheets, however, the new numeration was introduced: 5677, 5678, 5777, 5778. The sheets presenting the state from the 1880s were mainly used in the research. What is more, the later *Messtischblätter* editions were used to complete lacking Polish maps from the late '20s. Some of the sheets have basic legends located beyond the frame and in some cases the sheets covering the area of Poland have descriptions in two languages (German and Polish).

Two collections of the Prussian topographic maps used in the research share some common features, but also differ in some aspects. These are: the time of production, as *Urmesstischblätter* maps present the state from the first half of the 19th century, whereas *Messtischblätter* maps emerged in 1876. Both map types were based on geodetic surveys (on previously measured triangulation networks), however, inaccurate methods were still used on older maps. Thus, *Messtischblätter* maps showed higher measurement precision. The change in presenting the landform, apart from changes in the printing method and the use of colors, was also important. At the beginning, slopes of the land were presented through the method of hachuring, as contour lines were introduced in the second half of the 19th century. However, one should focus on the elements common to both types of maps, namely the same scale and invariable division into sheets. Maps were also characterized by a particular way of coding space used in Prussian topography. Separate legends for both series were produced and published in several editions.

2.2.3. Map of the Military Geographical Institute (Polish Wojskowy Instytut Geograficzny = WIG)

The late 1920s was another time period studied by the researchers. At that time, the Military Geographical Institute (English for: *Wojskowy Instytut Geograficzny = WIG*) was the institution responsible for the production and publication of maps. The institution was established after Poland had regained independence in 1918. Initially, its role was to collect maps produced at the time of partitions and copy them for the army. Then, the realization of triangulation work and the preparation of the new concept of topographic maps, used later as a basis for detailed maps made at the scale of 1:25,000, began. Maps were intended for the army, but they were also of great importance to economic and social needs [20].

The map was distinguished by the new system of topographic symbols (188 separations and 70 literal abbreviations). The map shared some common features with the Prussian *Messtischblätter* map in terms of the way of coding. The "Borowa Góra" coordinate system, along with a quasi-stereographic *WIG* projection, was adopted for the scale discussed. Those were monochromatic maps that sometimes used the brown color for the land form. The map produced by means of lithographic printing covered approximately 50% of the country's area [20].

According to indices, the modern area of Gliwice constitutes four sections of the map. Only two sheets covering the south-eastern and south-western part, made, respectively, in 1926 and 1933, were prepared. The remaining part of the area was a part of Germany at that time, therefore, two sheets presenting the state from 1928 and 1929 were completed with the *Messtischblätter* map. The WIG map had the basic legend with separations for roads, railroads and forests beyond the frame at the bottom of the sheet (Figure 3).

2.2.4. The Topographic Map in the 1942 Coordinate System

After WW2, when Poland became a zone under the strong influence of the Soviet Union, cartography was divided into the civil and the military. Furthermore, maps were censored and access to them was strictly limited. In 1952, the Conference of Geodetic Service of USSR and People's Republic Countries was held. At the conference, the guidelines on map production, such as the 1942 coordinate system based on the Krasowski ellipsoid, and the division into sheets according to the International Map of the World 1:1,000,000, were adopted. States that participated in the conference, including Poland, were obliged to prepare a basic topographic map at the scale of 1:25,000 [21].

Maps used in the research, made according to the Gauss–Krüger coordinate system, come from the '60 s (EPSG 3330). They were published to meet economic needs and lacked coordinates. Their content, compared to military editions, was also limited and, in terms of topographic objects and land form, masked by some tricks (the information about confidentiality of the data was placed on the map's frame). Only after 1990 were confidentiality regulations lifted, and civil maps were finally published in this coordinate system [20].

The area researched encompasses two sheets (number 3 and 5) of the former Gliwice county. The maps lacked the information about the system of the symbols used (the legend), however, as the content was limited, and one could distinguish the quality and order of the material presented [22].

2.2.5. The Topographic Map in the 1965 Coordinate System

The system adopted in 1968 was a result of the fact that the use of the 1942 system had been limited in civil cartography. Observing the global situation, i.e., the development of satellite programs, and running operations during the Cold War, the authorities in Moscow ordered the Eastern Bloc countries to prepare an independent system for producing economic maps. The above circumstances resulted in the preparation and publication of topographic maps at the scale of 1:25,000 and 1:50,000 since the late 1970 [21].

The Krasowski ellipsoid was the basic point of reference, but the system was not uniform for the entire country, as zones 1–4 were distinguished (prepared in the quasi-stereographic projection), and zone 5 was distinguished as well (made in the Gauss–Krüger coordinate system). The sheets had a different cut and division into sections in relation to the 1942 system. Moreover, maps in the 1965 system did not have the cartographic grid, so the topographic grid became the basis for the division into sections [20,21].

The current area of Gliwice is depicted on two map sheets in the 1965 system with 531.11 and 521.34 emblems in zone 5 (EPSG 2175), demonstrating the state of space from 1986. Furthermore, the limited content of the map was still censored and, in terms of graphic parameters, it was similar to military maps. An extended legend was placed on each sheet.

2.2.6. OpenStreetMap (OSM)

Originally a project of the Internet community aimed at creating a free map of the entire globe since 2004, it was adopted as the updated version to be used [23] (EPSG 3857). The OSM base map, as a topographic map, was used to create this article because it was up-to-date and compatible with the web technology used in the web programming stage. The service provides many opportunities, such as visible legend (the "i" button on the website). Downloadable OSM data are saved in a PostgreSQL relational database, without spatial extensions, in the WGS84 coordinate system. Roads constitute 28% of the data [24]. It is possible to display data in the 21-level zoom. The largest available scale: 1:500 (zoom 20), the smallest: 1:500,000,000 (zoom 0).

2.3. *Selection of Objects from the Legend*

Given that map series have several editions with symbol explanations, the query of maps and their legends in terms of the aspects studied (major roads) made the authors adopt basic legend editions (Figure 3). The authors adopted the explanations of symbols from map marginalia (if they occurred, e.g., WIG maps, the 1965 system) or, for *Urmesstischblätter* maps, the first edition of the legend, as it was adequate for the content of the sheets collected. Furthermore, the later, usually more comprehensive editions of legends did not demonstrate any changes in the way major road categories were distinguished. *Messtischblätter* maps' legends included highways in later editions.

In the research, legends constituted a starting point and a point of reference that allowed the authors to identify given road categories on maps. However, it was the analysis of every single sheet that provided the information about the category of roads that occurred there. On the maps analyzed, the category of the same road stretches would frequently change, depending on the time period the map came from. It could result from different rules of categorization or technical guidelines used for making maps, as well as from road rebuilding/expansion. Using legends, the authors followed the changes in the evolution of major road corridors in Gliwice.

Figure 3 presents some fragments of legends that show the ways of presenting roads of the highest category for the six maps analyzed. All of the legends depict roads, starting from the highest category. One can observe a similar way of coding these objects in the editions researched, namely the use of bold black contour on both sides of the line. Additionally, in full-color editions, the extra filling color was added (e.g., *Urmesstischblätter*, *OSM*). Explanations of symbols for the topographic map in the 1942 coordinate system constituted an exception among the materials collected. They came from the edition that was published many years after the map, as earlier legends, reserved for military map publications, had not been available.

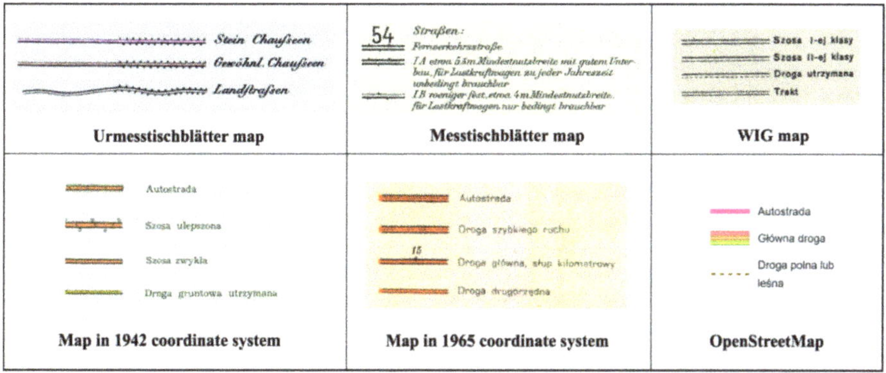

Figure 3. Ways of presenting the highest category of roads in the legends of the six maps.

2.4. Gathering and Processing Data

After six time periods demonstrating spatial structures with the road network of the city had been selected for further processing, the authors applied the scheme they had developed in previous studies [10]. In this stage of research, the Qgis 3.12 program was used. Maps were registered to the current WGS-84 coordinate system (EPSG 4326). The registration was made by means of affine transformation based on 4 ground control points [3]. Due to incomplete cartometricity, the georeferencing of the oldest map required particular care. In order to achieve optimum results (low RMSE with the appropriate proportion of a sheet), the larger number of ground control points was adopted. Selected fragments of *Urmesstischblätter* sheets that were located in Gliwice or nearby were also registered. The registration of *Urmesstischblätter* maps was the subject of the previous research [17,25].

Then, according to the scheme adopted, stretches of roads of the highest category, occurring on given sheets, were vectorized. On the OSM map, suitable fragments of linear objects were selected. Vectorization of roads according to the trace was carried out, starting from more recent to older maps and highlighting objects previously selected on the basis of legends. Roads from each period were saved in separate layers in the .shp format and finally six levels presenting the system of major roads in given time periods were gained [10].

2.5. Web Map Programming

Roads representing the evolution of European road corridors (the A1 and A4 highways) for the six maps analyzed, prepared in the process of vectorization, constituted a basis for the web map. In the process of programming the interactive web map, the Leaflet library was used. As a JavaScript library, Leaflet is efficient, useful and open-source [9,10,26]. It was also widely used in similar map versions.

Leaflet supports the following extensions: GeoJSON, TopoJSON and KML, therefore, vectors of roads saved as .shp extension files (and other, i.e., .dbf, .shx, .cpg, .prj) had to be converted to GeoJSON files [27]. According to Horbiński and Lorek [10], it is recommended to use this extension for files of small size and representing a relatively small area. Furthermore, GeoJSON is a frequently adopted standard in preparing layers of interactive web maps [28]. Necessary conversion, like in previous stages, was done in QGIS 3.12.

Having converted layers, researchers had to create the interactive web map on the basis of the previous assumptions [29]. According to the assumptions, the interactive web map presenting the evolution of major roads in Gliwice and construction of the junction of the European corridors (the A1 and A4 highways) is going to consist of the main map, overview map, layers and scale (Figure 4). In addition to the layers presenting the state of the highest class roads, the authors added layers of the boundaries of the Gliwice poviat and layers with the most important road intersections in a specific period of time.

To design the interactive web map (main map), the basic functions of Leaflet and two plugins were used (Leaflet Panel Layers [30] and leaflet-graphicscale [31]). The first one allowed the user to switch road layers on and off in the six stages analyzed. The other one represented map scales graphically. The overview map is an additional element. The overview map is an interactive map presenting the entire course of the roads. Created as a separate object, it does not have any links with the main map. The interactive web map (Figure 4) is available here: http://kartografia.amu.edu.pl/Motorway/index.html, on the server of Adam Mickiewicz University in Poznań.

Figure 4. Interactive web map of the development of the European freeway junction A1/A4.

3. Results and Discussion

Over a span of nearly two centuries, one can observe changes in the way major roads run. On the interactive web map, the occurrence of the junctions of these roads in given time periods were marked with the sight, starting with the situation when they connected in the center of Gliwice (the state from 1828). According to Antrop [32], it was one of the models typical of the settlement network of that time. The situation changed on the maps published nearly 100 years later. A crossing of major roads occurred in areas moved northwards from the city center. The change was linked with the urbanization and suburbanization process and intensified traffic that forced the construction of bypasses and moving the traffic away from the center [12]. A construction of the freeway on the outskirts of the city and moving the main road junction to the south-east was the next stage.

The level of industrialization and urbanization of the city and its neighborhood in given time periods determined the expansion of road network, including thoroughfares of the highest category [33]. The current state of the road network shows that mechanisms pushing the development were not only of the bottom-up character and were not resulting only from local needs in terms of economic development [34]. Considering this part of Europe, it becomes clearly visible that the city of Gliwice is located on the crossing of two corridors that were developing gradually in European states to finally make a network of the most significant roads on the continent [35]. Currently, both freeways are a part of the Trans-European Transport Network that makes a coherent transport network on the entire continent. The idea of the network emerged and was adopted in the 1990s as a result of international conferences (decisions of the European Parliament and the Council of the European Union) [36].

The A4 freeway is an extension of the east–west freeway running through Germany (*Autobahn 4*). Such roads were built there in the 1930s as a reply to a visible increase in the number of cars [37]. A short stretch of freeway was also constructed in Poland at that time. The A1 freeway ("Amber Highway") was constructed in the 1960s as a part of the project whose aim was to connect the north with the south of Europe. The name and course of the freeway refer to the historical Amber Road connecting the Baltic with the Mediterranean Sea. On the border with the Czech Republic, it connects with the D1 freeway (*dálnice D1*), dating back to the 1930s [38].

The "Gliwice Sośnica" junction was constructed between 2008 and 2010 and connected three thoroughfares on three separate levels. The A4 highway runs the lowest, the route 44 runs above and the A4 highway is the highest. Building was planned in a place where the Route 44 and the A4 highway crossed in 2005. Building one of the largest European freeway junctions in Gliwice ended in late 2009 and since September 2011 all sidings have functioned [39]. In the junction, two trans-European transport corridors were linked. The A4 constitutes a part of the international E40 road, leading from Ostend to Kiev, crossing the Polish–German border in Jędrzychowice and the Polish–Ukrainian border in Korczowa in eastern Poland. The M10 freeway leading to Lviv is supposed to be a continuation of the corridor on the Ukrainian side. The A1 highway constitutes a part of the international E75 north–south road. This corridor connects Norway with Greece and in Poland, when finished, is going to run from Gdańsk through Katowice to the border with the Czech Republic [40].

Table 1 contains information about the changes in the total length of major roads. The results come from calculating vector data presented on the maps from particular time periods, considering also the margin of 1 km around Gliwice. When comparing the data from the table, one needs to note that they result from the way particular stretches of roads were classified on different maps, which means that some of the roads may have been classified differently on particular maps, their rebuilding or modernization being actually not considered. Nevertheless, the above data juxtaposed with the interactive map present the gist of the changes occurring.

Among the six time periods researched, the greatest increase in major roads took place in the 19th century, reflecting changes occurring as a result of the industrial revolution (and also other factors). Out of all the periods studied, this was the longest one (1828–1880). At that time, the number of major roads increased more than twice. The processes initiated then continued for the following decades, as demonstrated on the map from 1928. A significant change on the Messtischblätter map was that the roads connecting north-west and south-west with south-east became officially recognized as major roads.

Another major change occurred between 1928 and 1960. The maps depicted the state before and after WW2. Generally, the road network became more dense and the increase in length of major roads (by approximately 40%) was connected with rebuilding the area after the war. Moreover, some projects are already being carried out to create new connections, such as constructing a bypass in the northern part to avoid passing through the city center.

The slightest changes are demonstrated on the map from 1960 and 1986. That was the shortest period of all. The web map shows that there was a balance between the number of roads that were constructed and those that disappeared. When calculating the total length of major roads for both map series, it is necessary to consider the issue of censorship and changes in the way particular road stretches were classified.

A highly significant change, the greatest one since the transformation of the 19th century, took place between the last two periods. The construction of freeways significantly impacted the road landscape of the region. New long corridors that run through Gliwice, connecting west with east and north with south, are the most relevant roads. Furthermore, they are completed by the northern bypass constructed in the previous decades, along with several road stretches marked as major, connected with freeways by means of junctions.

Table 1. The length of the highest category of road networks.

Year	1828	1880	1928	1960	1986	2020
Road Network (kilometers)	29.99	69.29	87.48	147.87	147.69	105.73

In order to collect data on changes occurring in space, historical maps are used and compiled with modern cartographic sources by means of new technologies [41–43]. Reconstructing the structure of historical road networks and the interactive way of presenting data help the authors develop the methodologies for retrospective studies [44–46]. The use of maps from many periods makes it possible to follow the course of changes taking place [47] in terms of the road network and the location of main junctions.

The approach adopted in the research develops previous studies by adding the new way of data presentation. Thematic visualizations in the form of web maps increase the availability of the information about the historical states of spaces. The web map may constitute a research material concerning historical cartography for all intranet users, including non-experts. Users receive the finished product from properly selected (in terms of scale and time periods), obtained and processed archival cartographic sources.

Each stage, from selecting the archival map to converting its content to the interactive form, is connected with work on the methodology of using historical maps in the research [17,48,49] or proper registration [25,50,51]. The access to historical maps (particularly to the editions from the early 19th century) is frequently difficult and the opportunity to publish archival materials is often limited. It is also important to collect data on the circumstances in which individual archival maps were published. Adopting a uniform scale for all the cartographic materials obtained allows one to compare data with similar accuracy, considering the specificity of each series (e.g., the subjectivity of topographers making maps in the first half of the 19th century [1] or censorship of the maps from the second half of the 20th century) [20].

According to the cartography cube theory [52], adding interactivity to spatial data obtained from the scans of archival maps boosts the potential of these maps for retrospective studies (obtaining new knowledge about the structure of the environment in the past). Such actions raise the status of paper maps in the cube from the level of data presentation to the level of research.

4. Conclusions

In comparison with other sources of information about space, such as descriptions, photos, drafts or projects, the map demonstrates a given area in a comprehensive way, along with its elements (objects), spatial layout, and the relations between them. Individual states of the evolving space were presented on the topographic maps used. The juxtaposition between selected objects in the form of vector layers of major roads from the periods following shows that the current area of Gliwice is characterized by the transformation of postindustrial areas' infrastructure and the change in space use in terms of the construction of thoroughfares and international freeway junctions [25,32,53].

Summing up the way of demonstrating changes that occurred in space in the example of the European freeway A1/A4 junction, the authors corroborate the hypothesis from this article's introduction. The interactive map that was designed proves that there are opportunities to reconstruct historical states of space, by means of modern multimedia technologies and cartographic sources. Creating interactive thematic web maps that show transformations of selected objects or spatial relations is possible thanks to the use of the approach presented. In addition, there is a wide range of opportunities provided by multimedia cartography in terms of presenting the data collected. The road layers placed on the interactive web map can be extended by adding other available multimedia, such as photos or even videos [25]. According to the authors, the use of the interactive web map is the most comprehensive method of all technologies used by modern cartography. Thanks to the interactive web map, the user has all the data collected in one place. Transparency in the reception of information is also guaranteed, as layers can be turned on and off. The interactive web map presented by the authors creates no information overflow effect [54,55], which may result in its better use [23]. Spatial data collected from different cartographic publications (from the first half of the 19th century till the present) consider the most significant aspects of changes in the road network of the area analyzed in a detailed and user-friendly way.

The authors considered the use of other archival cartographic publications that could make the research more accurate (by providing shorter time periods and more detailed maps), however, the adopted scale of 1:25,000 and time spans of 30–50 years, resulting from the previously executed query, provided the optimal juxtaposition of sources that allowed the authors to present spatial changes in the network of the most significant roads.

Moreover, the authors note the opportunity to increase the interactivity of the map created by introducing additional information with a mouse over or a click on the layers. In their interactive web map, as an example of placing information in a pop-up, they included information about the source map for each of the road layers. Opportunities to add new functions, linked with the development of the library that was used in creating the interactive web map or resulting from technological advancement (i.e., improvement of the software, updated search engines, implementation of the new W3C standards) were also considered [10]. Choosing layers independently and comparing them with one another would also be an interesting option. The layer selection order sets the display order, which allows the user to compare older layers to more recent layers and vice versa.

The above method of data presentation optimizes the possibility of interpreting selected thematic issues, among others in the field of transformation of thoroughfares. The aspect of evolution of the main thoroughfares and the interactive way of presenting it may be used for researching other significant places in Europe, e.g., to compare the level of land transformation resulting from the expansion of freeway junctions.

Author Contributions: Conceptualization, D.L.; Data curation, T.H.; Formal analysis, T.H.; Methodology, D.L.; Project administration, D.L.; Resources, D.L.; Software, T.H.; Visualization, T.H.; Writing—original draft, D.L. and T.H.; Writing—review & editing, D.L. and T.H. All authors have read and agreed to the published version of the manuscript.

Funding: This research received no external funding.

Acknowledgments: This paper is the result of research on visualization methods carried out within statutory research in the Department of Cartography and Geomatics, Faculty of Geographical and Geological Sciences, Adam Mickiewicz University in Poznań, in Poland.

Conflicts of Interest: The authors declare no conflict of interest.

References

1. Lorek, D.; Medyńska-Gulij, B. Scope of information in the legends of topographical maps in the nineteenth century—Urmesstischblätter. *Cartogr. J.* **2020**, *57*, 113–129. [CrossRef]
2. Stams, W.; Geschichtskartographie. *Lexikon der Kartographie und Geomatik*; Bollmann, J., Koch, W.G., Eds.; Spektrum Akademischer Verlag: Heidelberg, Germany, 2001; pp. 327–329.

3. Cybulski, P.; Wielebski, Ł.; Medyńska-Gulij, B.; Lorek, D.; Horbiński, T. Spatial visualization of quantitative landscape changes in an industrial region between 1827 and 1883. Case study Katowice, southern Poland. *J. Maps* **2020**, *16*, 77–85. [CrossRef]
4. Halik, Ł.; Medynska-Gulij, B. The differentiation of point symbols using selected visual variables in the mobile augmented reality system. *Cartogr. J.* **2017**, *54*, 147–156. [CrossRef]
5. Halik, Ł. Challenges in converting Polish topographic database of built-up areas into 3D virtual reality geovisualization. *Cartogr. J.* **2018**, *55*, 391–399. [CrossRef]
6. Edler, D.; Keil, J.; Wiedenlübbert, T.; Sossna, M.; Kühne, O.; Dickmann, F. Immersive VR experience of redeveloped post-industrial sites: The example of "Zeche Holland" in Bochum-Wattenscheid. *KN J. Cartogr. Geogr. Inf.* **2019**, *69*, 267–284. [CrossRef]
7. Smaczyński, M.; Medyńska-Gulij, B. Low aerial imagery—An assessment of georeferencing errors and the potential for use in environmental inventory. *Geod. Cartogr.* **2017**, *66*, 89–104.
8. Calka, B.; Cahan, B. Interactive map of refugee movement in Europe. *Geod. Cartogr.* **2016**, *65*, 139–148. [CrossRef]
9. Horbiński, T.; Medyńska, B. Geovisualisation as a process of creating complementary visualisations: Static two-dimensional, surface three-dimensional, and interactive. *Geod. Cartogr.* **2017**, *66*, 45–58. [CrossRef]
10. Horbiński, T.; Lorek, D. The use of Leaflet and GeoJSON files for creating the interactive web map of the preindustrial state of the natural environment. *J. Spat. Sci.* **2020**. [CrossRef]
11. Gwosdz, K. *Ewolucja Rangi Miejscowości w Konurbacji Przemysłowej. Przypadek Górnego Śląska*; Wydawnictwo Instytutu Geografii i Gospodarki Przestrzennej: Kraków, Poland, 2004; pp. 1–216.
12. Stankiewicz, B. *Dziedzictwo Kulturowe Przemysłu i Struktur Osadniczych w Obszarze Aglomeracji Górnośląskiej*; Wydawnictwo Politechniki Śląskiej: Gliwice, Poland, 2014; pp. 1–379.
13. Runge, J. Przeobrażenia funkcjonalno-przestrzenne miast tradycyjnego regionu społeczno-ekonomicznego—Wymiar teoretyczny. *Studia Miej.* **2018**, *32*, 21–33. [CrossRef]
14. Degner, H. Geschichtliche Entwicklung der amtlichen Preussischen Gradabteilungsblätter. *Mitt. Der Reichsamt Für Landesaufn.* **1930**, *2*, 85–99.
15. Zögner, L.; Zögner, G. *Preussens amtliche Kartenwerke im 18. und 19. Jahrhundert. Ausstellung und Katalog*; Institut für Angewandte Geodäsie, Aussenstelle Berlin: Berlin, Germany, 1981.
16. Engelmann, G. Kartengeschichte und kartenbearbeitung der preussischen urmesstischblätter. In *Festschrift für Wilhelm Bonacker*; Voggenreiter: Bad Godesberg, Germany, 1969.
17. Lorek, D. *Information Potential of Topographic Maps Urmesstischblätter Dated to 1822–33 from the Area of Wielkopolska, Western Poland*; Zakład Kartografii i Geomatyki UAM: Poznan, Poland, 2011; pp. 1–132.
18. Kraus, G. 150 jahre preussische messtischblätter. *Z. Für Vermess.* **1969**, *94*, 125–135.
19. Baumgart, G. *Gelände und Kartenkunde*; Verlag von E.S. Mittler und Sohn: Berlin, Germany, 1944.
20. Siwek, J. Mapy topograficzne. In *Wprowadzenie do Kartografii i Topografii*; Pasławski, J., Ed.; Nowa Era: Wrocław, Poland, 2006; pp. 235–289.
21. Krukowski, M.; Łoboda, A. Podstawy matematyczne współczesnych polskich map topograficznych. In *Dawne Mapy Topograficzne w Badaniach Geograficzno-Historycznych*; Czerny, A., Ed.; Uniwersytet Marii Curie-Skłodowskiej: Lublin, Poland, 2015; pp. 103–124.
22. Bertin, J. *Semiology of Graphics: Diagrams, Networks, Maps*; The University of Wisconsin Press: Madison, WI, USA, 1983.
23. Horbiński, T.; Cybulski, P.; Medyńska-Gulij, B. Graphic design and placement of buttons in mobile map application. *Cartogr. J.* **2020**. [CrossRef]
24. Nowak-DaCosta, J.; Bielecka, B.; Całka, B. Jakość danych OpenStreetMap—Analiza informacji o budynkach na terenie Siedlecczyzny. *Rocz. Geomatyki* **2016**, *14*, 201–211.
25. Lorek, D.; Dickmann, F.; Medyńska-Gulij, B.; Hannemann, N.; Cybulski, P.; Wielebski, Ł.; Horbiński, T. *The Cultural and Landscape Development of Polish and German Industrial Centres*; Bogucki Wydawnictwo Naukowe: Poznań, Poland, 2018; pp. 1–96.
26. Edler, D.; Vetter, M. The simplicity of modern audiovisual web cartography: An example with the open-source Javascript Library leaflet.js. *Kn J. Cartogr. Geogr. Inf.* **2019**, *69*, 51–62. [CrossRef]
27. Donohue, R.G.; Sack, C.M.; Roth, R.E. Time series proportional symbol map with leafet and jquery. *Cartogr. Perspect.* **2013**, *76*, 43–66.

28. Maiellaro, N.; Varasano, A. One-Page Multimedia Interactive Map. *Isprs Int. J. Geo-Inf.* **2017**, *6*, 34. [CrossRef]
29. Medyńska-Gulij, B. Cartographic sign as a core of multimedia map prepared by non-cartographers in free map services. *Geod. Cartogr.* **2015**, *63*, 55–64. [CrossRef]
30. stefanocudini/leaflet-panel-layers. Available online: https://github.com/stefanocudini/leaflet-panel-layers (accessed on 15 May 2020).
31. nerik/leaflet-graphicscale. Available online: https://github.com/nerik/leaflet-graphicscale (accessed on 15 May 2020).
32. Antrop, M. Landscape change and the urbanization process in Europe. *Landsc. Urban Plann.* **2004**, *67*, 9–26. [CrossRef]
33. Antrop, M. Changing patterns in the urbanized countryside of Western Europe. *Landsc. Ecol.* **2000**, *15*, 257–270. [CrossRef]
34. Bogdański, M. Influence of the A2 motorway on the economic development at local level. In *Bulletin of Geography. Socio-Economic Series*; Szymańska, D., Chodkowska-Miszczuk, J., Eds.; Nicolaus Copernicus University: Toruń, Poland, 2016; pp. 49–59.
35. Spiekermanm, K.; Wegener, M. Accessibility and spatial development in Europe. In *Scienze Regionali*; Franzo Angeli Edizioni: Milan, Italy, 2006; pp. 15–46.
36. europa. Available online: www.ec.europa.eu/transport (accessed on 15 May 2020).
37. autobahn. Available online: http://www.autobahn-online.de/a4geschichte.html (accessed on 15 May 2020).
38. ceskedalnice. Available online: http://www.ceskedalnice.cz (accessed on 15 May 2020).
39. gddkia. Available online: https://www.gddkia.gov.pl (accessed on 15 May 2020).
40. siskom. Available online: http://www.siskom.waw.pl/a1.htm (accessed on 15 May 2020).
41. Dickmann, F.; Kollecker, C. *Interaktive Kartographie—Eine praxisorientierte Einführung in die Methoden des digitalen Kartendesigns*; Materialien zur Raumordnung: Bochum, Germany, 2013.
42. Kaim, D.; Kozak, J.; Ostafin, K.; Dobosz, M.; Ostapowicz, K.; Kolecka, N.; Gimmi, U. Uncertainty in historical land-use reconstructions with topographic maps. *Quaest. Geogr.* **2013**, *33*, 51–59. [CrossRef]
43. Vervust, S.; Boùùaert, M.C.; De Baets, B.; Van de Weghe, N.; De Maeyer, P. A Study of the local geometric accuracy of count de Ferraris's Carte de cabinet (1770s) using differential distortion analysis. *Cartogr. J.* **2018**, *55*, 16–35. [CrossRef]
44. Helmfrid, S. Företal. In *Kartlagt Land—Kartor Som Källa till Förståelsen av de Areella Näringarnas Geografi Och Historia*; Jansson, U., Ed.; Kungliga Skogs och Lantbruksakademien: Stockholm, Sweden, 2007; pp. 5–8.
45. Karsvall, O. Retrogressiv metod: En översikt med exempel från historisk geografi och agrarhistoria. *Hist. Tidskr.* **2013**, *133*, 2–26.
46. Statuto, D.; Cillis, G.; Picuno, P. Using historical maps within a GIS to analyze two centuries of rural landscape changes in Southern Italy. *Land* **2017**, *6*, 65. [CrossRef]
47. Jongepier, I.; Soens, T.; Temmerman, S.; Missiaen, T. Assessing the planimetric accuracy of historical maps (sixteenth to nineteenth centuries): New methods and potential for coastal landscape reconstruction. *Cartogr. J.* **2016**, *53*, 114–132. [CrossRef]
48. Buczek, K. *History of Polish Cartography from the 15th to the 18th Century*; Zakład Narodowy im Ossolińskich: Wrocław, Poland, 1966.
49. Scharfe, W. *Abriss der Kartographie Brandenburgs 1771-1821*; Walter de Gruyter: Berlin, Germany, 1972.
50. Brovelli, M.A.; Minghini, M. Georeferencing old maps: A polynomial-based approach for Como historical cadastres. *e-Perimetron* **2012**, *7*, 97–110.
51. Affek, A. Kalibracja map historycznych z zastosowaniem GIS. In *Źródła kartograficzne w badaniach krajobrazu kulturowego*; Plit, J., Nita, J., Eds.; Komisja Krajobrazu Kulturowego PTG: Sosnowiec, Poland, 2012; Volume 16, pp. 48–62.
52. MacEachren, A.M. Visualization in modern cartography: Setting the agenda. In *Visualization in Modern Cartography*; MacEachren, A.M., Taylor, D.R.F., Eds.; Pergamon Press: London, UK, 1994; pp. 55–70.
53. Jepsen, M.R.; Kuemmerle, T.; Müller, D.; Erb, K.; Verburg, P.H.; Haberl, H.; Vesterager, J.P.; Andrič, M.; Antrop, M.; Austrheim, G.; et al. Transitions in European land-management regimes between 1800 and 2010. *Land Use Policy* **2015**, *49*, 53–64. [CrossRef]

54. Horbiński, T.; Cybulski, P. Similarities of global web mapping services functionality in the context of responsive web design. *Geod. Cartogr.* **2018**, *67*, 159–177. [CrossRef]
55. Cybulski, P.; Horbiński, T. User experience in using graphical user interfaces of web maps. *ISPRS Int. J. Geo-Inf* **2020**, *9*, 412. [CrossRef]

International Journal of
Geo-Information

Article

User Experience in Using Graphical User Interfaces of Web Maps

Paweł Cybulski * and Tymoteusz Horbiński

Department of Cartography and Geomatics, Adam Mickiewicz University, 61-712 Poznań, Poland; tymoteusz.horbinski@amu.edu.pl

* Correspondence: p.cybulski@amu.edu.pl

Received: 21 May 2020; Accepted: 26 June 2020; Published: 27 June 2020

Abstract: The purpose of this article is to show the differences in users' experience when performing an interactive task with GUI buttons arrangement based on Google Maps and OpenStreetMap in a simulation environment. The graphical user interface is part of an interactive multimedia map, and the interaction experience depends mainly on it. For this reason, we performed an eye-tracking experiment with users to examine how people experience interaction through the GUI. Based on the results related to eye movement, we presented several valuable recommendations for the design of interactive multimedia maps. For better GUI efficiency, it is suitable to group buttons with similar functions in screen corners. Users first analyze corners and only then search for the desired button. The frequency of using a given web map does not translate into generally better performance while using any GUI. Users perform more efficiently if they work with the preferred GUI.

Keywords: multimedia cartography; web map; graphical user interface; eye-tracking; UI/UX

1. Introduction

The importance of web maps in the development of the so-called "Multimedia Cartography" was noticed by researchers several years ago [1]. Also, the Commission on Maps and the Internet of the International Cartographic Association presented the purpose of investigating the role of efficient integration multimedia and maps on the Internet. Multimedia, such as graphics, photos, or video, plays an essential role in the transmission of spatial information on the web [2,3]. In internet cartography, one of the fundamental factors is interactivity. Interaction is a human–map communication system [4]. This is the way in which the user manipulates the map (by changing the scale or panning movement) [5,6]. The interaction on the map mainly takes place using the graphical user interface (GUI). It consists of buttons that have specific functions and a symbolic icon [7,8]. The most popular interactive buttons include geolocation, searching, changing layers, and routing [9]. They are available on almost all global map services, such as Google Maps, Bing Maps, OpenStreetMap, Baidu Maps, or Yandex Maps. Kraak and Ormeling [10], in their manual, described GUI as a 'minimum requirement' for designing an interactive map.

Interactivity is part of the users' experience [11,12]. Most web cartographic products are interactive. However, there is a difference between the user interface (UI) and the user experience (UX). User interface refers to tools (usually GUI buttons) that allow communication with a digital map and data manipulation [13]. User experience is a broader term. This refers to the experience and user preferences of the two-way communication process request–result [14]. According to Norman [15], this experience is responsible for the success of a given product. Thus, UI design is a process aimed at implementing tools that allow interaction, e.g., using a Leaflet.js library or dedicated API [16,17]. On the contrary, UX design involves developing the interaction result and the communication process itself so that the experience is satisfying for the user [18].

From a pragmatic point of view, assessment of both UI and UX seems relevant [19]. Objective evaluation of GUI tools, such as are currently used on online maps, is possible through efficiency parameters. Efficiency includes, among others, the time for proper competition and a spatial task using a given interactive tool [20–22]. Therefore, the GUI buttons themselves have an essential impact on the efficiency of a web map. GUI research considers the placement, number, and graphics of buttons in the first view as influential factors [4,23–25]. This is especially true for global online map services like Google Maps or OpenStreetMap that have the same interactive tools with various arrangements, numbers of buttons, and graphics [26]. As a result, the use of various map services is associated with a slightly different experience of interactivity. As a consequence, it leads to the separation of users who prefer a given web map [27].

Due to the diverse UX of interaction on different web mapping services, the need for measuring these differences arises. Eye-tracking is one of the methods that enable understanding the UX when working with multimedia and the interactive product [28,29]. It provides many indicators related to the movement of the gaze in space and time, such as time to the first fixation, fixation count, fixation duration, saccadic amplitude [30,31]. Fixations are places were gaze maintains relatively constant, and saccades are a quick movement of the gaze [32]. This provides an objective way of measuring the users' perceptual (visual) experience called the visual strategy [33]. It provides mental attention data [34]. This type of research is crucial, since some interactions, such as navigation, sometimes do not have a tool in the form of a button, and the user can still experience it.

The research topic presented in this way raises the question about differences in UX when interacting with a web map. Questions that arise include: does a different number of corners with buttons affect the user's view path? Are there measurable differences in UX between two GUIs with the same buttons but arranged differently? In particular, we aimed to define the difference in the users' visual experience using eye-movement metrics, such as the total number of fixations (FC), number of fixations on the specific GUI button (FCB), percentage of participants' attention needed for button identification (ATT), saccadic amplitude (SA), and time to the first fixation (FFT) in three interactive tasks. Participants performed a task in two different GUIs based on Google Maps or OpenStreetMap. The tasks consisted of finding and using the appropriate button that caused the desired interactions (geolocation, searching, finding the route). Eye movement measurements defined users' visual experience. An additional goal was to determine the efficiency of individual interfaces to compare UI/UX. This was measured with the use of time to the first click on the appropriate button in each task (TC). This study shows that different arrangement of the same interactive tools may have a different effect on the user experience and efficiency of web map.

2. Related Work

The number and placement of buttons play an important role in interface design. Wang [27] drew attention to this, especially during the opening page. He noticed that if the start page appears to be disordered and does not support the user's habits, it negatively affects its efficiency. Interface disorder is closely related to the location of the button because, as Wang noted, the search button plays an essential role in the interface layout. On the other hand, Nivala et al. [35] noticed that interactive web maps are meant for the general public and may not always meet all users' needs. Nivala, in conclusion, drew attention to the need to design the GUI in an orderly manner. Hegarty et al. [36] also referred to the issue of the importance of the interface in experiencing interaction. Newman et al. [37] came to similar conclusions about the simplicity of the user interface. Based on the respondents' responses, they redesigned the interface to suit preferences (e.g., they changed buttons location, removed navigation arrows, and changed the way of choosing layers).

Some researchers [7] have drawn attention to the GUI differences arising not only from the placement of individual buttons but also that each map provider has a different graphic style of buttons. Even the same interactive functions, such as wayfinding, may work differently, e.g., by adding waypoints manually or typing the next location. However, as noted by Horbiński et al. [9], UX and preferences may be different from what map providers propose. Haklay [13] explained this by the fact that user expectations may be due to the various devices on which they use web maps.

Assessment of the GUI and methods of assessing UX constitute a research gap in cartography. Although there are studies on the use of eye-tracking in interface studies, only some refer to cartography [38]. Çöltekin et al. [22] analysis is one of the few studies in the cartographic field. In their research, they focused on two differently designed interfaces. They measured effectiveness, efficiency, and overall satisfaction during three tasks. The experiment also used eye movement recording. The results gave interesting conclusions about user experience. It turned out that despite the more accurate execution of tasks with the Natlas interface, the study participants preferred the Carto.net interface. These types of conclusions also appear not only concerning for the interface but also for the methods of cartographic visualization [39]. Eye-tracking results discovered usability problems associated with individual buttons on both GUIs. According to Çöltekin, data from eye-tracking also provide information on users' expectations.

There are significant similarities in UX while using different interfaces of web map navigation in zooming tasks [40]. The eye-tracking analysis helped to find differences and similarities in UX while using four methods of zooming interaction: pan zoom, rectangle zoom, double click, and wheel zoom. Manson's research shows that most users performed better with rectangle zoom, even though they sometimes felt frustration using it. Pan zoom and click zoom were rarely preferred, and users were not satisfied using them.

According to research on the effectiveness of user interfaces, eye-tracking methodology combined with questionnaires provides a crucial setting for examining user experience. The presented studies motivated us to research the assessment of users' visual experience while performing different interactive tasks based on Google Maps or OpenStreetMap interface.

3. Methodology

The research methods used include web map and GUI design, efficiency and eyeball movement data acquisition, and data processing. We also presented the participants' pool and experiment procedures.

3.1. Web Maps

The study used a web map with an OpenStreetMap map designed with Leaflet.js library. We selected OSM because it is a free to use under an open license world map with geodata stored in a database that is accessible through JavaScript. The main factor in choosing the OpenStreetMap GUI is the compatibility of this web map with the Leaflet.js environment. In addition, the popularity of OSM is very large on a global scale, which was confirmed by search results of browsers such as Google, Bing, or Yahoo. The map view setting was φ: 52.17° N (latitude) and λ: 3.43° W (longitude) with zoom level 16. The area of the map was placed into the <body> section of HTML (Hypertext Markup Language) structure as <div id="map"></div>. Parameters of the map were set to 100% height and width in cascading style sheets (CSS). We added the map using the L.tileLayer function, which enables us to render the tile-based map in real-time [41]. Figure 1 presents the OpenStreetMap opening view developed in Leaflet.js.

Figure 1. Opening view of OpenStreetMap prepared in Leaflet.js.

3.2. *Graphical User Interface*

In GUI design, we used the placement of buttons according to Google Maps and OpenStreetMap, which are two of the most popular and recognizable web maps in the world. For the study, we adopted a simplified version of the interface, consisting of the six most significant buttons that enable interaction with the map: geolocation, search, route, change layers, and zoom buttons (plus and minus) [42]. According to Horbiński and Cybulski [7], these buttons are on every global web map. Figure 2 shows the location of the buttons on a computer monitor according to selected web maps. Button positioning (exact location) was possible with the use of CSS code. We used absolute positioning instead of relative without using JavaScript code. In our case, the map as an area of the website had no relationship with other elements. Thanks to this, we identified the map area *<div id="map">* as both *<body>* and *<html>*. Based on this assumption, we treated all buttons between the *<div id="map"> </div>* tags as independent elements. That is why we decided to use absolute positioning. The simplicity of CSS positioning dictates not using JavaScript coding in this case.

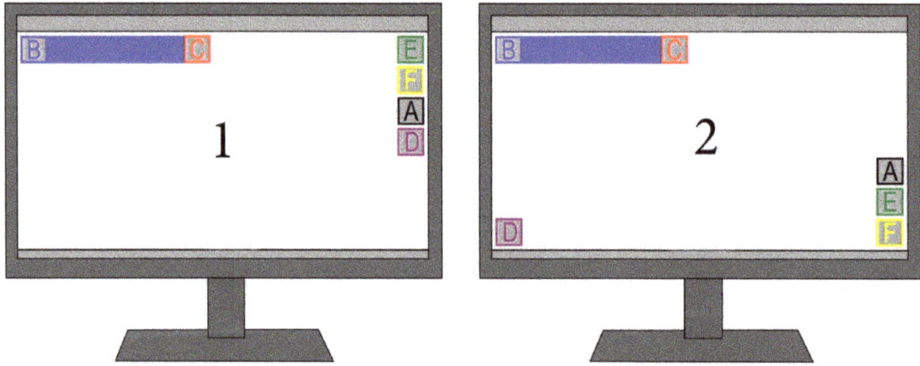

Figure 2. Buttons placement on the computer monitor according to (**1**) OpenStreetMap and (**2**) Google Maps. Letters indicate buttons: (A) geolocation, (B) search, (C) route, (D) change layers, (E) zoom in, and (F) zoom out.

Because each GUI has its unique graphic style, we decided to unify button graphics and dimensions. We proposed black and white 60 × 60 px buttons with a 2 px frame. Figure 3 shows all buttons along with the labels that refer to Figure 2. First, we prepared the concept of the layout, and then we inserted button graphics along with interactive functions.

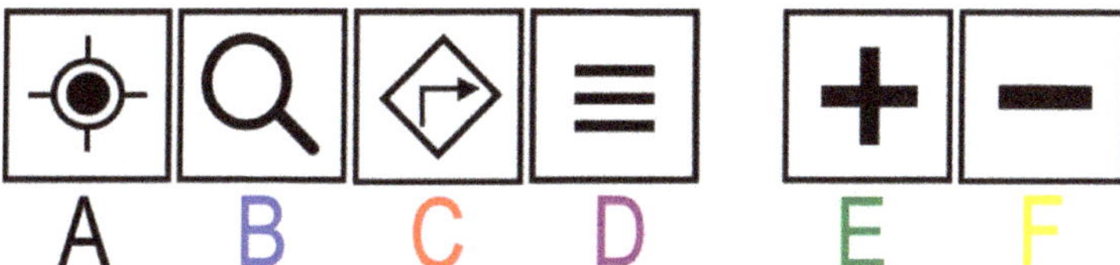

Figure 3. All buttons that we used in the graphical user interface (GUI) design. Labels refer to the description of Figure 2.

Buttons with specific graphics were placed in the GUI, as shown in Figure 4, following the arrangements of layouts according to OpenStreetMap and Google Maps. We believe that naming GUIs based on the names of web mapping services is necessary. The point is to draw attention to the fact that it is a practical solution used by map providers.

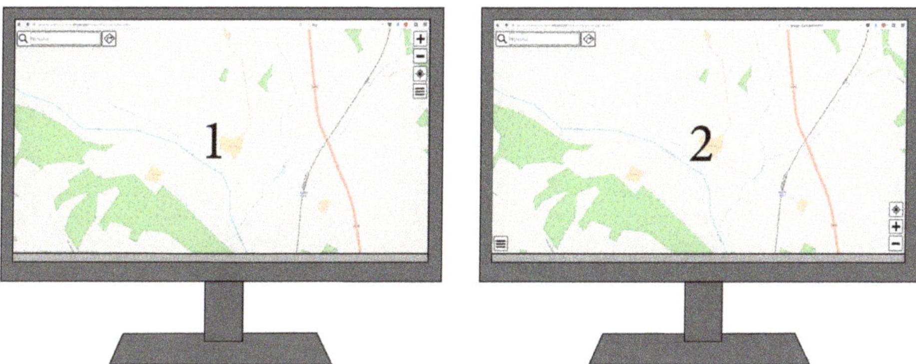

Figure 4. Web map and the unified GUI design according to (**1**) OpenStreetMap and (**2**) Google Maps.

For each button, we wrote a JavaScript code that allowed correct interaction with the map. The leaflet-control-geocoder.js (https://github.com/perliedman/leaflet-control-geocoder) plugin is responsible for geolocation, but it does not work correctly on desktop computers. For this reason, we simplified the plugin in such a way that the press of the button caused the view change to a specific location. The leaflet-search.js (https://github.com/stefanocudini/leaflet-search) plugin, which uses the OpenStreetMap spatial database, is responsible for the search. A circle with a diameter of 5 px marks the found point. We supplemented the plugin with the auto-complete function. We used the leaflet-routing-machine.js plugin (https://github.com/perliedman/leaflet-routing-machine) for route search. However, we had to simplify its operation by removing navigation options, using multiple markers, route length, and time in JavaScript code. We added a code that opens only after clicking a decision window, enabling the addition of more locations. Another coded interactive function was change of layers. In this case, we did not use an additional plugin but only the Leafleat.js library. We needed a second map background (Mapnik) for changing layers. Thanks to this, the layers were changed using radio switching (baseLayers). For zooming, we used the leaflet-control-zoombar.js plugin (https://github.com/elrobis/L.Control.ZoomBar) so that we could change the position of the zoom out and zoom in buttons. The possibility of zooming and dragging was blocked due to the study of using the geolocation button. We used Boolean values available in the L.map function of Leaflet.js:

zoomControl: *false*,
scrollWheelZoom: *false*,
dragging: *false*,
doubleClickZoom: *false*.

3.3. Participants and Experimental Process

In this study, we adopted two research groups of 20 participants each. One group performed survey tasks on a GUI based on OpenStreetMap and the other based on Google Maps. For both study scenarios, participants were randomly selected from a group of students from Adam Mickiewicz University in Poznan from a geodetic and cartographic course. Participants took part in the survey voluntarily and had the opportunity to opt-out of continuing at any time. In the participants' group performing tasks based on the Google Maps GUI, there were 16 men and 4 women. Nine people were in the 18–21 age range, ten in the 22–25 age range, and one was in the age group of 26 and over. We determined the experience in using Google Maps based on the frequency of use. Five people said they use it every day, eleven people once a week, three people once a month, and one person does not use it at all. In the same way, we defined the experience in using OpenStreetMap. Two people use it once a week, eleven once a month, and seven people do not use it at all. Four people use other web mapping services once a month and the other sixteen do not use them at all. There were 12 men and 8 women in the OpenStreetMap-based GUI group. Fourteen people were in the 18–21 age range, five people in the 22–25 age range, and one person in the age group of 26 and over. One person used Google Maps every day, nine use it once a week, seven people once a month, and three people said they did not use it at all. As for OSM, one person used it every day, eight people used it once a month, and eleven did not use it at all. One person uses other web mapping services once a month, and rest of the participants group do not use at all.

We experimented on a desktop computer with Windows 10 with the Firefox browser. We displayed web maps on a 21.5-inch screen with a resolution of 1920 × 1080 px. To capture gaze, we used Tobii eye-tracker X2-60 with a sampling frequency of 60 Hz. We used the Tobii Fixation Filter with a velocity threshold of 35 pixels/windows and distance threshold 35 pixels [43]. We used these parameters to classify raw eye movement data into fixations and saccades.

We asked participants in the questionnaire about the frequency of using Google Maps, OpenStreetMap, and other web maps. There were four possible answers for each question: I use daily (rank 4), once a week (rank 3), once a month or less (rank 2), I don't use it (rank 1). We used ranks to categorize participants' experience. Therefore, we were able to check if there were correlations between the frequency of using the specific GUI and the time of task completion.

Each group had three tasks to perform using GUI buttons. We implemented all web maps, instructions, tasks to solve, and questionnaires (with questions about age, gender, and frequency of using Google Maps, OpenStreetMap, and other web maps) in Tobii Studio 3.4. Before the participants started the task, it was necessary to calibrate the gaze with the eye-tracking device. The first task was to 'find your location using geolocation button'. The participant had to point to and click on the geolocation button. This action took him to the current location. However, to complete the task, the participant had to press the F10 button. The second task was to 'find the Żywiec city using the search button'. After clicking the appropriate search button, the user entered the particular city name and then confirmed it. The F10 button finished the task. The third task was to 'find a route between Leżajsk and Jasło using the route button'. After clicking the route button, a window appeared in which the participant entered the start and destination and then confirmed his choice. As in previous tasks, the F10 button ended the task. Each participant performed the tasks in this order with no time limit.

3.4. Efficiency and Eye Movement Metrics

Based on the web maps and questionnaires, we obtained time data on the solution of each task and the eye movement recordings. Because all users performed the individual tasks correctly, we used parameters related to the time of task execution to assess efficiency. Therefore, the first time parameter was the time taken to solve the task, evaluated from displaying the map to the first click on the geolocation button (TC). It was measured for each task separately. For the second task, it was time to the first click on the search button, and for the third task, the time to the first click on the route button. We used the time to the first click because participants also had to enter the name of the place, and typing names is not directly related to GUI efficiency—it relates to participants' ability to write speed on a computer. Mouse events, such as time to the first click, are often a part of web map assessment [44].

An essential part of UX is eye movement assessment. Therefore, the first UX assessment parameter was the total number of fixations during the task (FC). This tells you how frequently the participant stopped his gaze on the map screen or any GUI element. A visualization that is visually more demanding often has more fixations in overall [45]. The reason for the higher number of fixations may be users' lower spatial abilities [46]. However, in the study, we adopted homogeneous groups, so this factor does not influence the number of fixations. The next parameter is the number of fixations that appeared on the button used to complete the task (FCB). To obtain this parameter, it was necessary to specify the so-called areas of interest (AOI) [47]. The ratio of FCB to FC allows for determining what percentage of attention the participant needed to finding and identifying the specific button, which is another parameter (ATT). Saccadic amplitude is another parameter that defines the user's visual experience (SA). It is the angular distance of eye movement between fixations [48]. Some studies claimed that the shorter the SA, the less effective the visual scanning [49]. The last parameter defining visual UX is time to the first fixation (FFT) on the specific button.

4. Results

4.1. GUI Efficiency

GUI efficiency was determined based on TC for two interfaces independently. The time needed to complete individual tasks was lower when using the OpenStreetMap-based GUI. For the first task, TC median was 8.5 s for the OpenStreetMap-based GUI and 11.7 s for the Google Maps-based GUI. However, the Mann–Whitney test did not show statistically significant differences ($p > 0.05$). A similar situation, but with a smaller difference, was observed with the second task, in which TC median resulted in 1.8 s for Google Maps-based GUI and 1.6 s for OpenStreetMap-based GUI ($p > 0.05$). The third task presents similar efficiency measured with TC—median 3.0 s for Google Maps-based GUI and 4.2 s for OpenStreetMap-based GUI ($p > 0.05$).

As the TC results of individual participants show, the first task was the most time-consuming in both interfaces (Figures 5 and 6). The Mann–Whitney test confirms the statistical significance of results in both GUIs (TC in Task 1 > TC in Task 2 $p < 0.05$; TC in Task 1 > TC in Task 3 $p < 0.05$). As for the differences in TC between tasks 2 and 3, only the OpenStreetMap-based GUI has a statistically significant difference (TC in Task 2 < TC in Task 3 $p < 0.05$).

We used the Spearman correlation test to study the relationship between TC and participants' preferences. We found several statistically significant correlations ($p < 0.05$). As for the first task, which was to find your location using the geolocation button, there was a substantial relationship between the frequency of using Google Maps and TC on Google Maps-based GUI ($r = -0.54$). It means that the more often participants use Google Maps, the faster they perform the first task while using the Google Maps-based GUI. In the second task, the frequency of using OpenStreetMap correlated with TC while using the OpenStreetMap-based GUI ($r = 0.51$). This means that the more often participants used OpenStreetMap, the more time they needed to complete the task with the OpenStreetMap-based GUI. In the third task, there was a similar correlation as in the first task. The more often participants used Google Maps, the faster they performed tasks based on the Google Maps-based GUI ($r = -0.56$). It means that the more often participants use Google Maps, the faster they perform the third task using the Google Maps-based GUI. The correlation between the frequency of using OpenStreetMap and the time to complete the third task based on the Google Maps-based GUI is similar (r = −0.68).

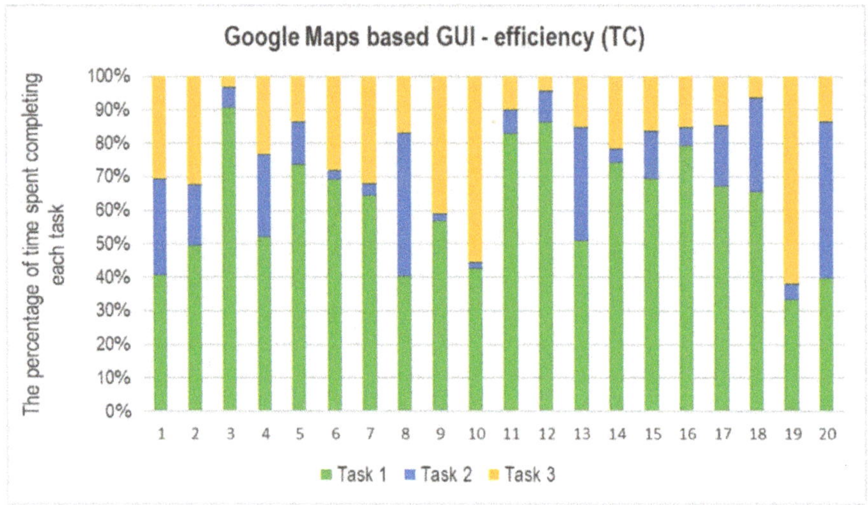

Figure 5. The percentage of time needed for completing tasks (TC) by each participant on Google Maps-based GUI.

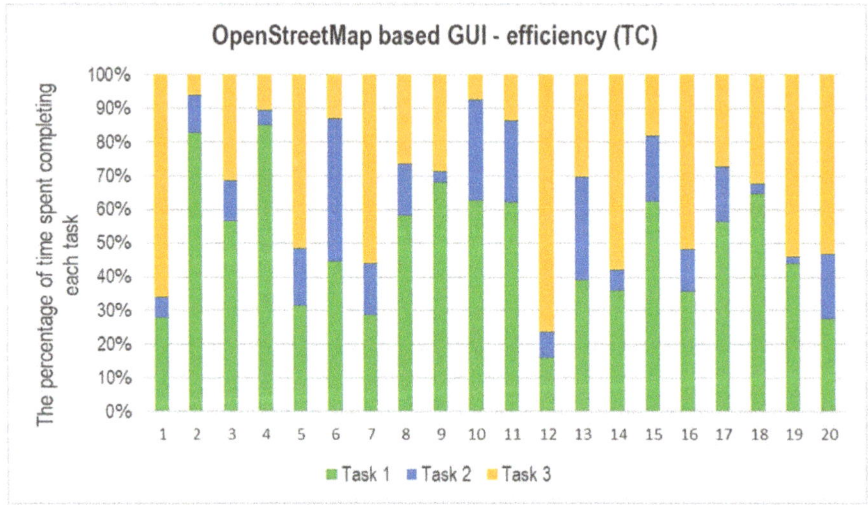

Figure 6. The percentage of time needed for completing tasks (TC) by each participant on OpenStreetMap-based GUI.

4.2. Users' Visual Experience

Despite the similarities in the effectiveness of both tested GUIs, the study participants showed differences in experiencing interactions. These differences were determined based on eye movement analysis. The first parameter is the number of fixations (FC). In the first task, the median was different for the Google Maps-based GUI—36 fixations—and for the OpenStreetMap-based GUI—38. Although the Mann–Whitney test did not show statistically significant differences ($p > 0.05$), the spatial distribution of fixations indicates different places of the participants' gaze concentration. This mainly applies to the first task (Figure 7). Google Maps-based GUI users searched for the geolocation button in three corners of the screen, while OpenStreetMap-based GUI users only searched for it in two, as shown by heatmaps. In the second task, the median FC on the Google Maps-based GUI was 34, while on the OpenStreetMap-based GUI, it was only 28. Although the Mann–Whitney test did not show statistical significance, the recorded distribution of fixations shows that, as in the first task, Google Maps-based GUI users were looking for a 'search' button in three corners of the screen. In the third task, the median FC is very similar. The Google Maps-based GUI median was 51 fixations, and the OpenStreetMaps-based GUI median had 54 fixations. Here, too, Google Maps-based GUI users searched for a 'route' button in three corners. The main similarity in UX when working with the interface is that the participants visually analyze all the corners where the buttons are, regardless of the task performed. The Google Maps-based GUI requires participants to analyze one corner more, compared to OpenStreetMaps-based GUI.

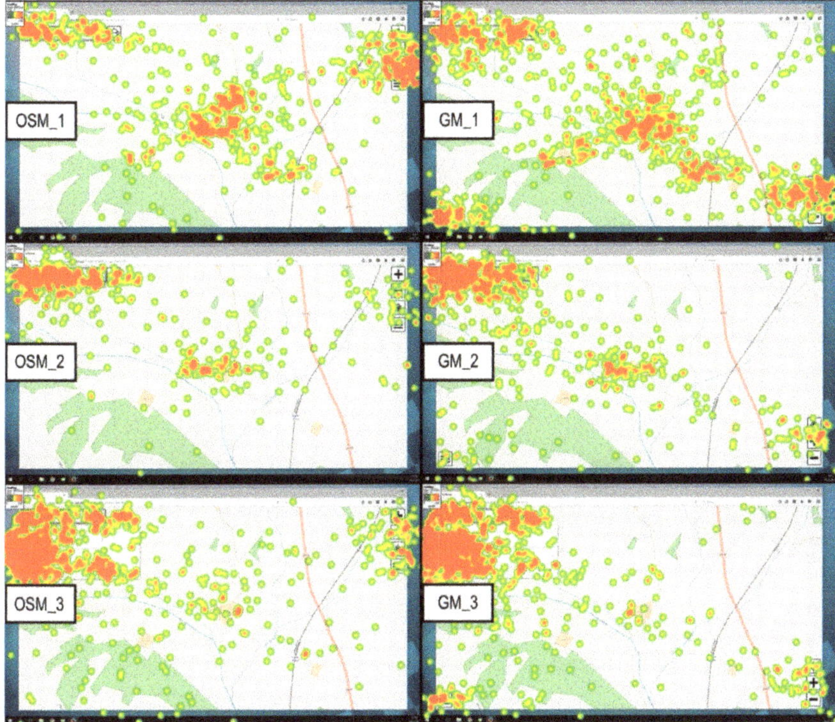

Figure 7. Six heatmaps presenting the spatial distribution of fixations in each task on both GUIs. OSM (from 1 to 3) stands for the OpenStreetMap-based GUI, and GM (from 1 to 3) stands for the Google Map-based GUI.

The number of fixations on the searched button (FCB) is another visual UX assessment parameter along with the percentage of attention paid to it (ATT). In the first task, the average FCB for both GUIs was 3. However, concerning FC, it gives 4.8% attention to the geolocation button in the Google Maps-based GUI and 9.3% attention while using the OpenStreetMap-based GUI. In the second task, the average ATT is at a similar level—35.5% (13 FCB) when using the Google Maps-based GUI and 40.2% (11 FCB) while using the OpenStreetMaps-based GUI. The longer interaction process associated with typing names caused an increased ATT level. In the third task, FCB for two interfaces was 2, which translated into 4.5% ATT using the Google Maps-based GUI and 3.5% ATT using the OpenStreetMap-based GUI, respectively. This task also requires a more sustained interaction. While the route search extends the name entry field, we did not obtain fixations from the area in which the participant entered the name.

The median saccadic amplitude (SA) in the first task on the Google Maps-based GUI was 6.50°, while on the OpenStreetMap-based GUI, it was 6.16°. In the second task, SA on the Google Maps-based GUI was 5.73°, and that on the OpenStreetMap-based GUI median was 5.53°. In the third task, SA on the Google Maps-based GUI was 4.59°, and on the OpenStreetMap median it was 4.90° on average. We did not found statistically significant differences using the Mann–Whitney test. Spearman's correlation showed that the more FC, the shorter SA ($r = -0.47$; $p < 0.05$) in the first task while using the Google Maps-based GUI. We found statistically significant differences ($p < 0.05$) with the Mann–Whitney test in FFT during the first task. For the Google Maps-based GUI, the FFT median was 7.8 s, and for OpenStreetMap-based GUI, FFT was 4.6 s. In the other two tasks, the Mann–Whitney test showed no statistical significance; however, the FFT on Google Maps-based GUI was slightly longer (for Task 2, the median was 0.5 s, and for Task 3 it was 1.2 s) comparing to OpenStreetMap-based GUI (Task 2—median was 0.4 s and Task 3—0.8 s).

5. Discussion

Based on the test results presented, we confirm Wang's [27] conclusions that different GUIs may cause different user experience. This is especially evident in the visual experience called the user's visual strategy described by eye movement metrics [30,33]. Depending on the GUI, the spatial distribution of fixations confirms this conclusion. In each task, participants observed three corners in the Google Maps-based GUI. However, in the OpenStreetMap-based GUI, only two corners were observed. This agrees with the claim of some researchers about the crucial button placement in the GUI layout [9,37]. More corners of the screen with buttons can translate into a longer cognitive process. However, as the correlation results show, the frequency of using a given map service has a high impact on the efficiency of the GUI. Wang [27] claimed that the preference of individual map services leads to the separation of users. However, the frequency of use (preference) of a given web map does not translate into generally better performance while using any GUI. Users perform more efficiently if they work with their preferred GUI. Using a preferred GUI of a multimedia map reinforces user habits [7]. However, participants who preferred OpenStreetMap were less efficient while using the OpenStreetMap-based GUI. It can also be a tip for designers of web multimedia maps to introduce GUI changes gradually, as Google does [50]. As noted by Horbiński and Cybulski [7], the GUI buttons of web mapping services are characterized only by pictograms. However, as noted by Muehlenhaus [51], button symbols are built on certain conventions that influence UX and efficiency.

Using the GUI for specific tasks on the map generates specific visual experiences. As the results in Figure 7 show, the participants investigate all the corners where the buttons occur. In this context, Nivala et al. [35] suggested that the interface on the map should be simple. Users examine all corners but not all buttons. According to this observation, we conclude that users analyze the GUI layout in search of the target button. They do not analyze all buttons in turn. On this basis, we can also recommend a guide for interactive map designers. The idea is to design groups of buttons with similar functions and place them together. In our opinion, this is a factor that contributes to increasing the efficiency of a web map, which is part of multimedia cartography [52]. This is also consistent with the principle presented by Shneiderman and Plaisant [53]—striving for consistency.

6. Conclusions

Map interactivity is a crucial element of multimedia cartography on the internet. The methodology of evaluating the visual user experience presented in the study has provided interesting conclusions about the performance of the graphical user interface. The results show the differences in experiencing interaction while using two different GUIs. The main difference in eye movement between Google Maps-based GUI and OpenStreetMap-based GUI is the visual analysis of a higher number of corners with buttons. There are also differences in time parameters such as the time to first fixation, fixation count, and saccadic amplitude. However, not all of the presented relationships showed statistical significance. This may be due to a small research sample. We think that increasing the number of participants could have resulted in more definite differences. This is true, especially since some test results showed $p < 0.10$.

Like any study, we rely on experimental simplifications. The two GUIs compared prove the importance of button placement in the UX. Of course, we see the possibility of developing our research on GUIs based on other web mapping services, e.g., Baidu Maps, Yandex Maps, Bing Maps, Map Quest, and HERE Maps. We can also include the GUI arrangement according to user preferences in future analysis. There are several significant lessons to be learned from extending this comparison. Firstly, if we examine the higher the number of GUIs based on different web maps, the more meaningful the results will be. Secondly, this can lead to more recommendations of proper GUI design that are desirable in GUI design. Third, it could help to exclude the least effective solutions. Fourthly, it would help in better understanding the UX when using the GUI. Among other things, one could answer the question of which solutions increase distraction and which are intuitive and supportive for users.

Eye-tracking combined, with a questionnaire, proved to be an effective method to obtain data on GUI efficiency and users' UX. In future studies, we see the possibility of adaptation of this methodology to study the GUI of web maps on mobile devices such as smartphones. It is also possible to examine UX while performing tasks with an animated map interface. This type of interface has additional buttons (sliders) responsible for changing the time [54,55]. It is also possible to use this methodology to study GUI on maps in augmented reality [56].

The use of eye-tracking brought surprising conclusions regarding UX. The participants of the study observe all the corners with buttons in each of the tasks. More corners with buttons generally result in a longer and more complex view path. This translates to eye-tracking metrics that define differences in UX. A larger number of corners with buttons also slightly reduces the performance of the web map. Studies have shown that the arrangement of GUI buttons according to the OpenStreetMap layout was more efficient when performing tasks. We noticed that participants who preferred Google Maps were better when working on an interface based on Google Maps. This allows us to believe that users who work with the web map interface are largely guided by their habits. However, the most problematic task was to find the geolocation button in both GUIs.

Observing all the corners (not all buttons) to find the appropriate button prompts us to recommend arranging buttons with similar interactive functionality together (searching, routing). Another recommendation is to arrange the buttons in the lowest number of screen corners. Since the participant visually examines all of the screen corners, the more corners a display uses, the more complex the scanning path.

Author Contributions: Conceptualization, Tymoteusz Horbiński; Formal analysis, Paweł Cybulski; Methodology, Paweł Cybulski; Resources, Tymoteusz Horbiński; Writing—original draft, Tymoteusz Horbiński; Writing—review and editing, Paweł Cybulski. All authors have read and agreed to the published version of the manuscript.

Funding: This research received no external funding.

Acknowledgments: The paper is the result of research on visualization methods carried out within statutory research in the Department of Cartography and Geomatics, Faculty of Geographical and Geological Sciences, Adam Mickiewicz University in Poznań, in Poland.

Conflicts of Interest: The authors declare no conflict of interest.

References

1. Cartwright, W.; Peterson, M.P.; Gartner, G. *Multimedia Cartography*, 2nd ed.; Springer: Berlin, Germany, 2007; ISBN 978-3540366508.
2. Medyńska-Gulij, B. Pragmatical Cartography in Google Maps API [in German]. *Kart. Nach.* **2012**, *5*, 250–255. [CrossRef]

3. Medyńska-Gulij, B.; Lorek, D.; Hannemann, N.; Cybulski, P.; Wielebski, Ł.; Horbiński, T.; Dickmann, F. Cartographic reconstruction of the landscape development of the Upper Silesian industrial area (Poland) and the Ruhr area (Germany) [in German]. *KN—J. Cartogr. Geogr. Inf.* **2019**, *63*, 131–142. [CrossRef]
4. Roth, R.E. Interactivity and Cartography: A Contemporary Perspective on User Interface and User Experience Design from Geospatial Professionals. *Cartographica* **2015**, *50*, 94. [CrossRef]
5. Roth, R.E. Interactive maps: What we know and what we need to know. *J. Spat. Inf. Sci.* **2013**, *6*, 59–115. [CrossRef]
6. Całka, B.; Cahan, B.B. Interactive map of refugee movement in Europe. *Geod. Cartogr.* **2016**, *65*, 139–148. [CrossRef]
7. Horbiński, T.; Cybulski, P. Similarities of global web mapping services functionality in the context of responsive web design. *Geod. Cartogr.* **2018**, *67*, 159–177. [CrossRef]
8. Panchaud, N.H.; Hurni, L. Integrating Cartographic Knowledge Within a Geoportal: Interactions and Feedback in the User Interface. *Cartogr. Perspect.* **2018**, *89*, 5–24. [CrossRef]
9. Horbiński, T.; Cybulski, P.; Medyńska-Gulij, B. Graphic Design and Button Placement for Mobile Map Applications. *Cartogr. J.* **2020**. published online. [CrossRef]
10. Kraak, M.-J.; Ormeling, F. *Cartography: Visualization of Spatial Data*, 3rd ed.; Routledge: London, UK, 2011; ISBN 978-0273722793.
11. Roth, R.E. User Interface and User Experience (UI/UX) Design. In *The Geographic Information Science & Technology Body of Knowledge*; (2nd Quarter 2017 Edition); Wilson, J.P., Ed.; University Consortium Geographic Information Science: Washington, DC, USA, 2017. [CrossRef]
12. Medyńska-Gulij, B.; Myszczuk, M. The geovisualization window of the temporal and spatial variability for Volunteered Geographic Information System. *Geod. Cartogr.* **2012**, *61*, 31–45. [CrossRef]
13. Haklay, M.M. *Interacting with Geospatial Technologies*; Wiley-Blackwell: Hoboken, NJ, USA, 2010; ISBN 978-0470998243.
14. Garrett, J.J. *The Elements of User Experience: User-cantered Design for the Web and Beyond*, 2nd ed.; Pearson Education: Berkeley, CA, USA, 2010; ISBN 978-0321683687.
15. Norman, D.A. *The Design of Everyday Things: Revised and Expanded Edition*; The MIT Press: Cambridge, MA, USA, 2013; ISBN 978-0262525671.
16. Horbiński, T.; Lorek, D. The use of Leaflet and GeoJSON for creating the interactive web map of the preindustrial state of the natural environment. *J. Spat. Sci.* **2020**. published online. [CrossRef]
17. Roth, R.E.; Donohue, R.G.; Sack, C.M.; Wallace, T.R.; Buckingham, T.M.A. A Process for Keeping Pace with Evolving Web Mapping Technologies. *Cartogr. Perspect.* **2014**, *78*, 25–52. [CrossRef]
18. Sundar, S.S.; Bellur, S.; Oh, J.; Jia, H.; Kim, H.S. Theoretical importance of contingency in human-computer interaction: Effects of message interactivity on user engagement. *Commun. Res.* **2016**, *43*, 595–625. [CrossRef]
19. Medyńska-Gulij, B. Map compiling, map reading and cartographic design in "Pragmatic pyramid of thematic mapping". *Quaest. Geogr.* **2010**, *29*, 57–63. [CrossRef]
20. Gkonos, C.; Enescu, I.I.; Hurni, L. Spinning the wheel of design: Evaluating geoportal Graphical User Interface adaptations in terms of human-cantered design. *Int. J. Cartogr.* **2019**, *5*, 23–43. [CrossRef]
21. Cybulski, P.; Wielebski, Ł. Effectiveness of dynamic point symbols in quantitative mapping. *Cartogr. J.* **2019**, *56*, 146–160. [CrossRef]
22. Çöltekin, A.; Heil, B.; Garlandini, S.; Fabrikant, S.I. Evaluating the effectiveness of interactive map interface designs: A case study integrating usability metrics with eye-movement analysis. *Cartogr. Geogr. Inf. Sci.* **2009**, *36*, 5–17. [CrossRef]
23. Peterson, M.P. Maps and the Internet: What a Mess It Is and How To Fix It. *Cartogr. Perspect.* **2008**, *59*, 4–11. [CrossRef]
24. Cartwright, W.; Crampton, J.W.; Gartner, G.; Miller, S.; Mitchell, K.; Siekierska, E.; Wood, J. Geospatial Information Visualization User Interface Issues. *Cartogr. Geogr. Inf. Sci.* **2001**, *28*, 45–60. [CrossRef]
25. Vincent, K.; Roth, R.E.; Moore, S.A.; Huang, Q.; Lally, N.; Sack, C.; Nost, E.; Rosenfeld, H. Improving spatial decision making using interactive maps: An empirical study on interface complexity and decision complexity in the North American hazardous waste trade. *Envrion. Plan. B Urban Anal. City Sci.* **2019**, *46*, 1706–1723. [CrossRef]
26. Skopeliti, A.; Stamou, L. Online Map Services: Contemporary Cartography or a New Cartographic Culture. *ISPRS Int. J. Geo-Inf.* **2019**, *8*, 215. [CrossRef]

27. Wang, C. Usability evaluation of public web mapping sites. *Int. Arch. Photogramm. Remote Sens. Spat. Inf. Sci.* **2014**, *4*, 285–289. [CrossRef]
28. Bojko, A. *Eye Tracking the User Experience: A Practical Guide to Research*; Rosenfeld Media: New York, NY, USA, 2013; ISBN 978-1933820101.
29. Duchowski, A. *Eye Tracking Methodology: Theory and Practice*, 2nd ed.; Springer: London, UK, 2007; ISBN 978-1846286087.
30. Krassanakis, V.; Cybulski, P. A review on eye movement analysis in map reading process: The status of the last decade. *Geod. Cartogr.* **2019**, *68*, 191–209. [CrossRef]
31. Kiefer, P.; Giannopoulos, I.; Raubal, M.; Duchowski, A. Eye tracking for spatial research" Cognition, computation, challenges. *Spat. Cogn. Comput.* **2017**, *17*, 1–19. [CrossRef]
32. Ooms, K.; Krassanakis, V. Measuring the spatial noise of a low-cost eye tracker to enhance fixation detection. *J. Imaging* **2018**, *4*, 96. [CrossRef]
33. Cybulski, P. Spatial distance and cartographic background complexity in graduated point symbol map reading task. *Cartogr. Geogr. Inf. Sci.* **2020**, *47*, 244–260. [CrossRef]
34. Webb, N.; Renshaw, T. Eyetracking in HCI. In *Research Methods for Human Computer Interaction*; Cairns, P., Cox, A.L., Eds.; Cambridge University Press: Cambridge, UK, 2008; pp. 35–69. [CrossRef]
35. Nivala, A.M.; Brewster, S.; Sarjakoski, L. Usability Evaluation of Web Mapping Sites. *Cartogr. J.* **2008**, *45*, 129–138. [CrossRef]
36. Hegarty, M.; Smallman, H.S.; Stull, A.T.; Canham, M.S. Naïve cartography: How intuitions about display configuration can hurt performance. *Geovisualization* **2009**, *44*, 171–186. [CrossRef]
37. Newman, G.; Zimmerman, D.; Crall, A.; Laituri, M.; Graham, J.; Stapel, L. User-friendly web mapping: Lessons from a citizen science website. *Int. J. Geogr. Inf. Sci.* **2010**, *24*, 1851–1869. [CrossRef]
38. Goldberg, J.H.; Kotval, X.P. Computer interface evaluation using eye movements: Methods and constructs. *Int. J. Ind. Ergon.* **1999**, *24*, 631–645. [CrossRef]
39. Wielebski, Ł.; Medyńska-Gulij, B. Graphically supported evaluation of mapping techniques used in presenting spatial accessibility. *Cartogr. Geogr. Inf. Sci.* **2019**, *46*, 311–333. [CrossRef]
40. Manson, S.M.; Kne, L.; Dyke, K.R.; Shannon, J.; Eria, S. Using Eye-tracking and Mouse Metrics to Test Usability of Web Mapping Navigation. *Cartogr. Geogr. Inf. Sci.* **2012**, *39*, 48–60. [CrossRef]
41. Edler, D.; Vetter, M. The Simplicity of Modern Audiovisual Web Cartography: An Example with the Open-Source JavaScript Library leaflet.js. *KN—J. Cartogr. Geogr. Inf.* **2019**, *69*, 51–62. [CrossRef]
42. Mitchell, T. *Web Mapping Illustrated: Using Open Source GIS Toolkits*; O'Reilly Media: Sebastopol, CA, USA, 2005; ISBN 9780596554866.
43. Olsen, A. The Tobii I-VT Fixation Filter. Tobii Technology. 2012. Available online: https://www.tobiipro.com/siteassets/tobii-pro/learn-and-support/analyze/how-do-we-classify-eye-movements/tobii-pro-i-vt-fixation-filter.pdf (accessed on 20 March 2012).
44. Kraak, M.-J.; Brown, A. *Web Cartography: Developments and Prospects*; CRC Press: London, UK, 2005; ISBN 978-0203305768.
45. Popelka, S.; Vondrakova, A.; Hujnakova, P. Eye-tracking Evaluation of Weather Web Maps. *ISPRS Int. J. Geo-Inf.* **2019**, *8*, 256. [CrossRef]
46. Dong, W.; Zheng, L.; Liu, B.; Meng, L. Using Eye Tracking to Explore Differences in Map-Based Spatial Ability between Geographers and Non-Geographers. *ISPRS Int. J. Geo-Inf.* **2018**, *7*, 337. [CrossRef]
47. Hessels, R.S.; Kemner, C.; van den Boomen, C.; Hooge, I.T. The area-of-interest problem in eyetracking research: A noise-robust solution for face and sparse stimuli. *Behav. Res. Meth.* **2016**, *48*, 1694–1712. [CrossRef] [PubMed]
48. Göbel, F.; Kiefer, P.; Giannopoulos, I.; Duchowski, A.T.; Raubal, M. Improving map reading with gaze-adaptive legends. In Proceedings of the 2018 ACM Symposium on Eye Tracking Research & Applications, Warsaw, Poland, 14–17 June 2018; Volume 29, pp. 1–9. [CrossRef]
49. Phillips, M.H.; Edelman, J.A. The dependence of visual scanning performance on saccade, fixation, and perceptual metrics. *Vis. Res.* **2008**, *48*, 926–936. [CrossRef]
50. Horbiński, T. Progressive Evolution of Designing Internet Maps on the example of Google Maps. *Geod. Cartogr.* **2019**, *68*, 177–190. [CrossRef]
51. Muehlenhaus, I. *Web Cartography: Map Design for Interactive and Mobile Devices*; CRC Press: London, UK, 2013; ISBN 978-1439876220.

52. Lorek, D. Multimedia integration of cartographic source materials for researching and presenting phenomena from economic history. *Geod. Cartogr.* **2016**, *65*, 271–281. [CrossRef]
53. Shneiderman, B.; Plaisant, C. *Designing the User Interface: Strategies for Effective Human-Computer Interaction*, 5th ed.; Addison-Wesley: Boston, MA, USA, 2010; ISBN 978-0321537355.
54. Cybulski, P.; Medyńska-Gulij, B. Cartographic Redundancy in Reducing Change Blindness in Detecting Extreme Values in Spatio-Temporal Maps. *ISPRS Int. J. Geo-Inf.* **2018**, *7*, 8. [CrossRef]
55. Medyńska-Gulij, B.; Wielebski, Ł.; Halik, Ł.; Smaczyński, M. Complexity Level of People Gathering Presentation on an Animated Map—Objective Effectiveness Versus Expert Opinion. *ISPRS Int. J. Geo-Inf.* **2020**, *9*, 117. [CrossRef]
56. Halik, Ł.; Medyńska-Gulij, B. The differentiation of point symbols using selected visual variables in the mobile augmented reality system. *Cartogr. J.* **2017**, *54*, 147–156. [CrossRef]

International Journal of
Geo-Information

MDPI

Article

Web Map Effectiveness in the Responsive Context of the Graphical User Interface

Tymoteusz Horbiński *, Paweł Cybulski and Beata Medyńska-Gulij

Department of Cartography and Geomatics, Faculty of Geographical and Geological Sciences, Adam Mickiewicz University, 61-712 Poznań, Poland; p.cybulski@amu.edu.pl (P.C.); bmg@amu.edu.pl (B.M.-G.)
* Correspondence: tymoteusz.horbinski@amu.edu.pl; Tel.: +48-61-829-6307

Abstract: The main objective of this article was to determine the effectiveness of a web map GUI (Graphical User Interface) layout designed specifically for desktop monitors and smartphones. A suitable design of buttons for the graphical user interface is vital for the effectiveness of web maps. This article presents a study of three rules that prevail in GUI map design in terms of responsiveness, which was analyzed on two devices: a smartphone and a PC screen. The GUI effectiveness study, based on six variants of web maps, was conducted by means of eye-tracking on a group of 120 participants. An additional goal was to find an index (based on eye movements, mouse tracking, and time) that would be assessing the effectiveness of the GUI layout on both devices. The main motivation for conducting the research described in the article was the desire to find a synthetic measure based on more than one factor (time) in the context of determining the effectiveness of the GUI.

Keywords: web map; effectiveness; responsive GUI; multimedia cartography; GUI effectiveness index; eye tracking

Citation: Horbiński, T.; Cybulski, P.; Medyńska-Gulij, B. Web Map Effectiveness in the Responsive Context of the Graphical User Interface. *ISPRS Int. J. Geo-Inf.* **2021**, *10*, 134. https://doi.org/10.3390/ijgi10030134

Academic Editors: Wolfgang Kainz and Christos Chalkias

Received: 28 January 2021
Accepted: 1 March 2021
Published: 3 March 2021

1. Introduction

Web maps, such as Google Maps and OpenStreetMap, are globally known products. Their phenomenon is related to the globalization of cartography [1], as the same maps have never been used by people all around the world on a daily basis and for such diverse reasons (i.e., planning trips, car navigation, storytelling). The wide range of web portals publishing web maps creates competition that allows one to constantly improve their functionality, adapting maps for technological changes and the needs of users. The rapid development of and easy access to the Internet and GIS (Geographic Information System) constitute a direct cause of such a dynamic improvement of maps [2]; hence, new versions or updates occur in web cartography almost daily [3,4]. Technologies of the creation and publication of web maps are understood as an API (Application Programming Interface) interface, frameworks, libraries, services, etc. Open Web Platform Technologies are a collection of (free) technologies that allows one to surf the Internet, employing HTML, CSS, SVG, and XML network standards and the JavaScript programming language. Moreover, web maps can be displayed at different levels of detail and are quite easily updated [5,6]. Web maps are commonly used and, although they can be quickly and easily updated, the problem of providing quick interaction in terms of receiving information from a global map service remains. Meeting users' needs and preferences is related to the button layout (placement) in a GUI (Graphical User Interface) and constitutes one of the most significant factors that determine the simple usage of web maps [7].

The search for effective rules of web map design has lasted since the emergence of the Internet. This technological innovation was made publicly available in 1991, when the first version of the World Wide Web appeared, or in 1993, when the first search engine that handled GUI was invented [8]. Currently, the literature provides one with multiple terms defining or referring to the significance of employing the rules of multimedia cartographic design in the context of web maps, such as web-mapping platform [9], public web mapping sites [2,10], online map services [1], Internet map [6,11], and web map service [12].

The fundamental attributes of a web map are as follows:

Adaptability [13,14]—the map's ability to adapt to the system or software (responsiveness), or providing users with tools that allow them to change properties [15], e.g., the language of the map,

Interactivity [16,17]—the term is defined as a dialogue between a man and a map by means of the computer device [18–21],

Mobility [22,23]—the opportunity the map user is provided with to handle the web map when moving and being supported by navigation [24],

Multiscale [25,26]—the term that describes the employment of cartographic interaction to change scale along with the level of detail regarding the information presented on the map [17],

Being up-to-date [27,28]—real-time updates, providing users with the opportunity to react to events and processes taking place in space as they are intensifying, e.g., traffic density.

Web map design is based on general cartographic rules, such as the choice of the mapping method and cartographic sign design [29], and more detailed rules, e.g., cartographic sign as a core of multimedia [30]. However, regarding online cartography, basic elements related to the human–computer interaction, such as clicks, need to be taken into consideration [31]. Opportunities to work out maps are created by different APIs that generate cartographic (geographic) content and basic interactions [32,33] in the form of base maps. In multimedia cartography, JavaScript libraries with the open code and the opportunity to create web maps on the customer's side are used. OpenLayers, Leaflet, and D3 constitute the examples of libraries with open source code.

In multimedia cartography, it is necessary to use the rules of GUI, which are defined as a way of presenting information on the computer and interacting with the user. The program window with cartographic content as the largest element is a basis for graphical interface. Basic elements, referred to as widgets and designed in the concept stage of the creative process, are responsible for interactivity. In the process of web map design, preparing a comprehensive web map layout [34] that includes typical map elements and user's interactions (Figure 1) is of great significance. For interactions in the map content window, one uses buttons that allow one to augment, reduce, move and in other ways alter the map view. On the interactive map, different icons, such as the one of a magnifying glass, hand, or an arrow, can appear, as users intuitively associate them with specific navigation functions. The object search function can be handled by the edit box, in which the searched name is typed. Moreover, cartographic signatures are interactive thanks to events (e.g., mouseover, click), which activate additional information included in tooltips and pop-ups.

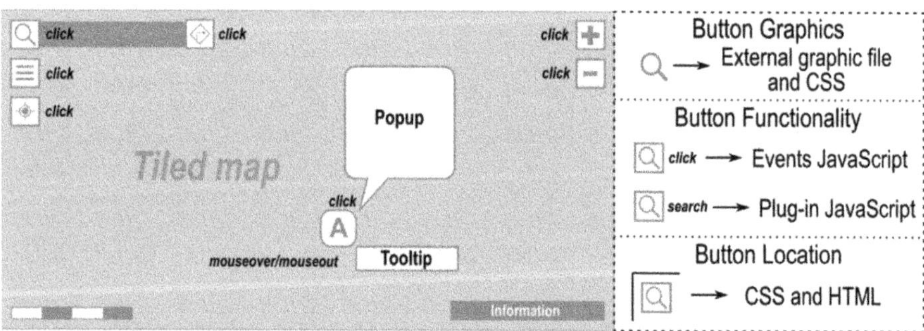

Figure 1. Web map GUI (Graphical User Interface) layout of buttons with interactive events.

GUI has become highly significant in software engineering and programming. In this paper, GUI denotes spots symbolized by buttons (with specific placement) for the user's interaction with a program or a web map. In this study, we use only buttons with symbolic icons as the GUI. It is a highly relevant element of each web map. GUI's functioning is based on associations and knowledge of users by the employment of simple symbols in button design. The same icons (symbols) have been used by the most popular web map, Google Maps, for many years. Google maps is widely used, which makes its symbols easily recognizable [7].

The problem of users' preferences is also related to the type of equipment on which a web map is displayed. At present, web maps and their GUI follow two rules on how the content adjusts to the device. The first one assumes a different button layout on the PC monitor and on the smartphone, e.g., Google Maps, whereas the second one assumes only the transformation of the existing web maps, which were originally designed for non-mobile devices, into mobile maps, such as OpenStreetMap [12]. The adjustment of the map size and GUI button placement to a particular display screen is defined as a responsive web map. In general terms, responsiveness is associated with the ability of a website to adjust to different devices and types of definition. Responsiveness consists in designing websites/web maps for mobile devices (with smaller display screens, such as smartphones) prior to extending the design process to devices with larger display screens, e.g., PC monitors [35]. So far, maps have not been analyzed in this respect.

The effectiveness of GUI of web maps, just as of any other product in multimedia cartography, can be studied by means of multiple techniques and methods. Questionnaires [36] and the usability method [2,37] are the two most important methods of measuring effectiveness in cartography. Along with technological advancement, the eye-tracking technique, which combines the questionnaire method with the usability method, extending the research by parameters resulting from the direct observation of human eye movement, has been worked out [38]. To meet the objective of this article, the previously suggested GUI effectiveness index, based on time parameters obtained thanks to the eye-tracking technique, both for mobile devices and PC screen, was employed.

2. Related Research

Studies of effectiveness regarding both cartographic products and the interface are well described in the literature. Goldberg and Kotval [39] noted that in the evaluation of the interface, the visual strategy of the user should be taken into account. They noticed also that a more effective interface results in a smaller number of fixations, whereas the less effective interface causes more fixations. Furthermore, they proved that higher effectiveness correlates with a shorter *scanpath* time, as opposed to interface elements laid out at random. Some authors identify an effective search for elements of the map with accuracy, whose criterion is determined by correct location in space and task completion time [38,40]. Sutcliffe et al. [40] showed that the effectiveness of an interface is affected by, among others, the knowledge of the system on which a task is performed and experience. Less experienced users demonstrated significantly more conservative strategies of interacting. Çöltekin et al. [38] compared two GUIs of multimedia and interactive web maps in tasks with users. In their conclusions, they showed that they had additionally used tools for eye movement registration, which enriched their study in terms of both quality and quantity, thus corroborating the hypothesis that one interface is faster and more accurate.

In computer simulations that use the map as the basis for spatial information, Saw and Butler [41] also employed the approach that consists in the comparison of interfaces. In their opinion, GUIs with buttons located in the bottom part and with map navigation functions were more effective during the work with users than the interface located on the left-hand side. Research on users' interactions with the interface was also conducted by Gołębiowska et al. [42]. They proved that during the exploration of a coordinated and multiple views (CMV) interface, novice users paid attention to elements explaining how specific tools worked. Moreover, it turned out that users would much more often use interactions by means of a mouse rather than buttons, such as the zoom in/out button. Furthermore, not all available interactions were used. The eye-tracking analysis helped researchers establish which interface elements were observed the most and how users' attention was changing during exploration.

3. Motivation, Aim, and Research Questions

To sum up the research described above, we would like to highlight that GUI effectiveness research, apart from correctness and task completion time, should also take the user's eye movement analysis into account. Hence, the willingness to find synthetic measurement based on more than one factor (time) in the context of determining GUI effectiveness constituted the main motivation for carrying out the research described in this article.

We ask the following questions to the discussed topic:

- Do the rules of web map design related to GUI have the same effectiveness for cartographic products that are displayed on devices of varying size?
- Does the way that interactive map-based tasks are performed depend on the interaction button layout?
- Should the rules of web map design for maps with the same cartographic content be the same for smartphones and PC monitors?

The main objective of the article was to determine the effectiveness of a web map GUI layout designed specifically for a desktop monitor and smartphone. An additional goal was to confront the existing principles of web map design with the rules of responsiveness. A second additional goal was to find an index (based on eye movements, mouse tracking, and time) that would be assessing the effectiveness of the GUI layout on both devices.

4. Methodology

An effectively designed GUI of web maps is characterized by the quick and correct location of the button with a map function searched for by the user. The map function, which is ascribed to the appropriate button and represented by the appropriate icon, gives the user the opportunity to interact with the map and obtain the information that it includes. In this article, effectiveness shall be understood as dependence between the time to first fixation, time of identification, and time of completion of the task (the so-called time to first mouse click). The fixation time is associated with the time to first fixation. A fixation is understood as focusing/fixing one's gaze on a given spot (i.e., the GUI button searched) for a longer time, which suggests that the person is paying attention to this spot. The identification time is related to identification, i.e., proper understanding of the icon that represents the searched map function located in the GUI button. Time to first mouse click is the time of proper interpretation of the icon and activation of the function by pushing the right button.

To achieve our objective, the author adopted four main stages of research:

- To select three variants of GUI (Section 4.1, Figure 2),
- To work out three variants of GUI (Section 4.2, Figure 2),
- To carry out the eye-tracking research with participants (Section 4.3),
- To formulate GUI effectiveness index (Section 4.4),
- To analyze and demonstrate the results (Section 5, Tables 1 and 2).

4.1. Choice of Three Gui Variants

One of the research simplifications in this article was to adopt the unified set of icons representing map functions, on the basis of which researchers decided to consider icons not because of their use by popular web maps but because they were understood by users best. Thus, the set of icons from the research by Horbiński et al. [7] was selected. In the research, a group of 100 respondents was questioned. The respondents were supposed to assign icons to specific web map functions. Icons that were matched correctly constitute the set used in this article.

Currently, there are two rules in terms of designing a web map GUI:

- To design two different web map GUIs for the PC monitor and the smartphone, e.g., Google Maps (Figure 2 Rule 3),
- To adjust the GUI of the web map designed for a PC monitor to a smartphone screen, e.g., OpenStreetMap (Figure 2 Rule 2).

The author of this article decided to verify the responsiveness rule by Marcotte [35] in the context of the GUI of the web map user. The rule consists of designing for mobile devices (smartphones) first and creating products for PC monitors afterward. Researchers decided to use the button layout variant worked out on the basis of the study by Horbiński et al. [7]. The GUI of the user is based on the choice of respondents, who decided on the map function layout on the smartphone screen. They could choose from 6 map functions, i.e., *Geolocation, Change layer, Search, Route, Default range map,* and *Measure*. Three map functions were selected by each respondent (*Geolocation, Search,* and *Route*). Referring directly to other currently existing web maps, researchers decided to include, apart from the three most frequently used functions, also the *Layer* button. It was necessary and dictated by the fact that exactly these four functions (*Geolocation, Layer, Route,* and *Search*) occurred in most mobile versions of web maps [11,12]. With the hypothetical variant of the web map GUI, designed exclusively for smartphones, the author of the article adjusted the variant to the PC monitor (decreasing the buttons) (Figure 2, Rule 1). Hence, he employed the responsiveness rule by Marcotte [35].

4.2. Creation of Three Gui Variants

All three variants were programmed for the needs of the research with the use of the Leaflet library (and coexisting plugins) (Figure 2) [43–45]. The responsiveness of variants was secured by media queries, fluid grid, and breakpoints that conditioned displaying elements created thanks to CSS coding. OpenStreetMap was used as a base map, as this global map with geodata stored in a database available through JavaScript could be used for free with the open license. The compatibility of the web map with the Leaflet.js environment constituted the main factor that determined the choice of GUI of OpenStreetMap. In addition, OpenStreetMap (OSM) was highly popular on a global scale, which was confirmed by search results by browsers such as Google, Bing, or Yahoo [46]. The view of the map was centered to be 52.17° N (latitude) and 3.43° W (longitude) with the zoom level of 16. The author emphasizes that GUI variants are being considered only at the first level of interactions, i.e., the activation of map functions occurs during the first interaction with the button (after pressing it). The base map is designed similarly to the topographic map. It contains points, lines, and areal objects. This includes anthropogenic and natural features. We did not change the map content. The OSM functionality contains interactions that enable e.g., legend preview or change layers; however, for the experimental simplification, we used only tools presented in Section 4.1.

4.3. Eye-Tracking Study with Respondents

A homogenous group of respondents with similar experience in working with web maps and from a similar age group participated in the research. Researchers assumed that each variant of tasks connected with the web map would be solved by 20 people. One hundred and twenty students aged 18–25 (66% men, 34% women) participated in the research. All the respondents declared that they used web maps (on smartphones and PC monitors). It is the most adequate research sample for such studies, as it consists of people that use mobile products in their daily life.

The research was conducted in a room with continuous lighting, with the use of the following equipment and software: Smartphone—Samsung Galaxy S7 (screen diagonal 5.1″, resolution 1920 × 1080), Monitor—LG Falatron E2260T-PN (screen diagonal 21.5″, resolution 1920 × 1080), Eye-tracker—Tobii X2-60, Software—Tobii Studio 3.4, Web browser—Mozilla Firefox, MDS—Mobile Device Stand. The same device for tracking the eye movement has been implemented on a desktop computer monitor and on a smartphone. For a smartphone, we used the Mobile Device Stand. In this solution, the mobile device is attached to a holder in a known location. Above the smartphone, an adjustable camera is placed, which enables recording the participants' interaction with the device. Eye tracking equipment is located below the smartphone so as to not interfere with the smooth use of the mobile device (Figure 3). In both solutions (PC and smartphone), the eye tracker is indirectly connected to the computer via a computing module. The overall accuracy corresponded to 24 pixels on the monitor and 72 pixels on the smartphone. This was based on the average distance between participants and the device. On a desktop monitor, it was 68.2 cm, and on a smartphone, it was 65.4 cm. For the detection of fixations and saccades, we used the velocity-based algorithm (I-VT). The velocity threshold was set to 2.1 px/ms.

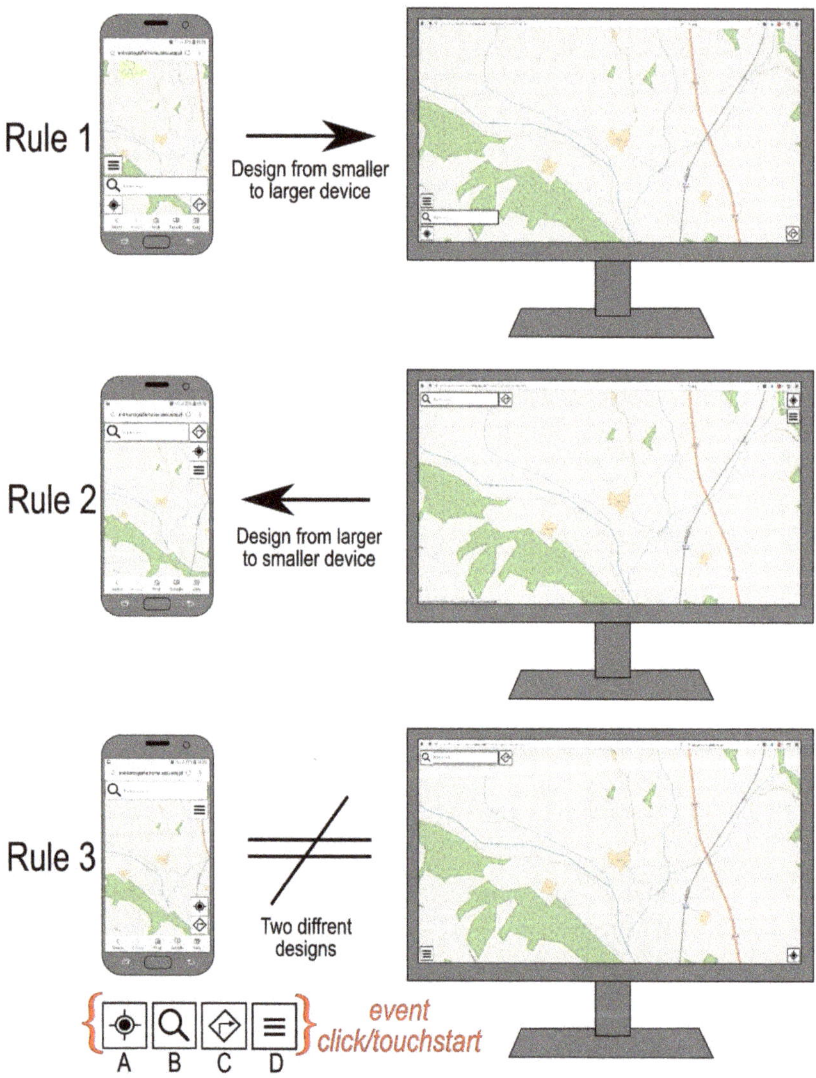

Figure 2. Variants of GUI (Graphical User Interface) design according to three rules (A—Geolocation, B—Search, C—Route, D—Change Layer).

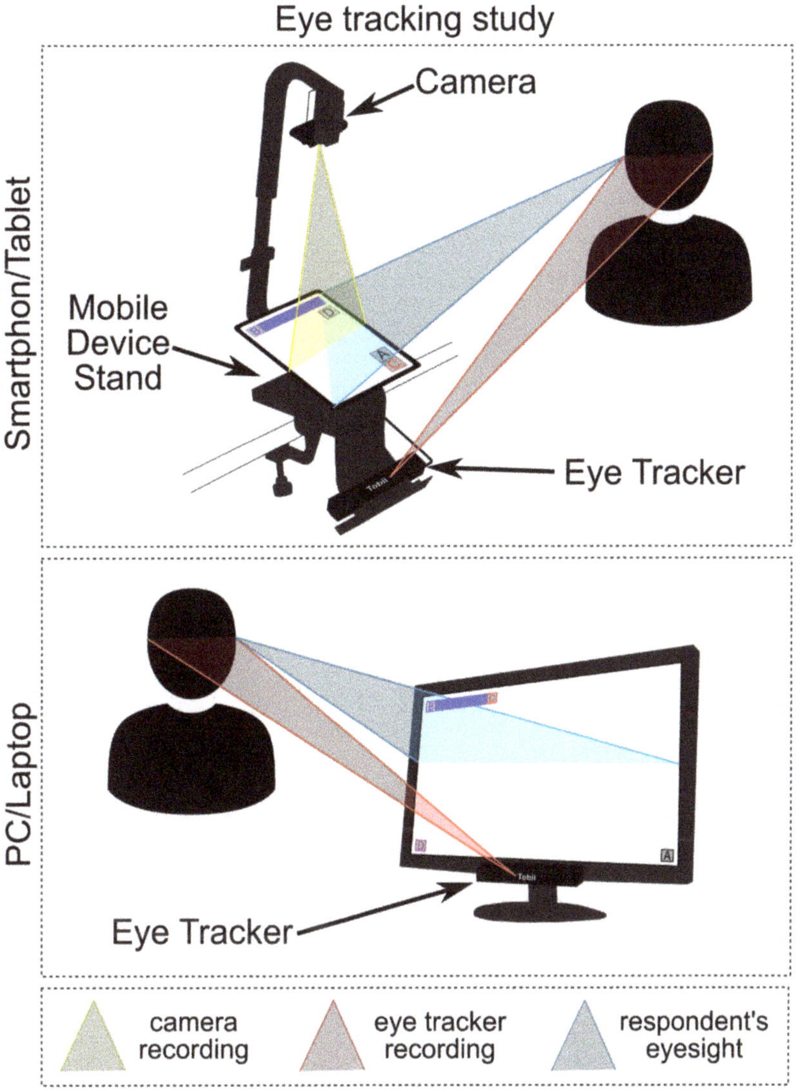

Figure 3. The method of conducting the eye-tracking study with a smartphone and PC screen.

Each respondent was instructed on how to use the equipment and the objective of the research, after which the equipment was calibrated. We performed a 5-point calibration, and during the test, we did not observe any anomalies. During the research, respondents had to complete three tasks that were supposed to verify the effectiveness of the web map button layout in analyzed variants, i.e., *Geolocation*, *Search*, and *Route*:

- Q1: Identify your location through geolocation (press *Geolocation*),
- Q2: Search for the town of Żywiec through the *Search* button (press *Search*),
- Q3: Determine the route connecting two towns: Jasło and Leżajsk (press *Route*).

Each group (20 people) completed three tasks for one variant of button layout (Figure 2) on one device (monitor or smartphone). Specific respondents were selected for groups on the basis of their characteristics to make groups as homogenous as possible. No time constraints were established in the research for completing the tasks. In tasks no. 2 and 3, town names with Polish diacritics were used to decrease the number of results shown. Although performing the tasks did not take too long, one needs to note that the entire research included introducing the respondent to the topic of the research, discussing the equipment, and explaining how specific elements worked (e.g., the issue of the mouse cursor, which was always invisible at the moment of beginning the task and the respondent had to move it from the left edge of the screen, was explained), the characteristics of the user, calibration and, finally, performing tasks by users.

4.4. GUI Effectiveness Index

As a part of the research, the GUI effectiveness index was suggested. The index uses the correlation between the time of fixing the gaze on the button (time to first fixation—T_{FF}), time of identifying the button (i.e., the difference between task completion time and identification time—IT), and task completion time (time to first mouse click—T_{FMC}). The importance of T_{FF} allows us to determine after what time the user consciously noticed the button. On the other hand, T_{FMC} tells us how quickly the user clicked on the button that he had noticed. IT, which denotes how much time the user needed for identifying the button from the moment he noticed the button until he interacted with it, also occurs in the index; however, it occurs indirectly. The index defines three activities (times) of the user on the map. Compared to previously used methods that considered only one measurement, the index turned out to be more comprehensive here.

The rate is considered in two variants; i.e., when (T_{FF} and IT $\neq$ 0):

$$T_{FF} <= IT \tag{1}$$

then:

$$EI_{GUI} = [((T_{FF}/IT)/T_{FMC}) * (T_{FF} + IT + T_{FMC})]/\text{value of } T_{FF} \tag{2}$$

and when:

$$T_{FF} > IT \tag{3}$$

then:

$$EI_{GUI} = [((IT/T_{FF})/T_{FMC}) * (T_{FF} + IT + T_{FMC})]/\text{value of IT.} \tag{4}$$

The analysis of the index in two variants is supposed to indicate the most effective GUI for web maps, both in terms of layout and the graphics used. Adopting only the first variant in the situation of very quick identification time and relatively longer first fixation time, the value of the index would not indicate high effectiveness. Multiplication by the sum of times eliminates the unit of measure (1/s) that we would receive only when dividing times by themselves. Final division by the value (without the unit of measure) of the first fixation time or identification time is the most significant element of the pattern. It has a highly relevant impact on the value of the index. Assuming that there are small differences between time to first fixation and identification time, the index would have the same value, regardless of the time value. Thus, division by the value of T_{FF} or IT favors a lower time value, which makes the rate value objective. Objectivity results mainly from the fact that when the value of both T_{FF} and IT is low, the index is high and may achieve the maximum value of 2. When the value for both times is high, the index is low. The index does not favor large disproportions between the times, i.e., when the value of T_{FF} is low and the one of IT is high or the other way around, the value of the index is low.

Correctness of the GUI index was tested by means of the program written in JavaScript (Figure 4), which was started along with the HTML code structure in each search engine. The *getRandomInt* function determined a random number from the range provided as function arguments (*min, max*) for the testing, thanks to which the function adopted random data, regardless of the testing person. The code presented in Figure 4 adopted the values in the range (0.3; 10), but any other range could be determined. The *index_EGUI* function calculated the GUI effectiveness index, considering two cases (T_{FF} <= IT | | IT <

T_{FF}), and all the values collected were included in the table in the key-value section. The size of the table (the number of verses in the table) was determined by the *val* attribute of the *index_EGUI* function.

```
function getRandomInt(min, max) {
    min = min;
    max = max;
return (Math.random() * (max - min)) + min;
}
function index_EGUI(val){
var dane = new Array(val);
    for ( i=0; i<val; i++) {
        var a = getRandomInt(0.3,10); // Time to first fixation
        var b = getRandomInt(0.3,10); // Identification time
        var c = a+b; // Time to first mouse click
        if (a<=b){
            var d = (((a / b) / c)*(a+b+c));
            var EGUI = d / a;
        }// TFF <= IT
        else{
            var d = (((b / a) / c)*(a+b+c));
            var EGUI = d / b;
        }// IT < TFF
        var obiekt = {Tz: a, Ti: b, Tr: c, ET: d, EGUI: EGUI};

        dane[i] = obiekt;

        document.write("<tr>" + "<td class='tz'>" +
        dane[i]["Tz"] + "</td>" + "<td class='ti'>" +
        dane[i]["Ti"] + "</td>" + "<td>" + dane[i]["Tr"] +
        "</td>" + "<td>" + dane[i]["ET"] + "</td>" +
        "<td class='jeden'>" + dane[i]["EGUI"] +"</td>"+"</tr>");
        // Presentation of the results in the form of a table
    }
}
index_EGUI(10);
```

Figure 4. Program for testing the GUI effectiveness index.

5. Results

The first step of the result analysis was to calculate the average T_{FF}, IT, and T_{FMC}. The average time value is presented in Table 1. Time results should be analyzed for specific tasks and devices. The pink color was used for marking the box with the shortest time and the blue color was used for the box with the longest time needed for individual tasks.

The lowest average T_{FF} both on the smartphone (0.66 s) and on the PC screen (0.34 s) was achieved for the task no. 2 (*Search*)—the button layout variant according to the first rule. The *Search* button in the layout variant following the first rule was identified on the smartphone the most quickly (2.61 s), whereas the *Route* button (Rule 1) was identified on the PC most quickly (1.33 s). Similar to T_{FF}, the average T_{FMC} was also the shortest for task no. 2 (rule 1), both on the smartphone (3.27 s) and on the PC (2.5 s) (Table 1).

Table 1. The average time to first fixation (T_{FF}), identification time (IT), and time to first mouse click (T_{FMC}) for individual tasks and button layout variants in GUI as suggested on the basis of three rules (Rule 1—Design from smaller to larger device; Rule 2—Design from larger to smaller device; Rule 3—Two different designs).

		T_{FF} [s]			IT[s]			T_{FMC} [s]		
	Rule	Q1 (σ)	Q2 (σ)	Q3 (σ)	Q1	Q2	Q3	Q1 (σ)	Q2 (σ)	Q3 (σ)
Smartphone	1	2.04 (1.65)	0.66 (0.7)	1.21 (1.07)	4.78	2.61	5.53	6.82 (3.76)	3.27 (1.5)	6.74 (6.86)
	2	1.73 (1.27)	0.86 (0.85)	0.93 (0.92)	4.17	3.06	4.66	5.9 (1.27)	3.92 (1.98)	5.59 (3.54)
	3	1.25 (1.54)	0.83 (0.73)	1.68 (1.51)	5.57	4.77	4.25	6.82 (2.67)	5.6 (6.05)	5.93 (1.57)
Desktop Monitor	1	5.96 (5.16)	0.34 (0.56)	3.15 (3.02)	5.72	2.16	1.33	11.68 (9.59)	2.5 (2.85)	4.48 (3.6)
	2	4.27 (2.96)	1.42 (3.36)	1.75 (2.06)	4.82	2.04	2.69	9.09 (6.38)	3.46 (4.44)	4.44 (3.47)
	3	8.23 (5.42)	1.29 (2.62)	5.26 (5.84)	5.12	3.17	2.11	13.35 (11.09)	4.46 (5.57)	7.37 (6.87)

One task, correctly and quickly completed by respondents on two devices, related to the *Search* button, does not determine the effectiveness of the GUI variant. Objective results for GUI can be achieved through the employment of the GUI effectiveness index suggested in the article.

According to the pattern suggested (Section 4.4), GUI effectiveness indices were calculated for individual tasks. All the results were juxtaposed in the table (Table 2). As in Table 1, the blue color was used to mark the lowest index and the pink color was used to mark the highest index. To evaluate GUI, the indices for individual tasks needed to be summed up (the Sum column). In the research, respondents completed three tasks, so the total value of the index can reach a maximum of 6 (for the individual device). In terms of responsiveness (for the smartphone or PC monitor), the variant following rule 2 (3.701) was the most effective GUI variant, which does not directly verify the responsiveness rule on websites (Rule 1—3.439) by Marcotte [35], according to which web maps should be designed for small mobile devices (smartphones) first and then adapted for large screen devices (PC monitor).

Table 2. GUI effectiveness index (EI_{GUI}) for individual tasks and variants of the GUI button layout.

		EI_{GUI}			
	Rule	Q1	Q2	Q3	Sum
Smartphone	1	0.418	0.766	0.362	1.546
	2	0.480	0.654	0.429	1.562
	3	0.359	0.419	0.471	1.249
Desktop Monitor	1	0.336	0.926	0.635	1.896
	2	0.415	0.980	0.743	2.139
	3	0.243	0.631	0.380	1.254

In Figure 5, 18 heatmaps were demonstrated for the spatial layout of fixations during performing tasks by individual users, both on the PC monitor and on the smartphone (compare to Figure 2). Due to the equipment limitation that made it impossible to place the smartphone in the same position on MDS (Mobile Device Stand) each time, it was impossible to generate summary heatmaps for all users for specific tasks (individual scenes for each research participant).

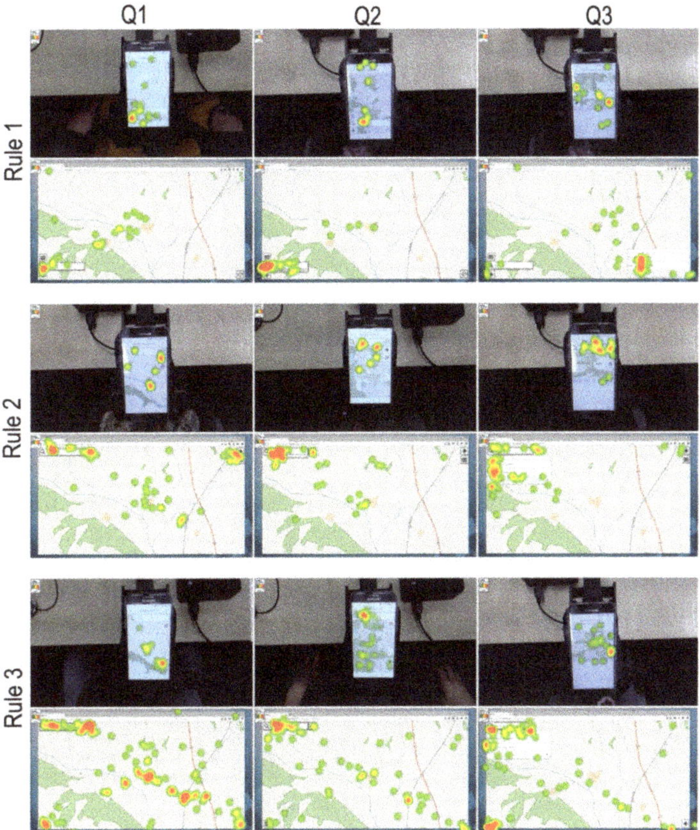

Figure 5. Eighteen heat maps presenting the spatial distribution of fixations in each task for the three rules on the computer monitor and on the smartphone (compare to Figure 2).

The spatial fixation layout of users performing tasks 1, 2, and 3 on the basis of Rule 3 on a PC monitor shows different gaze fixation spots. Such layout of fixation concentration directly decreases the index (Table 2) as well as affects the average task completion time (Table 1). One can also observe multiple indirect fixation clusters in the straight line between buttons (the diagonal of the screen). The GUI, based on Rule 3 (PC monitor), requires users to analyze one corner with buttons more compared to other rules. The results can be seen on the smartphone; however, the differences between the value of indices are smaller than those for the PC monitor.

Considering Rules 1 and 2, we do not note significant differences between the values of the index on a smartphone. The size of the screen used by the respondent on a smartphone may be of vital importance here, whereas for a PC computer monitor, the differences are significantly larger. The result could have been affected by placing buttons in the upper (Rule 2) and lower (Rule 1) part of the screen. Users' habit of using programs with an upper interface could have had an impact on the results in terms of GUI effectiveness for web map users.

6. Discussion

User–map interactions are crucial elements of both web and multimedia cartography. Different ways of designing web map GUI and their confrontation provided the researchers with interesting conclusions. In the evaluation of web map GUI, the effectiveness index suggested is key. In drawing conclusions and presenting data on the effectiveness of the map solutions studied, multiple authors used tables and specially designed diagrams [47,48]. The research considered the quickness and correctness of the answers delivered. These two parameters were associated with the effectiveness of the analyzed cartographic products. Subjective opinions of the respondents were also used [36]. In this analysis, to make both the results and the conclusions more accurate, the author suggested a GUI effectiveness index.

In this article, a group of 20 people completing tasks independently for each of the cases discussed was examined. The choice of such a research sample (the number and characteristics of respondents) was based on other multimedia cartography studies. According to Ware [49], the majority of studies are carried out on 12–20 respondents, whereas studies in which the learning effect may occur require more participants. Usually, the so-called "public users" participate in such studies [47]. These are often students, as they are easily available to researchers and relatively homogenous, particularly when specializing in the same field. A large group of students makes it possible to divide them into several teams, allowing respondents from different teams to evaluate different versions of mapping techniques that depict the same spatial data [50,51]. Multiple examples of studies conducted by other authors confirm such choice of research sample [47,52,53]. In effectiveness research, many authors used subjective opinions by users and, by means of timed task completion [47], checked the effectiveness of the multimedia cartographic presentation that employed VR (Virtual Reality), focusing mainly on talks with students and drawing conclusions on subjective opinions. Wielebski and Medyńska-Gulij [36], researching different variants of mapping methods, centered on determining effectiveness as quick and correct task completion. Such studies are most often carried out among university students and much less often among passers-by [54] or respondents that perform various tasks in space with mobile devices [55,56].

In their research, Nivala et al. [48] touched upon the role of the graphics of icons on web maps. They highlighted that the web map should be designed is such a way so that it considers the feelings of respondents, whereas interface and icons should be as simple as possible and easily understandable for the user at the same time. Not only did the authors use the most easily understandable (identifiable) icons in their research, they also tested the interface suggested by users' opinions. The interface was designed for mobile devices (smartphones) and then adapted for PC screens so that it still followed the responsiveness rule [35].

The objective of the article was to determine the effectiveness of web maps designed for mobile devices, which were confronted with the existing rules of web map design. The translation of the responsiveness rule by Marcotte [35] into web mapping, particularly into web maps, failed to prove the validity of this rule (Table 2). GUI is the most significant responsive element of each web map. People's habit of working on programs with the interface located in the upper part of the window translates into the effectiveness of the web map. It is worth mentioning that the employment of Rule 2 of GUI responsive design turned out to be the most effective. Analyzing the results of the index for individual tasks completed on a smartphone showed that each interface has one task that was completed more effectively. That may serve as the evidence that technological advancement should be heading toward the personalization of all the activities on smartphones. In terms of the results obtained and the differences between specific systems, solutions that personalize applications (through the choice of functionality) at the first activation should be favored, supported, and introduced more often. Naturally, it is just the first step in the context of individual GUI of web maps for each user.

On web maps, the possibility of changing a cartographic background is an important element of the interface [12,57]. The real content of the cartographic background is a vital part of map design. Research reports that the visual complexity of the map content is a factor that impacts effectiveness [58,59]. However, psychological studies confirm that in different conditions related to strong attention (when performing a specific map-based task), people tend to omit background information, focusing on task accomplishment [60]. Therefore, it is reasonable to infer that tasks performed by the participants in this study fit in the strong attention map-based tasks. This means that participants' performance may be similar with high probability when performing any other global web mapping service.

The GUI effectiveness index proposed by the authors requires in-depth studies in the future. Confirmation by further empirical research will ensure verification of the proposed GUI effectiveness index. Future research requires index verification, including other devices displaying a web map, different GUI layouts, various button graphics, or diverse functionality. The authors also believe that the developed index does not have to be solely and exclusively used to determine the effectiveness of the GUI of web maps. Further research in other areas may confirm its usefulness.

7. Conclusions

Three presented ways of designing button layout for map functions on smartphones indicate a responsive approach and, in this context, thanks to the index, one can clearly indicate features of each approach. The inclusion of responsiveness in the user–map interaction design seems legitimate, as the same products of multimedia cartography are used on display screens of different size, and the simultaneous designing for at least two devices becomes significant to the designer.

Answering the question asked in the article, it can be concluded that the value of the GUI effectiveness index differs not only for the rules of GUI design analyzed but also between the devices on which the web map is displayed. The fact that the layout of user–computer interaction buttons is of great importance to the times analyzed (T_{FF}, IT, and T_{FMC}) is one of the conclusions drawn. The way the task was performed in the context of web map GUI depends predominantly on three factors, which are graphics (symbols) representing the functions of a given button, the location of the button on screen, and the device itself. On the basis of the research, also, the fourth factor in the analysis of web map GUI, i.e., the number of corners with buttons, needs to be taken into account. As the results show (Table 1 and Figure 5), the number of corners with buttons not only affects the spatial fixation layout but also the GUI effectiveness index directly.

In answer to the first question related to the rules of web map design in terms of GUI, it is necessary to say that the GUI effectiveness index fails to provide the same value for the same cartographic products displayed on different devices (smartphone screen and PC monitor), which may be related to the difference in the size of their screens. It can also validate using less information for web maps (fewer buttons and reduced complexity of a base map) [12]. In answer to the second question, it needs to be said that the button layout plays a decisive role in the way tasks are performed (the more corners with buttons, the lower the index). Answering the third question, the highest value of the GUI effectiveness index for Rule 2 suggests that the web map should have the same button layout both on a PC monitor and smartphone screen. It may result from the previously mentioned habits of users, who are accustomed to buttons located in the upper part of the screen, and it also may corroborate the legitimacy of using the existing habits of users in the process of web map design [32].

The results related directly to individual buttons, particularly the *Search* button, make the statement by Horbiński and Cybulski [12] subject to further research. The statement referred to a habit of using buttons located in upper part of the screen. The authors of the article concluded that facilitation in the form of buttons located closer to the thumb on the smartphone screen might fail to bring the expected results, i.e., higher effectiveness of such map. The average times (Table 1) as well as EI_{GUI} (Table 2) for task 2 indicate that the location of the *Search* button in the lower part of the smartphone screen has a positive impact on the effectiveness of such solution. The fact of introducing the toolbar located at the bottom of the screen to Mozilla Firefox (for smartphones) in the latest updates (Android) may serve as the evidence that such solution works.

The interactions with the map presented in this study could be related with the map-using tasks presented by Keates [61]. Firstly, fundamental map activity is searching for places. However, users could perform random visual search, but more often, people are searching for a target place. On a paper map, one could search the name in the index. On a web map, the index of places is replaced by an interactive search button. Another fundamental map-using task is searching for route. This is the most common activity in navigation systems. In a web map, users are able to use the route search button, and the map is used to anticipate future action and verify current position. Maps can be used in more sophisticated ways. Keates use an example of a population map and chart comparison task. This would require extensive visual search in different patterns. A web map could support this type of map use with more complex interactive tools such as spatial queries [62].

The authors hope that the GUI effectiveness index will be useful in other studies of multimedia cartography. In the evaluation of effectiveness of the interface element or the map element, one can employ the index suggested, as it is numeral and objective. However, the interactive user–map action in which the analyzed interface element needs to be used by the respondent in the research is crucial. Indicating the effectiveness of cartographic symbols in web maps can be another example of how one can use the index, as cartographic content consists of various symbols depicting the same types of geographical objects.

Author Contributions: Conceptualization, Tymoteusz Horbiński and Beata Medyńska-Gulij; methodology, Tymoteusz Horbiński; formal analysis, Tymoteusz Horbiński and Paweł Cybulski; resources, Tymoteusz Horbiński; writing—original draft preparation, Tymoteusz Horbiński; writing—review and editing, Tymoteusz Horbiński, Paweł Cybulski and Beata Medyńska-Gulij. All authors have read and agreed to the published version of the manuscript.

Funding: This research received no external funding.

Acknowledgments: This paper is the result of research on visualization methods carried out within statutory research in the Department of Cartography and Geomatics, Faculty of Geographical and Geological Sciences, Adam Mickiewicz University Poznań, Poland.

Conflicts of Interest: The authors declare no conflict of interest.

References

1. Skopeliti, A.; Stamou, L. Online Map Services: Contemporary Cartography or a New Cartographic Culture. *ISPRS Int. J. Geo-Inf.* **2019**, *8*, 215. [CrossRef]
2. Wang, C. Usability evaluation of public web mapping sites. *Int. Arch. Photogramm. Remote Sens. Spat. Inf. Sci.* **2014**, *4*, 285–289. [CrossRef]
3. Haklay, M.; Singleton, A.; Parker, C. Web mapping 2.0: The Neogeography of the GeoWeb. *Geogr. Compass* **2008**, *2*, 2011–2039. [CrossRef]
4. Roth, R.E.; Harrower, M. Addressing Map Interface Usability: Learning from the Lakeshore Nature Preserve Interactive Map. *Cartogr. Perspect.* **2008**, *60*, 46–66. [CrossRef]
5. Dillemuth, J. Map Design Evaluation for Mobile Display. *Cartogr. Geogr. Inf. Sci.* **2005**, *32*, 285–301. [CrossRef]
6. Meng, L. Egocentric Design of Map-Based Mobile Services. *Cartogr. J.* **2005**, *42*, 5–13. [CrossRef]
7. Horbiński, T.; Cybulski, P.; Medyńska-Gulij, B. Graphic Design and Button Placement for Mobile Map Applications. *Cartogr. J.* **2020**, *57*, 196–208. [CrossRef]
8. Peterson, M.P. Maps and the Internet: What a Mess It Is and How to Fix It. *Cart. Perspect.* **2008**, *59*, 4–11. [CrossRef]
9. Wallace, T. A new map sign typology for the geoweb. In Proceedings of the 25th Annual International Cartographic Conference, Paris, France, 3–8 July 2011; ISBN 978-1-907075-05-6.
10. Mitchell, T. *Web Mapping Illustrated: Using Open Source GIS Toolkits*; O'Reilly Media: Sebastopol, CA, USA, 2005; ISBN 9780596554866.
11. Horbiński, T. Progressive Evolution of Designing Internet Maps on the example of Google Maps. *Geod. Cartogr.* **2019**, *68*, 177–190.

12. Horbiński, T.; Cybulski, P. Similarities of global web mapping services functionality in the context of responsive web design. *Geod. Cartogr.* **2018**, *67*, 159–177. [CrossRef]
13. Reichenbacher, T. Adaptive methods for mobile cartography. In Proceedings of the 21st International Cartographic Conference (ICC), Durban, South Africa, 10–16 August 2003; ISBN 0-958-46093-0.
14. Friedmannová, L.; Konečný, M.; Staněk, L. An adaptive cartographic visualization for support of the crisis management. In Proceedings of the AutoCarto 2006, the 16th International Research Symposium on Computer-Based Cartography, Vancouver, WA, USA, 26–28 June 2006; Available online: https://cartogis.org/autocarto/autocarto-2006/ (accessed on 2 March 2021).
15. Oppermann, R. Adaptive User Support: Ergonomic Design of Manually and Automatically Adaptable Software. In *Computers, Cognition, and Work*; Lawrence Erlbaum Associates: Hillsdale, NJ, USA, 1994; ISBN 978-0805816556.
16. Andrienko, G.L.; Andrienko, N.V. Interactive maps for visual data exploration. *Int. J. Geogr. Inf. Sci.* **2001**, *13*, 355–374. [CrossRef]
17. Roth, R.E. Interactive maps: What we know and what we need to know. *J. Spat. Inf. Sci.* **2013**, *6*, 59–115. [CrossRef]
18. Peterson, M.P. That interactive thing you do. *Cartogr. Perspect.* **1998**, *29*, 3–5. [CrossRef]
19. Cartwright, W. Extending the map metaphor using web delivered multimedia. *Int. J. Geogr. Inf. Sci.* **2001**, *13*, 335–353. [CrossRef]
20. Beaudouin-Lafon, M. Designing interaction, not interfaces. In *Advanced Visual Interfaces*; Association for Computing Machinery: New York, NY, USA, 2004; pp. 15–22.
21. Yi, J.S.; Kang, Y.A.; Stasko, J.T.; Jacko, J.A. Toward a deeper understanding of the role of interaction in information visualization. *IEEE Trans. Vis. Comput. Graph.* **2007**, *13*, 1224–1231. [CrossRef]
22. Clarke, K.C. Mobile mapping and geographic information systems. *Cartogr. Geogr. Inf. Sci.* **2004**, *31*, 131–136. [CrossRef]
23. Meng, L.; Reichenbacher, T.; Zipf, A. *Mapbased Mobile Services*; Springer: Berlin/Heidelberg, Germany, 2005; ISBN 978-3-540-26982-3.
24. Roth, R.E.; Young, S.; Nestel, C.; Sack, C.M.; Davidson, B.; Knoppke-Wetzel, V.; Ma, F.; Mead, R.; Rose, C.; Zhang, G. Global landscapes: Teaching globalization through responsive mobile map design. *Prof. Geogr.* **2018**, *70*, 395–411. [CrossRef]
25. Brewer, C.A.; Buttenfield, B.P. Framing guidelines for multi-scale map design using databases at multiple resolutions. *Cartogr. Geogr. Inf. Sci.* **2007**, *34*, 3–15. [CrossRef]
26. Roth, R.E.; Brewer, C.A.; Stryker, M.S. A typology of operators for maintaining legible map designs at multiple scales. *Cartogr. Perspect.* **2011**, *68*, 29–64. [CrossRef]
27. Boulos, M.N.K.; Burden, D. Web GIS in practice V: 3-D interactive and real-time mapping in Second Life. *Int. J. Health Geogr.* **2007**, *6*, 51. [CrossRef]
28. Goldsberry, K. Real-Time Traffic Maps for the Internet. Ph.D. Thesis, University of California, Santa Barbara, CA, USA, 2007.
29. Medyńska-Gulij, B. Map compiling, map reading, and cartographic design in "pragmatic pyramid of thematic mapping". *Quast. Geogr.* **2010**, *29*, 57–63. [CrossRef]
30. Medyńska-Gulij, B. Cartographic sign as a core of multimedia map prepared by non-cartographers in free map services. *Geod. Cartogr.* **2014**, *63*, 55–64. [CrossRef]
31. Seinfeld, S.; Feuchtner, T.; Maselli, A.; Müller, J. User Representations in Human-Computer Interaction. *Hum. Comput. Interact.* **2020**. [CrossRef]
32. Muehlenhaus, I. *Web Cartography: Map Design for Interactive and Mobile Devices*; CRC Press: London, UK, 2013; ISBN 978-1439876220.
33. Roth, R.E.; Donohue, R.G.; Sack, C.M.; Wallace, T.R.; Buckingham, T.M.A. A Process for Keeping Pace with Evolving Web Mapping Technologies. *Cartogr. Perspect.* **2015**, *78*, 25–52. [CrossRef]
34. Medyńska-Gulij, B. *Kartografia i Geomedia*; Wydawnictwo Naukowe PWN: Warszawa, Poland, 2021; ISBN 978-83-01-21554-5.
35. Marcotte, E. *Responsive Web Design*; A Book Apart: New York, NY, USA, 2010; ISBN 978-1937557188.
36. Wielebski, Ł.; Medyńska-Gulij, B. Graphically supported evaluation of mapping techniques used in presenting spatial accessibility. *Cartogr. Geogr. Inf. Sci.* **2019**, *46*, 311–333. [CrossRef]
37. Nivala, A.M.; Brewster, S.; Sarjakoski, L. Usability Evaluation of Web Mapping Sites. *Cartogr. J.* **2008**, *45*, 129–138. [CrossRef]
38. Çöltekin, A.; Heil, B.; Garlandini, S.; Fabrikant, S.I. Evaluating the effectiveness of interactive map interface designs: A case study integrating usability metrics with eye-movement analysis. *Cartogr. Geogr. Inf. Sci.* **2009**, *36*, 5–17. [CrossRef]
39. Goldberg, J.H.; Kotval, X.P. Computer interface evaluation using eye movements: Methods and constructs. *Int. J. Ind. Ergon.* **1999**, *24*, 631–645. [CrossRef]
40. Sutcliffe, A.G.; Ennis, M.; Hu, J. Evaluating the effectiveness of visual user interfaces for information retrieval. *Int. J. Hum. Comput. Stud.* **2000**, *53*, 741–763. [CrossRef]
41. Saw, J.; Butler, M. Exploring graphical user interfaces and interaction strategies in simulations. In *Hello! Where Are You in the Landscape of Educational Technology? Proceedings of the 25th Ascilite Conference*; Melbourne, Australia, 30 November–3 December 2008, ASCILITE: Tugun, Australia; Available online: http://www.ascilite.org.au/conferences/melbourne08/procs/saw.pdf (accessed on 2 March 2021).
42. Golebiowska, I.; Opach, T.; Rød, J.K. Breaking the Eyes: How Do Users Get Started with a Coordinated and Multiple View Geovisualization Tool? *Cartogr. J.* **2020**, *57*, 235–248. [CrossRef]
43. Edler, D.; Vetter, M. The simplicity of modern audiovisual web cartography: An example with the open-source Javascript Library leaflet.js. Kn. *J. Cartogr. Geogr. Inf.* **2019**, *69*, 51–62. [CrossRef]

44. Horbiński, T.; Lorek, D. The use of Leaflet and GeoJSON files for creating the interactive web map of the preindustrial state of the natural environment. *J. Spat. Sci.* **2020**. [CrossRef]
45. Lorek, D.; Horbiński, T. Interactive Web-Map of the European Freeway Junction A1/A4 Development with the Use of Archival Cartographic Sources. *ISPRS Int. J. Geo. Inf.* **2020**, *9*, 438. [CrossRef]
46. Cybulski, P.; Horbiński, T. User Experience in Using Graphical User Interfaces of Web Maps. *ISPRS Int. J. Geo Inf.* **2020**, *9*, 412. [CrossRef]
47. Medyńska-Gulij, B.; Zagata, K. Experts and Gamers on Immersion into Reconstructed Strongholds. *ISPRS Int. J. Geo Inf.* **2020**, *9*, 655. [CrossRef]
48. Nivala, A.-M.; Sarjakoski, L.T.; Sarjakoski, T. User-Centred Design and Development of a Mobile Map Service. W: ScanGIS'2005. In Proceedings of the 10th Scandinavian Research Conference on Geographical Information Sciences, Stockholm, Sweden, 13–15 June 2005; pp. 109–123.
49. Ware, C. *Information Visualization: Perception for Design*; Morgan Kaufmann: San Francisco, CA, USA, 2004; ISBN 978-0123814647.
50. Medyńska-Gulij, B. Point Symbols: Investigating Principles and Originality in Cartographic Design. *Cartogr. J.* **2008**, *45*, 62–67. [CrossRef]
51. Cybulski, P. Spatial distance and cartographic background complexity in graduated point symbol map-reading task. *Cartogr. Geogr. Inf. Sci.* **2020**, *47*, 244–260. [CrossRef]
52. Garlandini, S.; Fabrikant, S.I. Evaluating Effectiveness and Efficiency of Visual Variables for Geographic Information Visualization. In *Spatial Information Theory, Proceedings of the COSIT 2009, Aber Wrac'h, France, 21–25 September 2009*; Hornsby, K., Claramunt, C., Denis, M., Ligozat, G., Eds.; Springer: Berlin, Germany, 2009; pp. 195–211. ISBN 978-3-642-03832-7.
53. Brychtová, A.; Çöltekin, A. An Empirical User Study for Measuring the Influence of Colour Distance and Font Size in Map Reading Using Eye Tracking. *Cartogr. J.* **2014**, *53*, 202–212. [CrossRef]
54. Halik, Ł.; Medyńska-Gulij, B. The Differentiation of Point Symbols using Selected Visual Variables in the Mobile Augmented Reality System. *Cartogr. J.* **2016**, *54*, 147–156. [CrossRef]
55. Medynska-Gulij, B.; Halik, Ł.; Wielebski, Ł.; Dickmann, F. Mehrperspektivische Visualisierung von Informationen zum räumlichen Freizeitverhalten—Ein Smartphone-gestützter Ansatz zur Kartographie von Tourismusrouten. *J. Cartogr. Geogr. Inf.* **2015**, *65*, 323–329. [CrossRef]
56. Wielebski, Ł.; Medyńska-Gulij, B.; Halik, Ł.; Dickmann, F. Time, Spatial, and Descriptive Features of Pedestrian Tracks on Set of Visualizations. *ISPRS Int. J. Geo Inf.* **2020**, *9*, 348. [CrossRef]
57. Popelka, S.; Vondrakova, A.; Hujnakova, P. Eye-Tracking Evaluation of Weather Web Maps. *Int. J. Geo Inf.* **2019**, *8*, 29. [CrossRef]
58. Stachoň, Z.; Šašinka, Č.; Čeněk, J.; Angsüsser, S.; Kubiček, P.; Štěrba, Z.; Bilíková, M. Effect of size, shape and map background in cartographic visualization: Experimental study on Czech and Chinese population. *Int. J. Geo Inf.* **2018**, *7*, 427. [CrossRef]
59. Liao, H.; Wang, X.; Dong, W.; Meng, L. Measuring the influence of map label density on perceived complexity: A user study using eye tracking. *Cartogr. Geogr. Inf. Sci.* **2019**, *46*, 210–227. [CrossRef]
60. Rensink, R.; O'Regan, J.K.; Clark, J.J. To see or not to see: The need for attention to perceive changes in scenes. *Psychol. Sci.* **1997**, *8*, 368–373. [CrossRef]
61. Keates, J.S. *Understanding Maps*, 2nd ed.; Routledge: London, UK, 1996.
62. Robinson, A.C.; Demšar, U.; Moore, A.B.; Buckley, A.; Jiang, B.; Field, K.; Kraak, M.-J.; Camboim, S.P.; Sluter, C.R. Geospatial big data and cartography: Research challenges and opportunities for making maps that matter. *Int. J. Cartogr.* **2017**, 332–360. [CrossRef]

International Journal of
Geo-Information

Article

A Sightseeing Spot Recommendation System That Takes into Account the Visiting Frequency of Users

Yudai Kato and Kayoko Yamamoto * 

Graduate School of Informatics and Engineering, University of Electro-Communications, Tokyo 182-8585, Japan; k2030031@edu.cc.uec.ac.jp

* Correspondence: kayoko.yamamoto@uec.ac.jp; Tel.: +81-42-443-5728

Received: 18 May 2020; Accepted: 26 June 2020; Published: 27 June 2020

Abstract: The present study aimed to design, develop, operate and evaluate a sightseeing spot recommendation system that can efficiently and usefully support tourists while considering their visiting frequencies. This system was developed by integrating social networking services (SNSs), Web-geographic information systems (GIS) and recommendation systems. The system recommends sightseeing spots to users with different visiting frequencies, adopting two recommendation methods (knowledge-based recommendation and collaborative recommendation methods). Additionally, the system was operated for six weeks in Kamakura City, Kanagawa Prefecture, Japan, and the total number of users was 61. Based on the results of the web questionnaire survey, the usefulness of the system when sightseeing was high, and the recommendation function of sightseeing spots, which is an original function, received mainly good ratings. From the results of the access analysis of users' log data, the total number of sessions in this system was 329, 77% used mobile devices, and smartphones were used most frequently. Therefore, it is evident that the system was used by different types of devices just as it was designed for, and that the system was used according to the purpose of the present study, which is to support the sightseeing activities of users.

Keywords: sightseeing spot recommendation system; social networking service (SNS); web-geographic information systems (GIS); recommendation system; visiting frequency; sightseeing spot information

---

1. Introduction

Due to the advancement of information in recent years, anyone can easily send and receive information regardless of time and place, and obtain an abundance of various information from the internet. The same can be said with sightseeing information. Tourists must select necessary information among a significant amount of sightseeing information, and it is tremendously difficult to find information tailored to their purposes. Therefore, it is important to provide tourists with relevant sightseeing information using the internet.

At popular sightseeing sites, there are not only tourists who are visiting for the first time or who have only made a few visits, but there are also repeat travelers who have made several visits. Okamura et al. (2007) [1] identified the difference in sightseeing spots where tourists with different visiting frequencies visited. In this regard, taking up Kamakura City, Kanagawa Prefecture, Japan as an example, the enhancement of sightseeing support for repeat travelers was addressed as one of the issues on sightseeing (Kamakura City, 2016) [2]. Therefore, it is best to provide sightseeing spot information according to the visiting frequency of tourists. This is because it is difficult to use the same method to provide adequate and appropriate support for both tourists that visit for the first time and have limited knowledge and sense of locality, as well as tourists including repeat travelers who have knowledge and sense of locality concerning sightseeing spots.

Against the above backdrop, the present study aims to develop a sightseeing spot recommendation system that can efficiently and usefully support tourists, taking their visiting frequencies into account. Regarding the system developed in the present study, efficiency is related to the provision of relevant sightseeing information to tourists using the internet, and usefulness is related to the recommendation of sightseeing spots according to tourist preferences and visiting frequencies. This system is designed and developed by integrating social networking service (SNS), Web-geographic information systems (GIS) and recommendation system (Sections 3 and 4). SNS is used to gather, accumulate, evaluate and share sightseeing information. Web-GIS is used to visualize the sightseeing information on digital maps. The recommendation system is used to recommend sightseeing spots to users by taking the difference in the visiting frequencies into account. Additionally, the system will be used by various people from both inside and outside the operation target area during the operation period (Section 5), and improvement strategies will be submitted after the issues are identified by evaluating the system through a web questionnaire survey to users and an access analysis (Section 6).

Kamakura City, Kanagawa Prefecture was selected as the operation target area. The first reason for this selection is that there are many visits by tourists with different visiting frequencies. As mentioned above, Kamakura City is a well-known sightseeing area and the number of visits by repeat travelers has increased in proportion to the enhancements made to the sightseeing support for them. It has also received many first-timer tourists, making it a concentrated location by tourists with different visiting frequencies. The second reason is that Kamakura City has many sightseeing spots such as retail shops and restaurants for tourists in addition to historical buildings. Therefore, it is anticipated that the system can be used to recommend sightseeing spots according to tourist preferences and visiting frequencies.

2. Related Work

This system in the present study was developed by integrating multiple systems such as SNS, Web-GIS as well as the recommendation system into a single system. Therefore, the present study is related to three research fields including (1) studies related to sightseeing support systems, (2) studies related to sightseeing information system especially for repeat travelers, and (3) studies related to sightseeing recommendation systems.

Taking up the main preceding studies in recent years, regarding (1) studies related to sightseeing support systems, Anacleto et al. (2014) [3] presented PSiS Mobile, which is a mobile recommendation and planning application designed to support a tourist during his/her vacations. Brilhante et al. (2015) [4] proposed TripBuilder which is an unsupervised framework for planning personalized sightseeing tours in cities, using categorized points of interests (PoIs) from Wikipedia and albums of geo-referenced photos from Flickr. Zhou et al. (2016) [5] and Fujita et al. (2016) [6] developed navigation systems integrating SNS, Web-GIS and augmented reality (AR) to support sightseeing activities during normal times and evacuations during disasters. Based on these studies, Makino et al. (2019) [7] developed a system that visualizes spatiotemporal information in both real and virtual spaces to support sightseeing, integrating SNS, Web-GIS, Mixed Reality (MR) and gallery system as well as Wikitude, and connecting external social media. On the other hand, focusing on the language barrier while sightseeing, Yamamoto et al. (2018) [8] and Abe et al. (2019) [9] proposed sightseeing support systems using English and other nonlinguistic information including pictograms (symbols and marks). Referring to the above studies, Sasaki et al. (2019) [10] developed a sightseeing support system using AR and Pictograms.

For (2) studies related to sightseeing information systems, especially for repeat travelers, McKerche et al. (2014) [11] used global positioning systems (GPS) technology to compare and contrast the behavior patterns of first-time and repeat visitors. Masuda et al. (2012) [12] proposed a system to promote the creation of repeat travelers by providing only nearby information up to the sightseeing spots, and also provide a feeling of incompleteness when they are unable to find what they are looking for. Yorozu et al. (2015) [13] proposed a sightseeing planning support system which recommended hidden spots, especially for repeat travelers. Uchizono et al. (2016) [14] proposed a sightseeing recommendation system to increase repeat travelers, focusing on their experiences and interests, and the best season of sightseeing spots. Katayama et al. (2017) [15] developed an information providing system to promote the creation of repeat travelers by providing users with the information concerning hidden spots that are scarcely known but have been visited by others in the past, and have a relatively high satisfaction rate among tourists. Niibara et al. (2017) [16] proposed a system which provides sightseeing information especially for repeat travelers in response to their visit frequencies. Kang et al. (2018) [17] identified the spatial structure of the tourist attraction system by tourists' length of stay, employing anchor-point theory and social network analysis techniques with spatial statistics and using GIS. Uchida et al. (2019) [18] proposed KadaSola, which is a sightseeing support system for long stays to increase repeat travelers, classifying tourists by their attributes.

Regarding (3) studies related to sightseeing recommendation systems, Tarui (2011) [19] combined the collaborative filtering and content analysis methods to develop a system that recommends sightseeing spots from the travel history of users. Yu et al. (2012) [20] proposed a context-aware recommender system that provides personalized mobile travel planning services. Ikeda et al. (2014) [21] developed a social recommendation GIS that recommends sightseeing spots by means of the degree of similarity of sightseeing spots with individual preference information. Based on Ikeda et al. (2014) [21], Mizutani et al. (2017) [22] and Mukasa et al. [23] developed sightseeing spot recommendation systems that respectively take users' circumstances and priority conditions into account. Additionally, Kitayama et al. (2014) [24] developed a route recommender system that takes a user's visit duration at sightseeing locations into account. Gavalas et al. (2014) [25] proposed a mobile tourism recommendation system with context-awareness function. Li et al. (2017) [26] and Takahashi et al. (2017) [27] proposed a tourism course recommendation system using the data obtained from SNS. Aoki et al. (2019) [28] developed a recommendation system that interactively utilizes crowd information to support tour planning.

Regarding (1) studies related to sightseeing support systems, support is provided for sightseeing activities during normal times by means of the functions of submitting, viewing, recommendation, sightseeing planning assistance and navigation. Additionally, support is also provided for evacuation during emergencies by displaying support facilities (evacuation centers, water stations, etc.) on the digital map. Though the systems developed or proposed in (2) studies related to sightseeing information systems especially for repeat travelers promote the increase of repeat travelers, most of these do not provide sightseeing support for them. Regarding (3) studies related to sightseeing recommendation systems excluding Tarui (2011) [19], while sightseeing support through sightseeing spot recommendation by knowledge-based recommendation is conducted, the recommendation of appropriate sightseeing spots according to the preferences of users with different visiting frequencies has not been satisfactory. Additionally, with Tarui (2011) [19], though sightseeing information that suits the preferences of users based on their visiting history can be recommended adopting the collaborative filtering and content analysis methods, users are restricted as they must have experience visiting several sightseeing spots, and the travel history for specific sightseeing spots is not taken into consideration.

In comparison with the preceding studies mentioned above, the first original feature of the present study is that the system can be utilized by users with different travel frequencies including tourists that visit for the first time and have little knowledge and sense of locality as well as tourists who are repeat travelers and have an abundance of knowledge and sense of locality concerning sightseeing spots. The preceding studies did not develop the recommendation systems considering visiting frequency of users. The second original feature is that the system adopts both knowledge-based recommendation and collaborative recommendation as methods to recommend sightseeing spots according to the different preferences of users that occur due to the difference in visiting frequency. Thus, unlike the preceding studies, the system enables the recommendation of favorite sightseeing spots to each tourist according to his/her preferences and visiting frequency.

3. System Design

3.1. System Characteristics

As shown in Figure 1, this system is developed by integrating SNS, Web-GIS and the recommendation system. The purpose of this system is to support the sightseeing activities of users with different visiting frequencies, and by adopting both knowledge-based recommendations and collaborative recommendations as recommendation methods, provide appropriate sightseeing spot information that suits the preferences of such users. For users with low visiting frequency such as tourists visiting for the first time, recommendations will be made based on the required conditions entered for sightseeing spots adopting knowledge-based recommendation method. For users with high visiting frequency such as repeat travelers, recommendations will be made based on their information including evaluation history of sightseeing spots, sightseeing spot added to favorites and visiting history adopting collaborative recommendation method. Additionally, combining SNS and Web-GIS, the system enables users to display the evaluation data on digital maps, submit the comments and images for any location as well as create new sightseeing spot information, and easily gather and accumulate sightseeing spot information. As user information is simultaneously saved in the system, the longer the system is operated, the more support that caters to the preferences of users can be provided. In this way, the system can provide efficient support for sightseeing activities by recommending sightseeing spots that take the visiting frequency of users into account.

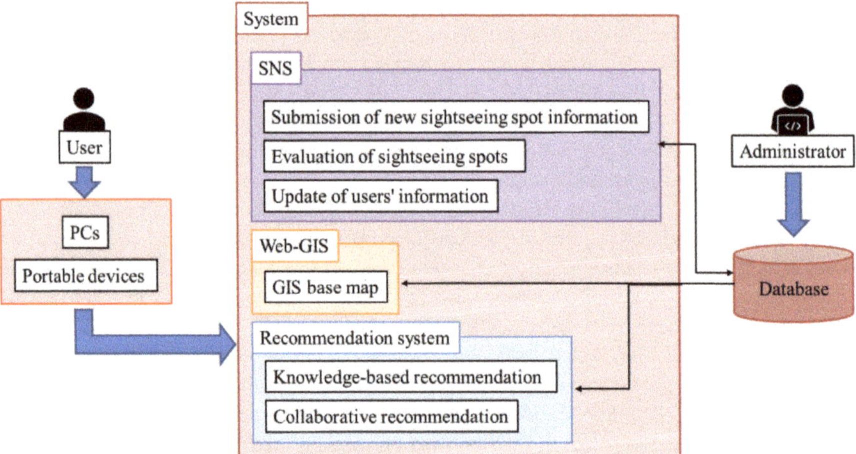

Figure 1. System design.

3.2. Target Devices

Though this system is expected to be accessed using PCs and portable devices, same functions can be used from any device as there is no difference in functions depending on the device used. The use from PCs, which are mainly indoors, is assumed to be the sightseeing planning assistance, by gathering sightseeing information, submitting new sightseeing spot information, and evaluating sightseeing spots already accumulated in the system. On the other hand, the main use from portable devices both indoors and outdoors is assumed to be the assistance of sightseeing activities by means of gathering sightseeing information, submitting new sightseeing spot information, and evaluating sightseeing spots already accumulated in the system.

3.3. System Operating Environment

This system is operated using the Web server, database server and the GIS server. The system operating environment is as shown in Figure 2. Heroku, which is a PaaS provided by the Salesforce company, was used for both the Web server and the database server. ArcGIS Online, which is provided by the ESRI, was used for the GIS server. Additionally, the web application developed with the system was implemented using PHP and JavaScript.

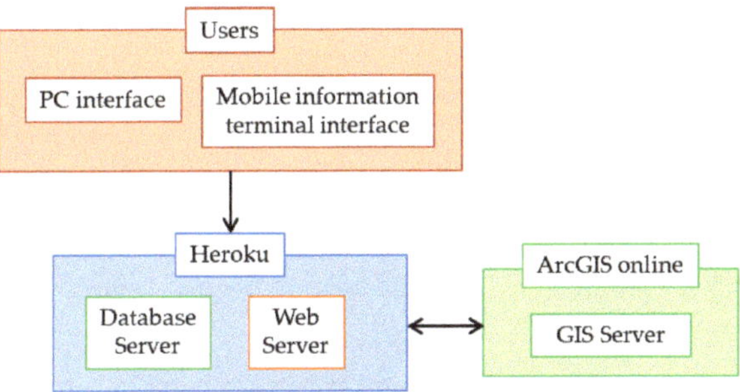

Figure 2. System operating environment.

3.4. Design of Each System

3.4.1. SNS

The main functions of SNS originally designed in this system are the submitting and viewing of sightseeing spot information. Additionally, using the designed SNS, submitted information, "favorite sightseeing spots" and "visited sightseeing spots" can be made public.

3.4.2. Web-GIS

Though there are many types of Web-GIS, this system uses ArcGIS API for JavaScript provided by the ESRI to display the location of recommended sightseeing spots, as it is convenient to access the websites without the installation of any software. Additionally, Leaflet, which is the JavaScript open-source map library using JavaScript, is used to display the recommendation results for sightseeing spots as well as the location of submitted information and sightseeing spots in the page for the detailed information of sightseeing spot.

3.4.3. Recommendation System

1. Selection of the recommendation methods

According to Jannach et al. (2011) [29] and Kamishima (2008) [30], there are three types of recommendation methods, including collaborative recommendation, content-based recommendation, and knowledge-based recommendation, that can make recommendations from vast information groups to match the preferences of users. This system is expected to be used by two types of users: users who visit a sightseeing spot for the first time, and users who have visited several times (repeat travelers). Therefore, the system will adopt the knowledge-based recommendation and collaborative recommendation methods. The knowledge-based recommendation method was selected, as it is the best method for users that visit a sightseeing spot for the first time and have little knowledge and sense of locality. The method can recommend appropriate sightseeing spots, by explicitly asking for preferences and creating preference data of users who receive recommendations. On the other hand, the collaborative recommendation method was selected, as it is the best method for users such as repeat travelers who have their own preferences based on their previous visits to the sightseeing spots.

The method can recommend appropriate sightseeing spots which suit the preferences of users, by referring to the utilization and preference information of users.

Additionally, the cold-start problem can be solved by adopting the knowledge-based recommendation method. Cold start is a problem where appropriate recommendations cannot be made due to insufficient past information. Since knowledge-based recommendations are conducted by creating preference data of users, appropriate recommendations can be made without any past information by explicitly asking for their preferences beforehand.

2. Knowledge-based recommendation system

Regarding the knowledge-based recommendation method, user profiles are created by having users evaluate the items set beforehand on a scale of 1 to 5. The created user profile is set as the user's characteristic vector. Regarding the evaluation data of sightseeing spots also, characteristic vectors for sightseeing spots are created by evaluating each item on a scale of 1 to 5. Based on the created characteristic vectors of users and sightseeing spots, a maximum of 10 sightseeing spots in descending order of similarity will be recommended by calculating the degree of similarity using Equation (1).

$$Sim_i = \frac{\sum_{j=1}^{n} U_j \times S_{ij}}{\sqrt{\sum_{j=1}^{n} \left(U_j\right)^2} \times \sqrt{\sum_{j=1}^{n} \left(S_{ij}\right)^2}} \quad (1)$$

Sim_i: Degree of Similarity
U_j: Preference information of user i
S_{ij}: Evaluation data of sightseeing spots

3. Collaborative recommendation system

According to Kamishima (2007) [30], collaborative filtering (collaborative recommendation in the present study) method can be divided into memory-based method and model-based method. The memory-based method of collaborative recommendation system was selected, as user information will be accumulated in the database, and the preference information of users will be expected to be insufficient right after the start of the operation of this system.

Additionally, according to Kamishima (2007) [30], the memory-based method can be divided into two types: user-based type and item-based type. The system will adopt item-based type of collaborative recommendation system for the same reason memory-based method was selected which is that the lack of preference data of users can be expected right after the start of the operation of the system. This method enables recommendations to be made based only on the preference data of users, by accumulating the evaluation data of sightseeing spots beforehand. Regarding the preference data of users, they can be gathered by registering the evaluation data of sightseeing spots as well as "favorite sightseeing spots" and "visited sightseeing spots" using the designed SNS in Section 3.4.1. Therefore, the system will use the item-based type to develop a memory-based collaborative recommendation system. More specifically, the Item-Based Neighborhood Model proposed by Aggrawal (2016) [31] will be used. First, the degree of similarity between sightseeing spots is calculated using Equation (2) from the evaluation data of sightseeing spots accumulated in the database of the system.

$$Sim(i,j) = \frac{\sum_{u \in U_i \cap U_j} \left\{\left(r_{ui} - \underline{r_u}\right) \times \left(r_{uj} - \underline{r_u}\right)\right\}}{\sqrt{\sum_{u \in U_i \cap U_j} \left(r_{ui} - \underline{r_u}\right)^2} \times \sqrt{\sum_{u \in U_i \cap U_j} \left(r_{uj} - \underline{r_u}\right)^2}} \quad (2)$$

$Sim(i,j)$: Degree of Similarity between sightseeing spot i and sightseeing spot j
r_{ui}: Evaluation data of sightseeing spot i by user u

$\underline{r_u}$: Average value of evaluation data of user u

At the same time, user profiles will be created based on the users' evaluation data of sightseeing spots registered in the system using the designed SNS in Section 3.4.1. Aside from the users' evaluation data, the evaluation values of sightseeing spots according to each category as calculated using Equation (3) are added to the evaluation values of sightseeing spots that belong to their respective categories, based on the data of "favorite sightseeing spots" and "visited sightseeing spots" registered by users.

$$C_i = 5 \times \frac{s_{ui}}{I_u} \quad (3)$$

C_i: Evaluation data of category i
I_u: Number of favorites and visits of user u
s_{ui}: Number of favorite sightseeing spots and visited sightseeing spots of user u

Next, for sightseeing spots that are not evaluated by users, the estimated evaluation values of users are calculated using Equation (4), based on the degree of similarity between sightseeing spots and user profiles, and up to 10 sightseeing spots will be recommended in the descending order of the estimated evaluation value.

$$p_{ut} = \frac{\sum_{j \in Q_{t(u)}} Sim(j,t) \times p_{uj}}{\sum_{j \in Q_{t(u)}} |Sim(j,t)|} \quad (4)$$

p_{ut}: Estimated evaluation values of sightseeing spots that have not been evaluated by users
$Q_{t(u)}$: Aggregation of sightseeing spots evaluated by user u

4. System Development

4.1. The Front-End of the System

This system will implement unique functions for users, which will be mentioned below, in response to the purpose of the present study, as mention in Section 1. In order to implement these several unique functions, the system was developed by integrating plural systems into a single system. Additionally, the system was operated targeting Japanese people and those who can understand Japanese, while selecting Kamakura City, Japan as the operation target area. Therefore, all websites included in the system are written in Japanese with English notations.

1. Viewing function of sightseeing spots

Users are transferred to the page for the viewing function of sightseeing spot information (Figure 3) from the "Kamakura area map" in the menu of the top page. This page allows users to search for sightseeing spots using "search from map" or "search by category". When using "search from map", a popup including the "name of sightseeing spot", "category" and "link to the page for the detailed information of sightseeing spot (Figure 4)" will be displayed, by clicking onto the markers on the digital map. Additionally, the markers are color-coded according to category. On the other hand, when using "search by category", sightseeing spots will be displayed in a list according to category. Users are transferred to the page for the detailed information of sightseeing spots by selecting a sightseeing spot from the list. Additionally, users who have created their own account and log in to the system can register their "favorite sightseeing spots" and "visited sightseeing spots", by clicking onto the buttons of "register favorites" and "register past visits" from the list.

In the page for the detailed information of sightseeing spot (Figure 4), the "name of sightseeing spot", "category", "postal code", "address", "link (external site of the sightseeing spot)", "details (sightseeing spot information)" and "number of comments on the sightseeing spot" as well as the "image of the sightseeing spot" and "map" are displayed. The buttons of "register favorites" and "register page visits" are displayed on this page, and users can use them to register their "favorite sightseeing spots" and "visited sightseeing spots". The link for each category transfers users to the list of sightseeing spots for the selected category, while the link in the comments on a sightseeing spot transfers users to the list of comments. Additionally, the link of "click here to update sightseeing spot information" transfers users to the page for the editing of sightseeing spot information. This page allows users to update information, excluding the "name of sightseeing spot", "category", "image" and "location information".

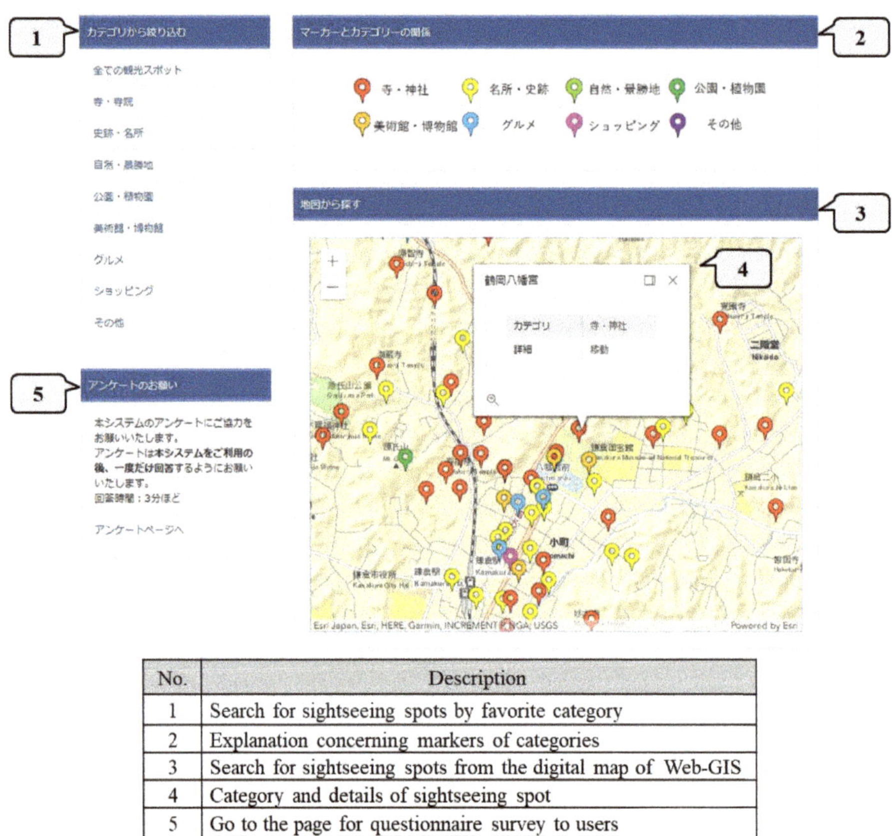

No.	Description
1	Search for sightseeing spots by favorite category
2	Explanation concerning markers of categories
3	Search for sightseeing spots from the digital map of Web-GIS
4	Category and details of sightseeing spot
5	Go to the page for questionnaire survey to users

Figure 3. Page for viewing function of sightseeing spots.

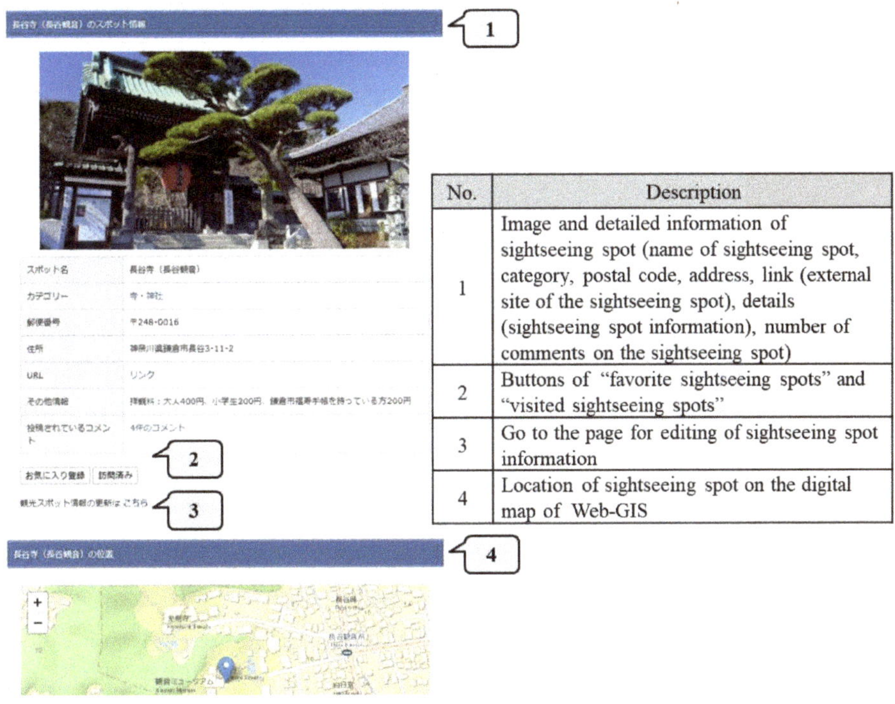

No.	Description
1	Image and detailed information of sightseeing spot (name of sightseeing spot, category, postal code, address, link (external site of the sightseeing spot), details (sightseeing spot information), number of comments on the sightseeing spot)
2	Buttons of "favorite sightseeing spots" and "visited sightseeing spots"
3	Go to the page for editing of sightseeing spot information
4	Location of sightseeing spot on the digital map of Web-GIS

Figure 4. Page for detailed information of sightseeing spot.

2. Submitting function of sightseeing spots

Users are transferred to the page for the submitting function of sightseeing spots (Figure 5) from the "submitting function" in the menu of the top page. In this page, information on a location can be submitted by entering the "name of sightseeing spot", "title", "category", "comment", "image" and "evaluation for each item", and clicking onto the location on the digital map or by acquiring the present location information. For each evaluation item, five items including the "satisfaction level", "access", "non-crowdedness", "landscape", and "accessibility for those with special need" must be evaluated on a scale of 1 to 5. Regarding the scale, "5" means the best and "1" means the worst. Additionally, in the page for this function, users can be transferred to the list and map of submitted information. Users can visit the page for the submitted information list from the page for the submitting function of sightseeing spots as well as the top page. This page displays the "title", "name of sightseeing spot", "submitter", "submitting date and time" and "image". Users can go to the page for the detailed information of sightseeing spot, by selecting one of the submitted information.

Regarding the map with submitted information, markers are displayed on the digital map based on the location information from the submitted information. By clicking onto a marker, a popup including the "title", "name of sightseeing spot", "category", "submitter", "submitting date and time" and "comments" is displayed. These markers are color-coded according to category. On the page for the detailed information of sightseeing spots, the "title", "target sightseeing spots", "category", "submitter" and "comments" as well as "submitted image" and "map" are displayed. Selecting a target sightseeing spot leads to the page for the detailed information of sightseeing spots, selecting a category leads to the page for the submitted information list of such category, and selecting a submitter leads to the My Page of the person who submitted the information. Additionally, only the administrator and the user who submitted the information can go to the page for the editing of sightseeing spot information to delete submitted information.

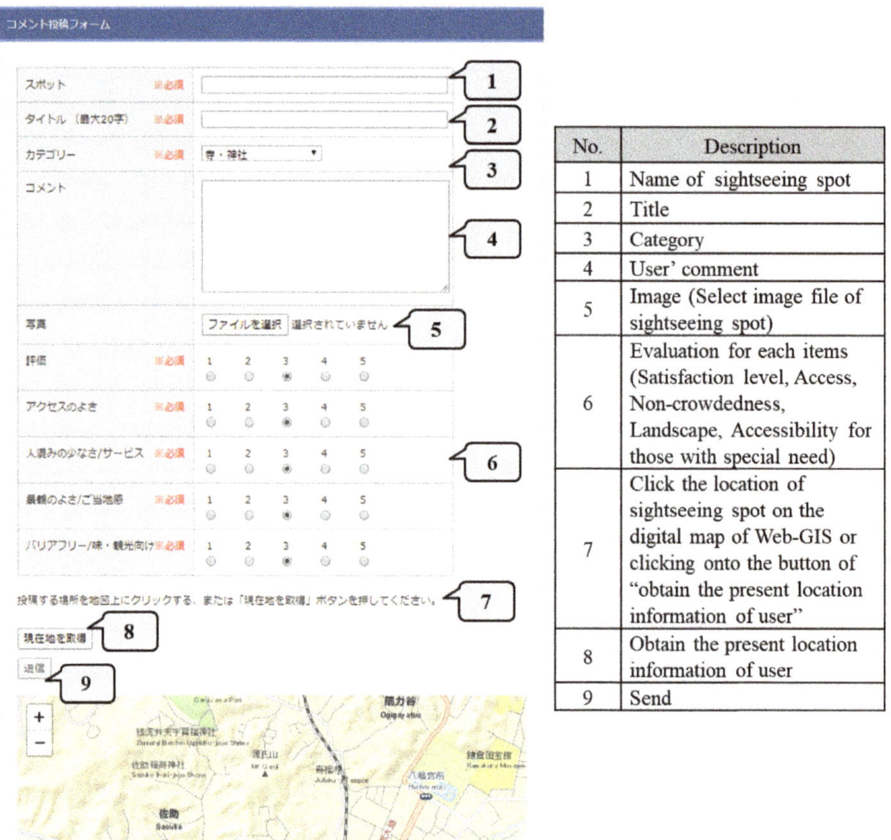

No.	Description
1	Name of sightseeing spot
2	Title
3	Category
4	User' comment
5	Image (Select image file of sightseeing spot)
6	Evaluation for each items (Satisfaction level, Access, Non-crowdedness, Landscape, Accessibility for those with special need)
7	Click the location of sightseeing spot on the digital map of Web-GIS or clicking onto the button of "obtain the present location information of user"
8	Obtain the present location information of user
9	Send

Figure 5. Page for submitting function of sightseeing spots.

3. Recommendation function of sightseeing spots

Users are transferred to the page for the recommendation function of the sightseeing spots adopting knowledge-based recommendations (Figure 6) from the "recommendation conditions" in the menu of the top page. By clicking onto the button of "send" after evaluating each items for sightseeing spots on a scale of 1 to 5, and selecting the range of recommendation results from "main station" and "distance (250m, 500m, 1km, or not specified) from the mains station", users can go to the page for the recommendation results. The center of recommendations can be set as users' present locations by using their present location information. Additionally, users can be transferred to the recommendation function of sightseeing spots adopting collaborative recommendations from the "recommended spots" in the menu of the top page. Sightseeing spots are recommended in the same way as the recommendation function of the sightseeing spots adopting knowledge-based recommendations.

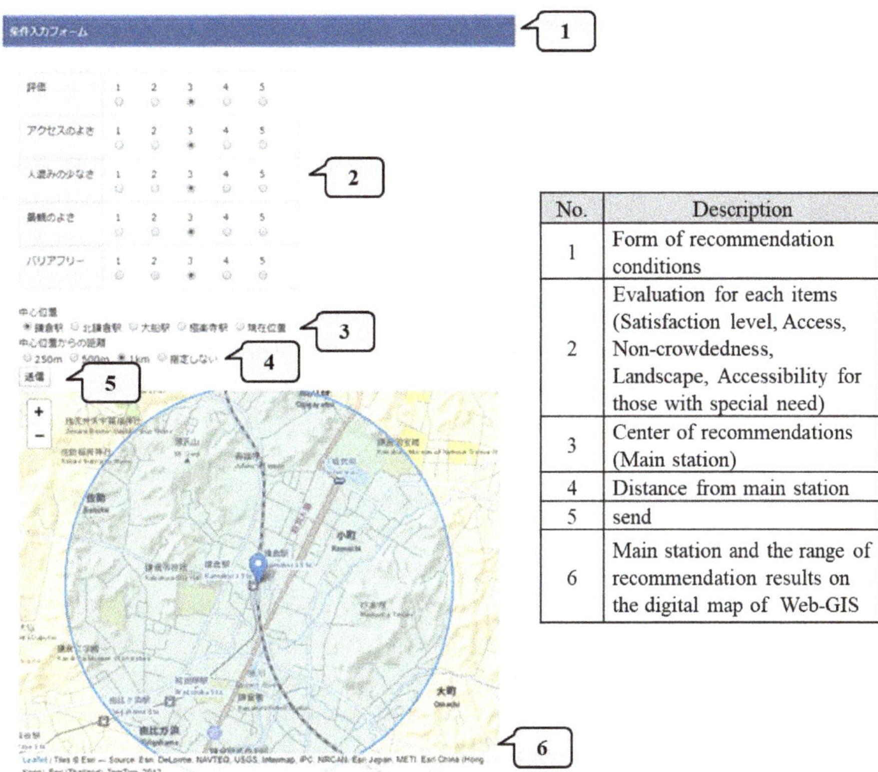

No.	Description
1	Form of recommendation conditions
2	Evaluation for each items (Satisfaction level, Access, Non-crowdedness, Landscape, Accessibility for those with special need)
3	Center of recommendations (Main station)
4	Distance from main station
5	send
6	Main station and the range of recommendation results on the digital map of Web-GIS

Figure 6. Page for recommendation function of sightseeing pots adopting knowledge-based recommendations.

Users can go to the page for the recommendation results (Figure 7) by clicking onto the button of "send" on the pages of "recommendation conditions" and "recommended spots". The page for the recommendation results displays a list of sightseeing spots (up to 10) that were recommended as well as the map with the locations. A popup containing the "name of sightseeing spot", "category", and the "link to the page for the detailed information of sightseeing spot" can be displayed, by clicking onto the markers on the digital map. These markers are color-coded according to category. The list of recommended sightseeing spots and the links in the popups lead to the link to the page for the detailed information of sightseeing spot.

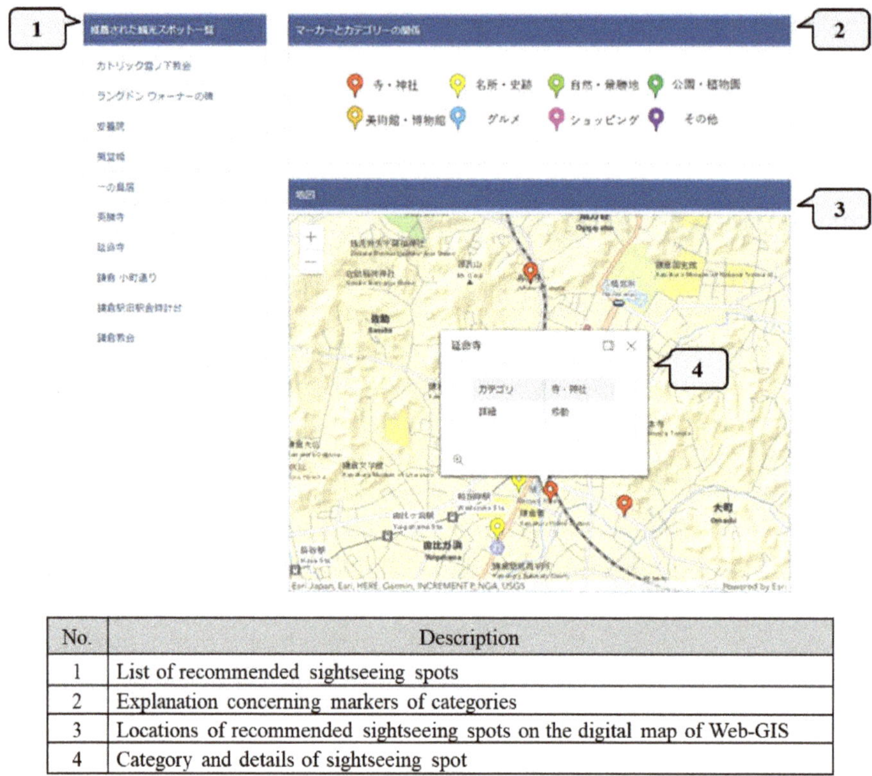

No.	Description
1	List of recommended sightseeing spots
2	Explanation concerning markers of categories
3	Locations of recommended sightseeing spots on the digital map of Web-GIS
4	Category and details of sightseeing spot

Figure 7. Page for recommendation results.

4.2. The Back-End of the System

1. Processing concerning the evaluation of sightseeing spots

In order to improve the accuracy of the recommendation function of sightseeing spots, users are asked to evaluate sightseeing spots using the designed SNS in Section 3.4.1. and new evaluation values are calculated with the back-end of this system.

2. Processing of knowledge-based recommendations

The back-end of the system is used for the process of calculating the degree of similarity adopted to recommend sightseeing spots to users, creating user profiles based on the preference information entered by users, and narrowing of sightseeing spots within the range of the recommendation results to be displayed. Users can receive recommendations, by entering their preference information and the range of recommendation results for sightseeing spots.

3. Processing of collaborative recommendations

The calculation process for the degree of similarity between sightseeing spots, the evaluation information of sightseeing spots obtained through the users' SNS, the degree of similarity between the user profile created from the registered information such as "favorite sightseeing spots" and "visited sightseeing spots" and evaluation data of sightseeing spots were used. The back-end of the system was used for the calculation process of the estimated evaluation values of unevaluated sightseeing spots, and the narrowing process of sightseeing spots within the range of the recommendation results entered by users. After the users have used each function of SNS, they can receive recommendations by entering the range of recommendation results for sightseeing spots.

4.3. System Interface

The interface of this system has two types: the PC and portable device screen of users, and the PC screen of the administrator. For the users' screen, a responsive design was selected and two types of interfaces were prepared according to the screen size of the devices used. For the administrator's page, users, sightseeing spot information, and submitted information can be managed. By utilizing Graphic User Interface (GUI), malicious users and inappropriate sightseeing spot information can be deleted without being affected by the information technology (IT) literacy of the administrators.

5. Operation

5.1. Sightseeing Spot Data

5.1.1. Data Gathering

The sightseeing spot information must be gathered beforehand in order to enable the use of the functions just after the start of the operation of this system. A total of 133 places of sightseeing spot data were gathered in Kamakura City, Kanagawa Prefecture that were evaluated in the travel review site of 4travel.jp [32] as well as in the Jalan Kankou Guide [33].

5.1.2. Data Processing

Though the sightseeing spot data gathered in the previous section were divided into six categories including "temple", "famous and historical site", "art museum and museum", "beach", "nature and scenic site" and "park and botanical garden", information on "restaurant" can be expected to be submitted during the operation of this system. Therefore, the categories were reconsidered based on the travel review site of 4travel.jp [32] as well as in the Jalan Kankou Guide [33]. As a result, eight categories including "temple", "famous and historical site", "nature and scenic site", "park and botanical garden", "art museum and museum", "food", "shopping" and "others" were put in place, and the 133 sightseeing spots were re-categorized. The "others" category includes hot springs, lodgings and events.

5.2. User Assumption

This system is expected to be used by various users such as those who are planning on visiting the operation target area for the first time, and those who have already visited the operation target area and are planning another visit. Since those who visit for the first time can be expected to have little knowledge and sense of locality, the knowledge-based recommendation method that does not require previous knowledge will be adopted as a tool to gather sightseeing spot information. For those who have visited the operation target area several times and already have knowledge and sense of locality, the collaborative recommendation method will be adopted as a tool to gather sightseeing spot information that suits their individual preferences based on their visiting history. Additionally, all users who are both indoors and on-site will be encouraged to submit evaluations for sightseeing spots that they have visited and sightseeing spot information that have not been registered.

5.3. Operation

5.3.1. Operation Overview

The operation of this system was conducted over the course of six weeks with people inside and outside the operation target area as subjects. The authors promoted the use of the system through the website, Twitter and Facebook of their labs. Users will register when using the system for the first time. Registration is completed by entering their account names, email addresses, genders, age groups, numbers of past visits to Kamakura City and passwords. The email address and password are required when logging in to the system. Users are automatically transferred to the top page only after the initial registration. After completing the registration process, users can use all the functions of the system only if they are logged in. While the "Kamakura area map" and other information concerning Kamakura area can be utilized without logging in, "favorite sightseeing spots" and "visited sightseeing spots" cannot be registered. The My Page can be used to change the information concerning users and their preferences which will enable them to receive recommendations of sightseeing spot that suit individual preferences.

5.3.2. Operation Results

The users of this system are shown in Table 1. There were a total of 61 users including 44 men and 17 women. There were no incentives for users, and those who wanted to use the system had to register when using the system for the first time, as mentioned in the previous section. There were more male users and males in their 20s made up the highest percentage. Regarding age groups, users in their 20s made up the highest percentage of 64% including both men and women. Users in their 50s made up 13%, users aged 60 and over made up 10%, users in their 10s made up 5%, users in their 30s made up 7%, and the lowest percentage was users in their 40s who only made up 2%. All users are Japanese, and their places of residence mainly concentrate in the Kanto region (the Tokyo metropolitan area). Additionally, as a result of the operation, more than half of the users were young people who are familiar with new technologies.

Table 1. Breakdown of system users and web questionnaire survey respondents.

Age Groups of Users	10–19	20–29	30–39	40–49	50–59	60–	Total
Number of Users	3	39	4	1	8	6	61
Number of Web Questionnaire Survey Respondents	1	36	4	1	6	5	53
Valid Response Rate (%)	33.3	92.3	100	100	75.0	83.3	86.9

Regarding the number of visits to the operation target area before using the system, based on the classification of tourists focusing on visiting frequency in Adachi et al. (2007) [34], those with zero visits are defined as "first-time visitors", 1–2 visits are "few-time visitors", 3–5 visits are "semi-repeat travelers", and 6–11 or more visits are "repeat travelers". According to this definition, 16% of users were first-time visitors, 25% were few-time visitors, 14% were semi-repeat travelers, and 45% were repeat travelers. In this way, the percentage of repeat travelers was high for both men and women, and there were also many first-time and few-time visitors. Therefore, it is evident that the system was used by people with different visiting frequencies including first-time and few-time visitors who have limited knowledge and sense of locality as well as semi-repeat travelers and repeat travelers who have sufficient knowledge and sense of locality concerning sightseeing spots.

Additionally, the number of new sightseeing spots submitted during the operation was 22, and the number of evaluations for sightseeing spots that were accumulated in the system before the operation was 45. Therefore, it can be expected that the number of submissions for new sightseeing spot information as well as the evaluations for accumulated sightseeing spots will increase, by conducting the operation of the system on a long-term basis.

6. Evaluation

In this section, first of all, the system developed in the present study will be evaluated based on the results of a web questionnaire survey to users and an access analysis of users' log data. Next, based on the results, improvement strategies for the system will be submitted.

6.1. Evaluation Based on the Web Questionnaire Survey

6.1.1. Overview of the Web Questionnaire Survey

According to the purpose of the present study, a web questionnaire survey was carried out in order to conduct an (1) evaluation concerning the system utilization as well as an (2) evaluation concerning the overall system and original functions. This questionnaire survey was carried out on the website one week after the operation commenced. The overview of the questionnaire survey is also shown in Table 1. As shown in Table 1, 53 people out of the 61 users responded which is an 87% valid response rate. The second evaluation focused on the original functions of this system used by users on their own initiative.

6.1.2. Evaluations Concerning the System Utilization

1. Evaluation concerning the compatibility with the information acquisition methods of sightseeing spots

Regarding the information acquisition methods for sightseeing spots (multiple answers allowed), 31% answered PCs, 57% answered portable devices, and 11% answered guidebooks. This result clearly shows that the methods used to acquire sightseeing information mostly involve the use of the internet from PCs or portable devices such as smartphones and not only printed media such as guidebooks. Therefore, it is evident that this system was effective in supporting sightseeing activities, as it enabled users to acquire sightseeing spot information using their PCs or portable devices.

2. Evaluation concerning the status of system utilization

Regarding the devices used to access the system, 21% answered PC while 79% answered smartphones, showing that the system is mostly accessed from smartphones. For the purpose of using the system (multiple answers allowed), 37% answered "Gather sightseeing information", 34% answered "Create travel plans", and 27% answered "On-site travel". Therefore, this results show that each purpose has the same level of demand. Based on the above, situations in which the system can be utilized were expanded by preparing an interface for portable devices in addition to one for PCs.

6.1.3. Evaluations Concerning the Overall System and Original Functions

1. Evaluations concerning the overall system and the submitting function of sightseeing spots

Regarding the usefulness of this system while sightseeing, 51% answered "I think so" and 47% answered "I somewhat think so", and 2% answered "I don't think so". Therefore, the system can be considered useful in supporting sightseeing activities. For the submitting function of sightseeing spot information, only 30% of those who responded to the questionnaire survey used the function. Users that had not used the function gave reasons such as "I didn't have any sightseeing spot information", "I didn't want to submit sightseeing spot information", "It takes time to enter information in the submitting form", and "The design of the page makes it difficult to submit". Therefore, it was revealed that improvements were necessary as the reasons provided that were related to the design of the page for the submitting function of sightseeing spot information.

2. Evaluations concerning the overall recommendation function of sightseeing spots

The evaluation results concerning the overall recommendation function of sightseeing spots, which is an original function in the system, are shown in Figure 8. Regarding the suitability of sightseeing spots recommended (10 spots), 98% answered "I think so" or "I somewhat think so". Therefore, it was appropriate to recommend several sightseeing spots. For the usefulness to display sightseeing spots recommended on digital map, and the usefulness to specify range of recommendation results for sightseeing spots, 96% answered "I think so" or "I somewhat think so". From these results, it can be considered effective to allow users to select a range when recommending sightseeing spots as well as display them on the digital map of Web-GIS.

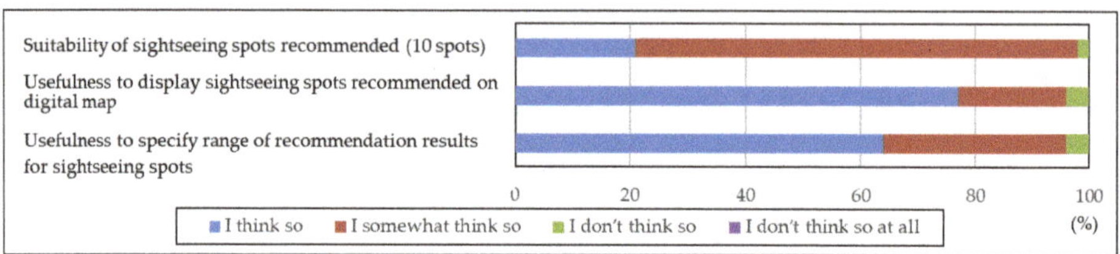

Figure 8. Evaluations concerning the overall recommendation function of sightseeing spots.

3. Evaluations concerning the recommendation function of sightseeing spots adopting knowledge-based recommendations

The evaluation results concerning the recommendation function of sightseeing spots adopting knowledge-based recommendations are shown in Figure 9. As mentioned in Section 6.1, when using this function, users must enter their recommendation conditions for sightseeing spots. The function was used by 72% of those who responded to the questionnaire survey. Regarding the compatibility of the recommended sightseeing spots with the preferences of users, 92% of those who used the function answered "I think so" or "I somewhat think so". For the suitability of recommendation conditions for sightseeing spots, 97% answered "I think so" or "I somewhat think so". Additionally, for the satisfaction rate of the sightseeing spots recommended, 90% answered "I think so" or "I somewhat think so".

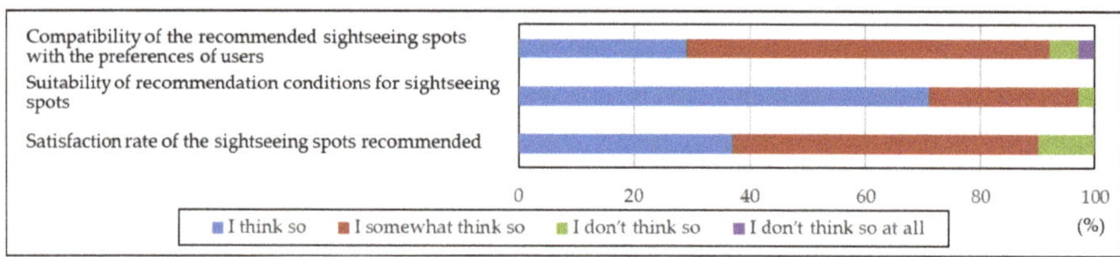

Figure 9. Evaluations concerning the recommendation function of sightseeing spots adopting knowledge-based recommendations.

From these results, sightseeing spots recommended by adopting the knowledge-based recommendation method, which is based on the preference information of users obtained by recommendation conditions selected by users, matched the preferences of users. On the other hand, users who did not use the function answered that "I couldn't find it". This may be because it was hard to find the buttons leading users to the page for the function, when using the system from smartphones.

4. Evaluations concerning the recommendation function of sightseeing spots adopting collaborative recommendations

The evaluation results concerning the recommendation function of sightseeing spots adopting collaborative recommendations are shown in Figure 10. This function was used by 81% of those who responded to the questionnaire survey. Regarding the compatibility of the recommended sightseeing spots with the preferences of users, 91% of those who used the function answered "I think so" or "I somewhat think so". For the Satisfaction rate of the sightseeing spots recommended, 95% answered ""I think so" or "I somewhat think so". Therefore, sightseeing spots recommended adopting the collaborative recommendation method, which is based on the preference information and evaluation history of users, and matched the preferences of users.

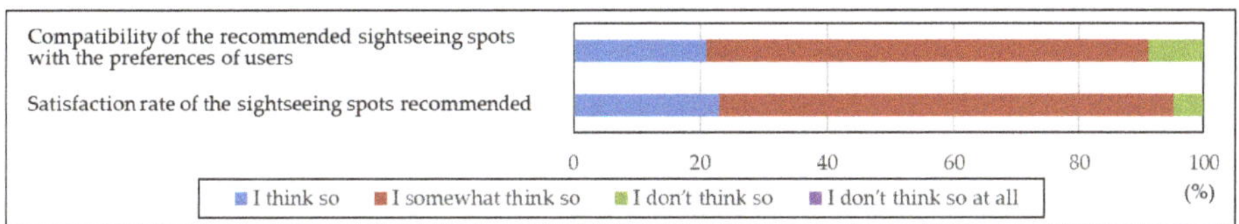

Figure 10. Evaluations concerning the recommendation function of sightseeing spots adopting collaborative recommendations.

6.2. Evaluations Based on the Access Analysis

An access analysis was conducted in the present study using the log data of users during the operation period. The present study used the Google Analytics of Google. The access log can be obtained by calling the Python Program, which contains the analysis code created with Google Analytics, from the HTML file read on each page within the website subject to the access analysis.

The total number of sessions in this system was 329. Regarding the devices used to access the system, 23% used PCs, 74% used smartphones, and 3% used PC tablets. The reason for this is that smartphones have been used most frequently as a convenient information acquisition method in recent years. Therefore, it can be considered effective to design the system to be used equally regardless of the type of device in order to eliminate the difference in obtaining information.

The top 10 visited pages are shown in Table 2. As evident from Table 2, the pages for the recommendation conditions and the recommendation function of sightseeing spots are frequently accessed. Therefore, the system was used in line with the purpose of the present study which was to support sightseeing activities of users with different visiting frequencies adopting different methods. Additionally, the pages for the "Kamakura area map" to view sightseeing spot information and submitting function of sightseeing spot information were also frequently accessed. However, as the access number of the page for the submission completion was low, it can be presumed that users visited the page for the submitting function of sightseeing spot information, but did not make any submissions. As mentioned in Section 6.1.3, the reason for this is considered to be related to the users and the design of the website.

Table 2. Number of visits for each page (top 10).

Rank	Page Name	Number of Visits	Percentage (%)
1	Top page	288	20.4
2	Page for "Kamakura area map"	103	7.3
3	Page for submitting function of sightseeing spot information	82	5.8
4	Page for creating of an account of user	70	5.0
5	Page for recommendation conditions	65	4.6
6	Page for recommendation function of sightseeing spots	61	4.3
7	Login page	43	3.0
8	Page for recommendation function of sightseeing spots adopting knowledge-based recommendations	40	2.8
9	Page for submitted information list	38	2.7
10	Page for recommendation results	31	2.2

6.3. Submission of Improvement Strategies

The tasks concerning this system submitted based on the results of the questionnaire survey and the access analysis are summarized below.

1. Web-page design

It is essential to create a web-page design that enables users to easily view sightseeing spot information, easily notice the information users are lacking, and easily and quickly make information updates. This will promote the use of other functions related to recommendation and submitting in addition to the information updates.

2. Recommendation function of sightseeing spots

It is necessary to implement the recommendation function according to category, and a function that allows for changes to the number of sightseeing spots to be displayed according to the range of recommendation results for sightseeing spots. This will improve the compatibility of the recommended sightseeing spots with the preferences of users as well as the visibility of recommendation results, enabling users to create a more detailed travel plan.

3. Submitting function of sightseeing spots

It is necessary to implement the functions such as submitting new sightseeing spot information, evaluating sightseeing spots already accumulated in the system, and submitting comments and images on different pages. Additionally, it is essential to submit from the digital map or the page for the detailed information of sightseeing spot, and visualize other users' reactions using the "like" button. These aspects will promote the use of the submitting function of sightseeing spot information to users. Furthermore, the accuracy of the recommendation results for sightseeing spots will be improved, as this will allow for the gathering and accumulation of more sightseeing spot information.

7. Conclusions

In the present study, a system was designed and developed (Sections 3 and 4), the operation was implemented (Section 5), and evaluations and the submission of improvement strategies were conducted (Section 6). In comparison with the systems in sightseeing spot services for tourists developed in the preceding studies mentioned in Section 2, the system developed in the present study can be utilized by users with different travel frequencies including tourists that visit for the first time and have little knowledge and sense of locality, as well as tourists who are repeat travelers and have an abundance of knowledge and sense of locality concerning sightseeing spots. Additionally, the system adopted both knowledge-based recommendation and collaborative recommendation as methods to recommend sightseeing spots according to the different preferences of users that occur due to the difference in visiting frequency.

The present study can be summarized in the following three points.

1. In the present study, a system was designed and developed by integrating SNS, Web-GIS and recommendation system in order to recommend sightseeing spots to users with different visiting frequencies. This system reduced the burden of gathering sightseeing information to recommend sightseeing spots to users with different visiting frequencies and enabled the gathering and accumulation of sightseeing spot information. Kamakura City, Kanagawa Prefecture, Japan was selected as the operation target area, and the operation and evaluations of the system were conducted.
2. The operation of the system was conducted over the course of six weeks with people inside and outside the operation target area as subjects, and the total number of users was 61. A web questionnaire survey was conducted for users. Based on the results of this questionnaire survey, it revealed that recommendation system for sightseeing spots adopting two recommendation methods is effective in supporting the sightseeing activities of users with different visiting frequencies.

3. The results of the access analysis made clear that the system was used in line with the purpose and design of the present study which was to enable the system to be used regardless of the type of device used, and support the sightseeing activities of users with different visiting frequencies adopting different methods. The total number of sessions in the system was 329. Regarding the devices used to access the system, 77% used mobile devices, and smartphones were used most frequently.

In regards to future research tasks, the improvement of the system according to the outcome in Section 6.3 as well as the improvement of its utilization significance by increasing performance records of the system in other urban sightseeing destinations can be raised.

Author Contributions: Yudai Kato design, develop and operate the sightseeing spot recommendation system that takes into account the visiting frequency of users in the present study. He also initially drafted the paper. Kayoko Yamamoto carried out background work, and evaluates the system. All authors contributed to write up and review, and approved the paper manuscript. All authors have read and agreed to the published version of the manuscript.

Funding: This research received no external funding.

Acknowledgments: In the operation of the sightseeing spot recommendation system and the web questionnaires survey of the present study, enormous cooperation was received from those mainly in the Kanto region such as Kanagawa Prefecture and Tokyo Metropolis. We would like to take this opportunity to gratefully acknowledge them.

Conflicts of Interest: The authors declare no conflict of interest.

References

1. Okamaura, K.; Fukushige, M. How to Promote Repeaters? Empirical Analysis of Tourist Survey Data in Kansai Region. *KISER Disc. Pap. Ser.* **2007**, *10*, 35.
2. Kamakura City. Issues on Sightseeing in Kamakura City. Available online: https://www.city.kamakura.kanagawa.jp/kankou/kankoujijouh28.html (accessed on 31 January 2020).
3. Anacleto, R.; Figueiredo, L.; Almeida, A.; Novais, P. Mobile Application to Provide Personalized Sightseeing Tours. *J. Netw. Comput. Appl.* **2014**, *41*, 56–64. [CrossRef]
4. Brilhante, I.R.; Macedo, J.A.; Nardini, F.M.; Perego, R.; Renso, C. On planning sightseeing tours with TripBuilder. *Inf. Process. Manag.* **2015**, *51*, 1–15. [CrossRef]
5. Zhou, J.; Yamamoto, K. Development of the System to Support Tourists' Excursion Behavior Using Augmented Reality. *Int. J. Adv. Comput. Sci. Appl.* **2016**, *7*, 197–209. [CrossRef]
6. Fujita, S.; Yamamoto, K. Development of Dynamic Real-Time Navigation System. *Int. J. Adv. Comput. Sci. Appl.* **2016**, *7*, 116–130. [CrossRef]
7. Makino, R.; Yamamoto, K. Spatiotemporal Information System Using Mixed Reality for Area-Based Learning and Sightseeing. In *Computational Urban Planning and Management for Smart Cities*; Geertman, S., Allan, A., Pettit, C., Stillwell, J., Eds.; Springer: Berlin, Germany, 2019; pp. 283–302.
8. Yamamoto, K. Navigation System for Foreign Tourists in Japan. *J. Environ. Sci. Eng.* **2018**, *10*, 521–541.
9. Abe, S.; Yoshitsugu, N.; Miki, D.; Yamamoto, K. An Information Retrieval System with Language-Barrier-Free Interfaces. *J. Inf. Syst. Soc. Jpn.* **2019**, *14*, 57–64.
10. Sasaki, R.; Yamamoto, K. A Sightseeing Support System Using Augmented Reality and Pictograms within Urban Tourist Areas in Japan. *Int. J. Geo-Inf.* **2019**, *8*, 381. [CrossRef]
11. McKercher, B.; Shoval, N.; Ng, E.; Birenboim, A. First and Repeat Visitor Behaviour: GPS Tracking and GIS Analysis in Hong Kong. *Int. J. Tour. Space Place Environ.* **2012**, *14*. [CrossRef]
12. Masuda, M.; Izumi, T.; Nakatani, Y. System that Promotes Repeat Tourists by Making Sightseeing Unfinished. *Hum. Interface* **2012**, *14*, 259–270.

13. Yorozu, N.; Abe, A.; Ichikawa, H.; Tomizawa, H. Development of sightseeing Planning Support System That Take into Account Repeat Travellers. In Proceedings of the 77th Annual Meeting of the Information Processing Society of Japan, Kyoto, Japan, 30 April–1 May 2015; pp. 871–872.
14. Uchizono, Y.; Takayama, T. A Guide System That Induce Travellers' Revisit to Onomichi City. In Proceedings of the 78th Annual Meeting of the Information Processing Society of Japan, Yokohama, Japan, 11–14 April 2016; pp. 619–620.
15. Katayama, S.; Isogawa, N.; Obuchi, M.; Nishiyama, Y.; Okoshi, T.; Yonezawa, T.; Nakazawa, H.; Takashio, K.; Tokuda, H. SpoTrip: Evaluation of Information Provision System for Hidden Spots to Promote the Increase Repeat Travellers. *IEICE Tech. Rep.* **2017**, *116*, 185–192.
16. Niibara, S.; Takayama, T. A System That Provides the appropriate Levels of Information for Tourists. In Proceedings of the 79th Annual Meeting of the Information Processing Society of Japan, Nagoya, Japan, 16–18 March 2017; pp. 781–782.
17. Kang, S.; Lee, G.; Kim, J.; Park, D. Identifying the Spatial Structure of the Tourist Attraction System in South Korea Using GIS and Network Analysis: An Application of Anchor-Point Theory. *J. Destin. Mark. Manag.* **2018**, *9*, 358–370. [CrossRef]
18. Uchida, K.; Izumi, R.; Kunieda, T.; Yamada, S.; Yonetani, K.; Gotoda, A.; Yaegashi, M. Development "KadaSola": A Sightseeing Support System for Long Stay. In Proceedings of the 81st Annual Meeting of the Information Processing Society of Japan, Fukuoka, Japan, 14–16 March 2019; pp. 841–842.
19. Tarui, Y. Recommendation System of Tourist Site Using Collaborative Filtering Method and Contents Analysis Method. *Bull. Dep. Manag. Inf. Sci. Jôbu Univ.* **2011**, *36*, 1–14.
20. Yu, C.C.; Chang, H.P. Towards Context-Aware Recommendation for Personalized Mobile Travel Planning. In *Context-Aware Systems and Applications*; Cong, P., Nguyen, V., Nguyen, M.H., Tung, T., Suzuki, J., Eds.; Springer: Berlin, Germany, 2012; pp. 121–130.
21. Ikeda, T.; Yamamoto, K. Development of Social Recommendation GIS for Tourist Spots. *Int. J. Adv. Comput. Sci. Appl.* **2014**, *5*, 8–21. [CrossRef]
22. Mizutani, Y.; Yamamoto, K. A Sightseeing Spot Recommendation System That takes into Account the Change in Circumstances of Users. *Int. J. Geo-Inf.* **2017**, *6*, 303. [CrossRef]
23. Mukasa, Y.; Yamamoto, K. A Sightseeing Spot Recommendation System for Urban Smart Tourism Based on Users' Priority Conditions. *J. Civ. Eng. Archit.* **2019**, *13*, 622–640. [CrossRef]
24. Kitayama, D.; Ozu, K.; Nakajima, S.; Sumiya, K. A Route Recommender System Based on the User's Visit Duration at Sightseeing Locations. In *Software Engineering Research, Management and Applications*; Lee, R., Ed.; Springer: Berlin, Germany, 2014; pp. 177–190.
25. Gavalas, D.; Konstantopoulos, C.; Mastakas, K.; Pantziou, G. Mobile Recommender Systems in Tourism. *J. Netw. Comput. Appl.* **2014**, *39*, 319–333. [CrossRef]
26. Li, S.; Takahashi, S.; Yamada, K.; Takagi, M.; Sasaki, J. A Proposal of the Tourism Course Recommendation System (TCRS) for Foreign Tourists by Using Photo Data in Social Network Service. In *New Trends in Intelligent Software Methodologies, Tools and Techniques*; Fujita, H., Selamat, A., Omatu, S., Eds.; IOS Press: Amsterdam, The Netherlands, 2017; pp. 331–338.
27. Takahashi, S.; Li, S.; Yamada, K.; Takagi, M.; Sasaki, J. Case Study of Tourism Course Recommendation System Using Data from Social Network Services. In *New Trends in Intelligent Software Methodologies, Tools and Techniques*; Fujita, H., Selamat, A., Omatu, S., Eds.; IOS Press: Amsterdam, The Netherlands, 2017; pp. 339–347.
28. Aoike, T.; Ho, B.; Hara, T.; Ota, J.; Kurata, Y. Utilising Crowd Information of Tourist Spots in an Interactive Tour Recommender System. In *Information and Communication Technologies in Tourism*; Pesonen, J., Neidhardt, J., Eds.; Springer: Berlin, Germany, 2019; pp. 27–39.
29. Jannach, D.; Zanker, M.; Felfernig, A.; Friedrich, G. *Recommender Systems: An Introduction*; Cambridge University Press: Cambridge, UK, 2011.
30. Kamishima, T. Algorithms for recommender systems (2). *Trans. Jpn. Soc. Artif. Intell.* **2008**, *23*, 89–103.
31. Aggrawal, C.C. *Recommender Systems: The Textbook*; Springer: Berlin, Germany, 2016; p. 498.
32. 4travel.jp. Available online: https://4travel.jp/ (accessed on 31 January 2020).

33. Jalan Kankou Guide. Available online: https://www.jalan.net/kankou/ (accessed on 31 January 2020).
34. Adachi, H.; Shioya, H. *Study Report: Study on the Process of Forming Repeat Travellers*; JTB Tourism Research & Consulting Co.: Tokyo, Japan, 2007; p. 20.

International Journal of
Geo-Information

Article

A Feasibility Study of Map-Based Dashboard for Spatiotemporal Knowledge Acquisition and Analysis

Chenyu Zuo [1], Linfang Ding [2,*] and Liqiu Meng [1]

1 Chair of Cartography, Technical University of Munich, 80333 Munich, Germany; chenyu.zuo@tum.de (C.Z.); liqiu.meng@tum.de (L.M.)
2 KRDB Research Centre for Knowledge and Data, Faculty of Computer Science, Free University of Bozen-Bolzano, 39100 Bozen-Bolzano, Italy
* Correspondence: linfang.ding@unibz.it

Received: 8 September 2020; Accepted: 22 October 2020; Published: 27 October 2020

Abstract: Map-based dashboards are among the most popular tools that support the viewing and understanding of a large amount of geo-data with complex relations. In spite of many existing design examples, little is known about their impacts on users and whether they match the information demand and expectations of target users. The authors first designed a novel map-based dashboard to support their target users' spatiotemporal knowledge acquisition and analysis, and then conducted an experiment to assess the feasibility of the proposed dashboard. The experiment consists of eye-tracking, benchmark tasks, and interviews. A total of 40 participants were recruited for the experiment. The results have verified the effectiveness and efficiency of the proposed map-based dashboard in supporting the given tasks. At the same time, the experiment has revealed a number of aspects for improvement related to the layout design, the labeling of multiple panels and the integration of visual analytical elements in map-based dashboards, as well as future user studies.

Keywords: dashboard; eye-tracking; spatiotemporal analysis; usability

---

1. Introduction

Interactive dashboard is a multimedia presentation style that concisely combines texts, images, charts, maps, videos, and gauges to allow users' instant perception. The interactions on dashboards, such as selecting, filtering, searching, arranging, or drilling down, would additionally empower users with the flexibility to view and explore information effectively [1]. Few [2] describes dashboard as "a visual display of the most important information needed to achieve one or more objectives, consolidated and arranged on a single screen so the information can be monitored at a glance". With the increasing amount of data available in a variety of domains, e.g., natural resources and urban infrastructures, the map-based dashboard with its dedicated components of maps and geovisualizations has become a popular tool that provides an at-a-glance overview of geospatial knowledge and supports stakeholders in making strategic decisions that lead to innovative businesses [3]. Map-based dashboard are designed to present a collection of data, and also to support the visual learning and analytical reasoning of geospatial knowledge [4]. For example, map-based dashboards are often designed to present heterogeneous georeferenced information to citizens and to encourage them to comprehend their living environments. We list several popular map-based city dashboards in Figure 1. In these city dashboards, maps are applied as the main visualization method to organize and show the information from a spatial perspective. The Dublin Dashboard (Figure 1a) is designed to display the census mapping. The spatial distribution, temporal trend, and detailed data values can be interactively retrieved via the dashboard. The Boston Dashboard (Figure 1b) shows train and bike station information on maps, giving users an overview of their locations and the distribution. In addition, the juxtaposition of the maps allow users to correlate the distributions of the trains and bike stations. The Galway Dashboard (Figure 1c) utilizes maps and charts to show various factors related to enterprise and industry. The geovisualizations in this dashboard are arranged like a waterfall on a web page so that users scroll down to check the factors one by one. In general, the map-based dashboards focus on a limited number of factors, showing the spatial and numerical distribution of the data in a self-explanatory way. The users can drill down to the data values, subsets, and a deeper understanding such as correlation and causality through simple interactions. In addition, acquiring knowledge from the dashboard has to be fast. Studies have reported that it takes less than two minutes to construct a single piece of knowledge from a dashboard [5–7]. It is important to note that some studies show that high-level spatiotemporal analysis is a growing need among dashboard users, such as spatial search, comparison, correlation analysis, prediction, and outlier detection [1,8]. The design of effective map-based dashboards with proper analytical functions is generally challenged by constraints such as at-a-glance display, limited viewing time, and limited user ability to understand complex geospatial data [9–11].

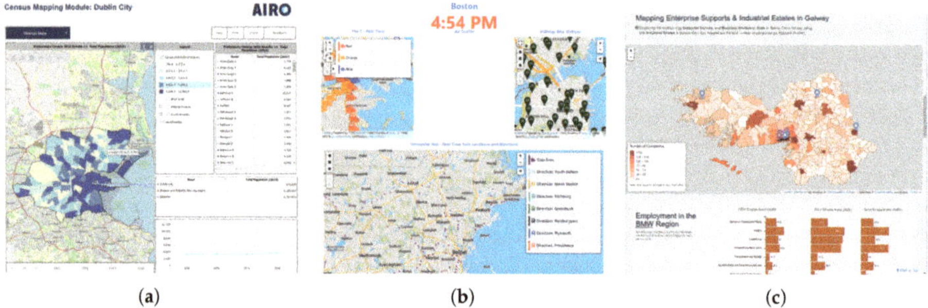

Figure 1. The screenshots of three city map-based dashboards. (**a**) Dublin dashboard [12]. (**b**) Boston dashboard [13]. (**c**) Galway dashboard [14].

Visual analytics is a booming field for the identification and understanding of complex data patterns by combining the machine's computing capability and human visual perception [15]. Effective visual interface design and complex interactive visualizations have been proposed to facilitate the visual analytical procedure. For example, Robinson et al. [16] (2015) have applied visual analytics methods to identify geo-located events patterns in social media data. They have designed an interface with multiple-linked views to visualize the temporal trend, spatial locations, keywords, and detailed media texts. In a follow up study, Pezanowski et al. [17] (2017) have designed an interface of multiple linked-views with a map, a table, and a matrix to support the correlation analysis among the detected events in social media. Li et al. [18] went a step further by combining different visualizations in each view to reveal significant occurrence patterns, i.e., co-, pre-, and post-occurrence patterns for pairs of locations. They designed map multiples with a timeline to show the spatiotemporal patterns, juxtaposed bar charts and radial charts to show the (re)occurrence patterns. To satisfy the increasing demand of dashboard users on overviewing spatiotemporal distribution of regional phenomena and their relationship, we propose to integrate the visual analytics approach with dashboards and set the emphasis on the design of analytical functions following the principle of understanding at a glance. For instance, data acquisition and analysis can be better supported by dashboards after applying the high interactivity characteristics. Users may solve simple tasks (identifying, locating, and distinguishing) and complex analytical tasks (cluster identification, ranking, comparing, associating, and correlating) [1,19] by quickly viewing and interacting with a map-based dashboard. Yalçın et al. [7] (2018) have proposed a dashboard with multiple-linked views to support novice users to identify tabular data patterns. They used maps and basic charts (bar chart, line chart, and pie chart) in each view to present a perspective of a dataset. Nazemi et al. [19] (2019) proposed a dashboard with the juxtaposition of visualizations such as a map, a chord diagram, small multiples, and a bar chart to allow users to perform analysis and comparison tasks.

Whether dashboards are useful and effective depends highly on user-centered evaluation [20]. User studies are commonly used to evaluate various types of cartographic visualizations. Numerous studies have focused on investigating the influence of specific design elements on interactive maps. Evaluating the user experience and usability of the geovisualizations in high-level spatial insights construction should be further studied [21–23]. Andrienko et al. [24] have proposed two types of analysis tasks, i.e., identification and comparison, to evaluate analytical geovisualizations. These two types of tasks differ in cognitive operations. The identification tasks focus on finding the characteristics of objects or locations. The comparison tasks require one to compare or summarize characteristics in different times or places. Yalçın et al. [7] have grouped the insights constructed by users into five categories: fact, min/max, correlation, distribution, and comparison. Bogucka et al. [25] proposed benchmark tasks that differ in query types, search output, and cognitive operations. In our work for the evaluation of the designed map-based dashboards, we propose three identification tasks and three reasoning tasks, each differing in search output, query type, dashboard interaction, and data uncertainty.

Popular methods for the evaluation of map-based dashboards include survey, interview, think-aloud, and eye-tracking [26]. Robinson et al. [16] have evaluated their map-based interface with a task-solving session and a survey on 25 domain experts. These two tasks were open-ended tasks, requiring the users to understand spatiotemporal patterns. In their survey of usability and utility, they applied the System Usability Scale method [27]. In a further study, Pezanowski et al. [17] have conducted an online usability and utility survey with 23 completed responses out of 327 participants. Many evaluations are done by interviewing the experts [18,28–31]. The expert interview gives self-reported outcomes, but the outcomes could be biased. McKenna et al. [32] and Yalçın et al. [7] conducted an evaluation on their dashboards by the think-aloud method during free exploration and a post-survey. The think-aloud session can reflect the usability in a natural usage scenario. Many studies use the eye-tracking method to externalize the knowledge construction procedure of users in viewing geovisualizations. In such studies, the authors collect and analyze eye movements when users are performing the predefined tasks. Hegarty et al. [33] found that users are more likely to be attracted by visually complex visualizations than simple ones. Opach et al. [34] have used the eye-tracking method to study the viewing strategy of users in obtaining insight on multi-component animated maps. They analyzed the order of response accuracy, fixation durations, dwells and transitions of the area of interests (AOIs). Bogucka et al. [25] accessed the feasibility of a space-time cube by analyzing the tasks' complete rate, duration, and search strategy from the eye movement data. Popelka et al. [35] have evaluated analytical maps by analyzing participants' attention, fixation sequence, and comparing the viewing points between correct and wrong answers. They formed a set of suggestions for the map-based interactive analytical application design.

In this study, we evaluated the effectiveness and efficiency of our proposed map-based dashboard by means of eye-tracking and interviews. The evaluated map-based dashboard is composed of multiple linked-views, and aims at supporting users in acquiring and analyzing geo-knowledge, such as spatial distribution, clusters, correlation, and temporal trend, at a glance. The dashboard is implemented as a web-based prototype so that users can interact with it during the experiment. Our experiment consisted of two major components. First, we designed six tasks and collect the eye movement data when the participants were performing the tasks. Next, we interviewed the participants about their attitudes towards the dashboard. We then analyzed the eye-tracking data and interview results using a variety methods to reflect the effectiveness and efficiency of the dashboard. The remainder of this paper is structured as follows: Section 2 introduces the design of the map-based dashboard. Section 3 describes the design of the experiment. In Section 4, we analyze the results of the experiment. Section 5 discusses the performance of the dashboard and the limitation of our experiment. In Section 6, we provide our conclusions from this study and offer outlooks for the future.

2. Design of the Dashboard

This section introduces the background for designing and implementing the dashboard, the test data, and the visual user interface of the proposed map-based dashboard.

2.1. Background

In a fast-developing society, stakeholders in many domains are updated with rapidly and constantly changing information about their surrounding economic environment. McKenna et al. [36] have studied the information needs of different stakeholders in an enterprise. The analysts need the most detailed information to understand how each factor changes at each location and time. The managers and directors need more general information, such as data distributions and trends. The chief executive officers (CEOs) need the most general information, and they care more about the future trend rather than historical events. Previous studies have suggested that dashboards serve as an effective tool for stakeholders in their decision-making procedure [1,37]. According to the information needs of stakeholders, dashboards are categorized into three main types according to their roles: operational, analytic, and strategic [2]. The operational dashboards aim to monitor the situation with a high temporal resolution and use dynamic visualizations to show the changes in detail. The analytical dashboards present patterns at a higher abstract level, and provide interactions for users to explore further information. The strategic dashboards provide an overview of the most general information and require fewer updates and interactions.

A map-based dashboard was designed in this study for stakeholders like leaders of small and medium-sized enterprises (SMEs) and citizens. These stakeholders need the overview information as well as certain analytical functions to understand and analyze the factors of the economic environment to make their decisions. They want to answer typical questions such as how high the population density in a city is, how much the average income of the citizens is, what is the trend of the economic-related factors in recent years, and how the economic development and transportation infrastructure correlate. Thus, we have designed our dashboard as an analytical map-based dashboard that not only shows the overview of the spatiotemporal patterns of multiple economic factors, but also supports users to fulfill their analytical tasks.

2.2. Test Data

Our test datasets are socioeconomic data at the municipality level provided by Yangtze River Delta Science Data Center (http://nnu.geodata.cn:8008/). The datasets originated from the census data. The datasets cover various topics, including gross domestic product (GDP), public investment, industrial output, population, and employment. In this study, we have selected four representative categories of socioeconomic environment, including enterprise, GDP, population, logistic, and have further identified 22 related factors in these four categories. The temporal coverage is from 2013 to 2015, where the data is available in most municipalities. Table A1 in the Appendix A shows the categories and factors used in this study.

We have chosen Province Jiangsu, China as the study area. Jiangsu is located in the east of China, at the lower reaches of the Yangtze River, which covers 107,200 km^2 and 98 municipalities. The sizes of the municipalities vary from 54 km^2 to 3059 km^2. With over 80 million people, Jiangsu is among the most densely populated and economically fastest developed regions in China. In recent years, the industry structure is undergoing rapid transformations in Jiangsu. Therefore, mastering the local economic conditions is very valuable for the stakeholders.

2.3. User Interface

We aim to show the users the overview of the spatiotemporal patterns of various economic factors, reveal their correlations, and allow users to compare the patterns at different levels of detail.

The interface of our dashboard consists of five panels, (A) the title panel, (B) the toolbar panel, (C) the spatial panel, (D) the temporal panel, and (E) the ranking panel.

Figure 2a shows the interface and its components. The title panel shows the topic of the dashboard. The toolbar panel allows users to reset maps, reset selections, and read explanations to various factors. The spatial panel presents the spatial distribution of multiple factors on maps. Each map shows a socioeconomic category, i.e., enterprise, GDP, population, and logistics. There are several layers within each map, and each layer shows one factor. The temporal panel shows the temporal trend of the data distribution along time. When a county is selected, its historical data is displayed as a bar chart in the temporal panel. The ranking panel presents the top five municipalities from the four economic categories respectively. The color scheme of these panels is kept consistent. Each category of factors is represented with a single-hue color scheme, where category enterprise is represented in red, GDP in purple, population in green, and logistic in blue.

Moreover, the interactions are shown in Figure 2b. The spatial, temporal, and ranking panels are linked, and serve together as an at-a-glance representation of the economic condition of the study area. The search function allows users to locate any county in the area. When a county is searched, the maps are zoomed in to this county and the temporal data of the searched county are shown. The factors can be selected by the switching function. Whenever a county or a factor is selected on one panel, this selection is applied to other panels. The four maps are sychronized in zoom level and central point. When users move one map, the others follow. Last but not least, users can reset the maps, reset the whole dashboard, and read the available factors and the their detailed information (see Table A1) by clicking the buttons on the toolbar.

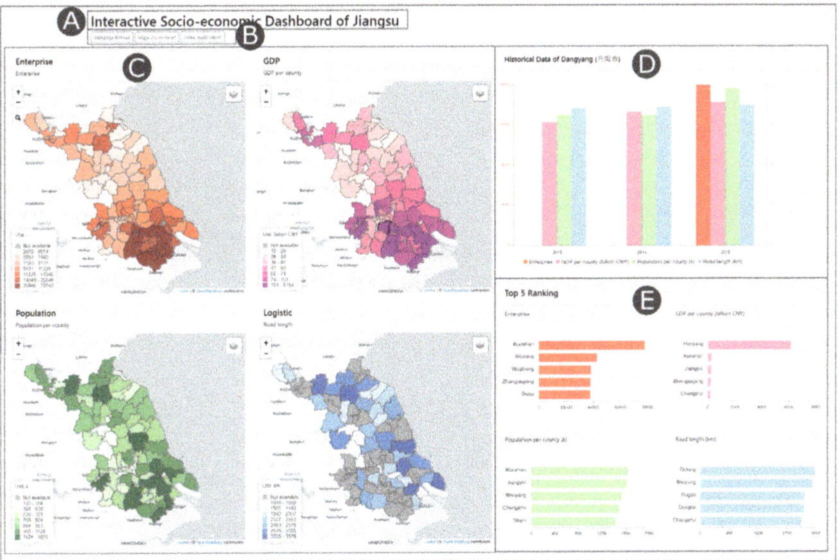

(**a**) The panels of the dashboard design. The panels are labeled as: Ⓐ the title panel, Ⓑ the toolbar panel, Ⓒ the spatial panel, Ⓓ the temporal panel, Ⓔ the ranking panel.

Figure 2. *Cont.*

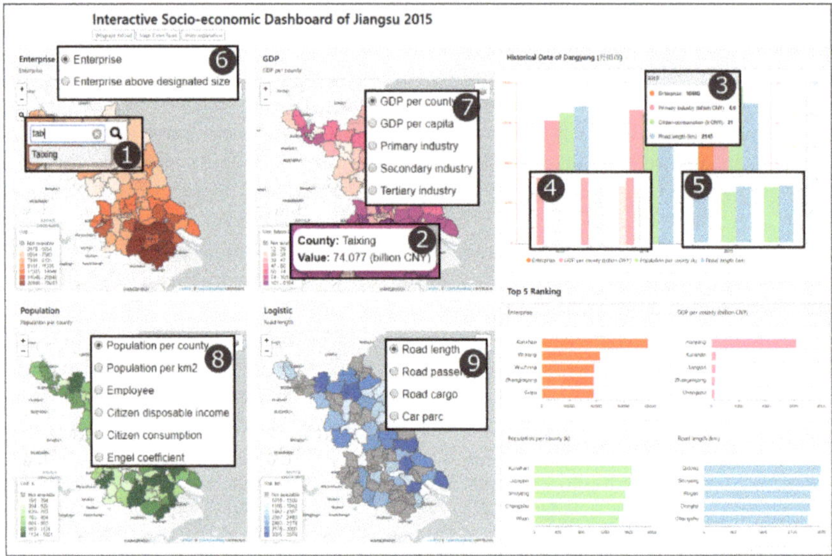

(**b**) The interactions of the dashboard. The functions are labeled as: ① searching, ② and ③ hovering, ④ and ⑤ highlighting and hiding, ⑥–⑨ switching context.

Figure 2. The interface and the interaction design of the map-based dashboard.

This dashboard design allows users to retrieve detailed information on demand. For example, users can retrieve the value of a factor of a certain municipality in a certain year. Furthermore, the dashboard supports users in obtaining high-level knowledge. For instance, users can learn the temporal trend of a factor, compare the spatiotemporal distributions of factors in different municipalities, compare the patterns of several factors of one specific municipality, or visually find correlation among factors. The dashboard interface was developed in JavaScript. The maps and charts were developed based on open source libraries such as Leaflet (https://leafletjs.com/) and ApexCharts.js (https://apexcharts.com/). We used Bootstrap (https://getbootstrap.com/) to arrange the layout of the dashboard. The interface can be browsed in various web browsers, such as Google Chrome, or Firefox.

3. Design of the Evaluation Experiment

This study aims to access the feasibility of the map-based dashboard for knowledge acquisition, especially with regard to spatiotemporal patterns and correlations in socioeconomic data. We have designed a qualitative study to collect and analyze the performance of the dashboard. More specifically, we have evaluated the effectiveness and efficiency of the dashboard by analyzing participants' gaze behavior and studying their attitudes towards the dashboard. To achieve this, we set up several benchmark tasks at several difficulty levels and designed an eye-tracking experiment to collect the participants' visual attention. Then we conducted an interview to collect feedback. In this section, we describe the design of these evaluation methods in detail.

3.1. Participants

We recruited 40 participants with the means of short introductions in classrooms, posters on the campus, and online advertisements. One of the participants dropped out of the experiment due to near-sightedness. The remaining 39 participants had normal or corrected-normal eyesight and completed the experiment. After the experiment, we found that the eye-tracking ratios of seven participants were less than 70%, and could not be considered. Thus, the analysis was based on the recoded eye movement data from the remaining 32 participants. Among the 32 participants, there were 17 females and 15 males. Their average age was 25.9, with a standard deviation of 2.36. The participants had diverse educational backgrounds: one participant had high school or equivalent degree, 17 participants had bachelors' degree, and 14 participants had masters' degree. They had various interactive dashboard usage experience. Twelve participants had used dashboards more than five times, five participants had used less than five times, eight participants had heard about it but not used, seven participants had never heard about it. In addition, none of the participants had ever lived in the study area. To clarify: in this paper the participants are the volunteers who took part in our experiment, the users refer to our target users of the designed dashboard.

3.2. Apparatus

We used a Gazepoint GP3 eye tracker, equipped with the software Gazepoint Analysis to collect eye movement data. The eye tracker has a 0.5–1 degree of visual angle accuracy and 60 Hz update rate. As Figure 3 shows, the eye tracker was placed under a monitor. We used two 2560 × 1440 resolution DELL monitors, one of which was for the participants to explore the dashboard, another for the controller of the experiment. The dashboard was running on a local server with Google Chrome as the browser. The participants were provided with a keyboard and a mouse to interact with the dashboard. The experiment environment was set up in the eye-tracking lab at the Technical University of Munich. The experiment lab was in a stable, quiet condition, and with scattering light during the experiment.

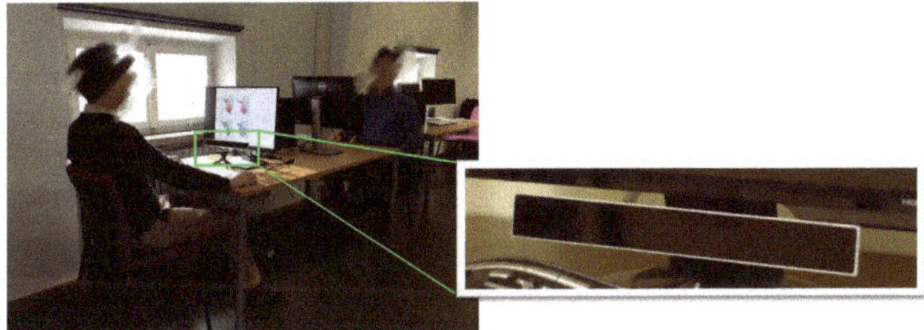

Figure 3. A picture showing the experiment environment. A participant (**left**) was doing the eye-tracking experiment with the Gaze Point GP3 eye tracker, while a controller (**right**) was observing the eye and mouse movements of the participant on a second screen.

3.3. Benchmark Tasks

Considering the tasks in [7,24], Zuo et al. proposed four benchmark tasks in dashboard usability testing in [38], which differ with regard to cognitive operations. However, the tasks were open-ended and caused great variations in participants' answers. In this study, we have proposed six close-ended benchmark tasks in different cognitive operations and dashboard interactions. These tasks belong to two types: (1) identify specific value(s) from the dashboard, and (2) compare or summarize high-level knowledge based on the facts found via the dashboard. For each type we proposed three specific tasks with increasing difficulty. The tasks were presented as statements that should be judged by the participants as being correct, wrong, or unknown based on their interactions with the dashboard. Table 1 describes the six statements and their associated answers.

The execution of these proposed tasks involves different search areas, periods, attributes, querying types, cognitive operations, dashboard interactions, and data availabilities. Table 2 outlines the complexity of each task in the aforementioned aspects. T1, T2, and T3 require participants to locate a value (or more) from the dashboard and make basic comparisons. The cognitive operation difficulty increases from T1 to T6. T1 requires participants to only find a value of an area. T2 requires participants to compare the values in an area. The cognitive complexity of T3 is higher than T1 and T2, because participants need to compare the values of multiple areas. T4, T5, and T6 require participants to summarize the patterns from multiple areas. T5 is slightly more complex than T4, because more search attributes and more dashboard interactions are involved. T6 is the most difficult task, because it requires participants to deduce the results from incomplete data, and the participants need to understand the economic concept "industrialized level".

Table 1. The six statements and their associated answers of the benchmark tasks.

Task Number	Statement	Answer
Task 1	In 2015, the Tertiary Industry value of *Qidong* is 85 billion Chinese Yuan (CNY).	Wrong
Task 2	The number of enterprises in *Jintan* increases from 2013 to 2015.	Unkown
Task 3	In 2015, among all the counties in Jiangsu, *Kunshan* has the largest number of enterprises.	Correct
Task 4	The south part of Jiangsu is economically stronger than the north part.	Correct
Task 5	In Jiangsu, the more employees in a county, the higher the citizens' disposable income is.	Wrong
Task 6	In Jiangsu, the longer the total length of the road of a county, the higher the industrialized level is.	Wrong

3.4. Experiment Tool

To guide participants through the hands-on part of the experiment, we developed an interactive experiment tool on the dashboard interface. As shown in Figure 4, the tool is in the left side of the dashboard with a dark background color to differentiate with the data visualization panels. It consists of eight items, including the first item of "Start free exploration", the six items of Statement 1–6, and the last item of "Finished!". When an item is clicked, the item expands with a concrete instruction of the step. When the item Start free exploration is clicked, it shows the instruction of "Please start tasks in 3:00 minutes!". The number is a real-time countdown clock to remind participants of the remaining time during the free exploration. Note that the six tasks were shown in a random order for each participant in the Statement items. Thus the influence of the order for the response time of tasks was be minimized. The item "Finished!" confirmed the completeness of tasks and informed the participants that they were free to move their bodies. The participants were asked to click the items following the order from top to bottom. Only one item could be clicked at one time. After clicking a new item, the dashboard on the right side is reset.

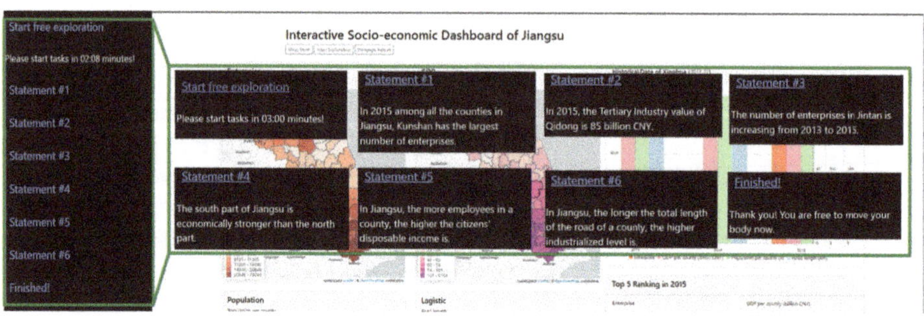

Figure 4. The integrated experiment tool with the step guides. The screenshot shows when the *"Start free exploration"* item is clicked. On the right side, it shows an example of the items being clicked.

Table 2. Benchmark tasks for the evaluation the map-based dashboard.

Tasks	T1	T2	T3	T4	T5	T6
Description	Find an attribute of a place.	Find the attribute temporal trend of a place.	Identify a place with the highest attribute value.	Summarize the spatial distribution of an attribute.	Compare the spatial distribution of two attributes.	Abstract the attributes. Compare the spatial distribution of two attributes.
Search Area	Single	Single	Multiple	Multiple	Multiple	Multiple
Search Time	Single	Multiple	Single	Single	Single	Single
Search Attribute	Single	Single	Single	Single	Multiple	Multiple
Query type	State	Change	Order	State	State	State
Cognitive Operation	Identification	Comparison	Identification, Comparison	Comparison, Summary	Comparison, Summary	Comparison, Summary, Deduction
Dashboard Interaction	Query, Switch content	Query, Highlight *	Switch content *	Switch content *	Switch content	Switch content *
Data Availability	High	High	High	High	High	Middle

* Owning to the default layer setting, the interactions are optional.

3.5. Procedure

The experiments were conducted from 26th November 2019 to 21st December 2019, in the Eye-tracking lab at the Technical University of Munich. The experiment was conducted in consecutive order, with one participant following the other. The participants were allowed to terminate the experiment at any time. The experiment consisted of six steps: pre-experiment, introduction, calibration, free-exploration, task-solving, and interview. In addition, all the steps were carried immediately after the other. In this section, we introduce each step in detail.

Pre-experiment. Before the experiment, each participant had several minutes to relax, because we found that many participants were too nervous or excited to start the experiment directly. When the experiment started, we first informed the participants about the data protection policy, the approximate duration, the experiment steps, and the data collection. If the participant agreed, we would proceed with the experiment. The participants were then asked to fill out a form with their personal information, including gender, age, education level, and dashboard usage experience.

Introduction. We conducted a standard introduction for the participants. The introduction included a short description of the factors and operation tutorial of the dashboard, namely data categories and factors, panels of the dashboard, and the experiment tool. The participants were allowed to ask usage-related or general questions in this step. Additional information that might have influenced the results of the experiment was not given.

Calibration. First, we asked the participants to find a comfortable position while keeping their eyes within the detection range of the eye-tracker. We informed the participants that they needed to hold the position during the calibration, free exploration, and the task-solving steps. We then repeatedly calibrated the eye tracker until it met the experimental requirement.

Free-exploration. During this step, the eye movements of the participants were tracked. Every participant was asked to explore the dashboard freely for three minutes. All the participants were allowed to view or interact with the dashboard freely. They began this step by clicking the Start free exploration item on the experiment tool. They could check the time with the countdown clock on the experiment tool, or the experiment controller would remind the participants when the time was up.

Task-solving. The task-solving step was also performed while the eye movements was being tracked. The participants solved the tasks following the order of Statement showing on the experiment tool. After clicking and reading the task item, they could interact with the dashboard and check the answers as correct, wrong, or unknown on a prepared sheet. After finishing a task, they were only allowed to proceed to the next task and could not return to or change any previous answers. The tasks did not have a time limit for completion.

Interview. In the last step, we interviewed each participant with four questions. First, we asked the participants to rate their confidence levels of the answers in the range of 1 (not confident at all)–10 (very confident). Second, we asked them to rate the difficulty level of using the dashboard between 1 (very hard)–10 (very easy). Third, we asked them to list the design elements that helped them during the completion of the tasks. Lastly, we asked them to list the design items or elements that were not easy to understand or interact with. During the discussion, the participants were also asked to describe more whenever necessary. The answers of the participants were recorded as written protocols.

3.6. Methods of Analysis

We analyzed the acquired eye-tracking data and the interview results to assess the performance of the designed map-based dashboard. We focused on analyzing five themes: the attraction of the panels, the effectiveness of the dashboard, the efficiency of the dashboard, the task-solving strategy of the participants, and their attitude towards the dashboard.

We explain these five themes in detail. (1) The attraction on the dashboard panels was well reflected by the visual fixation on the dashboard during the free-exploration stage. To show the attention distribution, we visualized the fixation positions in the first 90 seconds using heatmaps.

(2) The effectiveness is measured by the task-solving correctness. We compared the success rates in solving the benchmark tasks first among the different cognitive complexities, second among the increasing familiarity of the proposed dashboard, and finally among different participant groups. More specifically, we assumed that the familiarity of the proposed dashboard increases while the participants were carrying out the tasks. (3) We measured the efficiency using the response time of completing a task. The duration of each task-solving step was recorded between the starting click and the ending click. Similarly, the response time was also compared according to different cognitive complexities, the familiarity of the proposed dashboard, and different participant groups. In addition, the response time was compared between the successfully and unsuccessfully performed tasks. (4) The search strategies of the participants were measured. These were comprised of the search sequence, average dwell time, and the transition and return probabilities among the Areas of Interest (AOIs) in each task. According to [39], we listed the selected metrics of the eye movements in Table 3. We selected seven AOI areas on the dashboard shown in Figure 5, including Task, AOI Enterprise, AOI GDP, AOI History, AOI Population, AOI Logistic, and AOI Ranking. Since we focused on how the participants construct knowledge via multiple panels, the AOIs were selected based on the dashboard panels and their contents. The spatial panel was divided into four AOIs, as each map shows different categories of data. Moreover, these metrics were visually analyzed. The sequences of the fixations on the AOIs were visualized in sequence charts, with each fixation shown as a color block along the timeline. Based on the visualization method of transition states proposed in [40], we designed a dwell and transition chart to show the eye movement patterns between the AOIs. In this chart, each AOI is represented by a circle, and the radius stands for the average dwell. The transition between two AOIs is represented by a line, and the width stands for the transition probability. (5) The attitude of participants is reflected by their answers from the interview. The confidence rates and overall usability rates were qualitatively analyzed. The positive and negative design items listed by the participants were grouped in the dashboard panel, layout, interaction, and others.

Table 3. The selected eye movement metrics.

Metric	Description
Sequence	The order of fixation within the AOIs.
Dwell time	The sum of all the fixations and saccades within an AOI.
Transition	The movement from one AOI to another.
Return	It is a transition to an AOI itself, also known as revisit.
Transition probability	The probability of the fixation moving from one AOI to another AOI in a sequence.

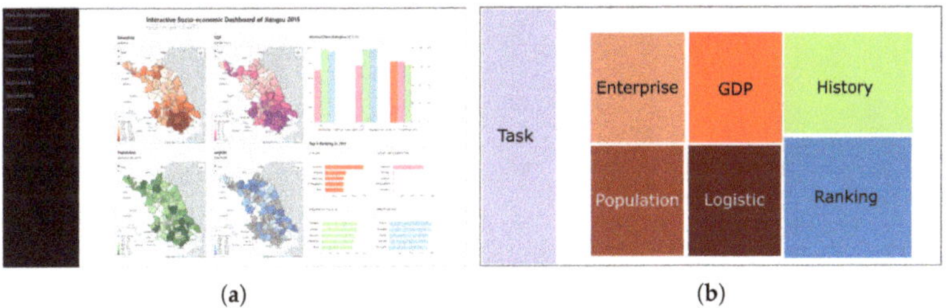

(a) (b)

Figure 5. The outline of the AOIs (**a**) and the color coding of each AOI (**b**).

4. Evaluation Results

This section describes the analysis results of the eye movement data and the interview feedback collected from the experiment. More specifically, we introduce the results in the following five aspects: (1) The participants' fixations in free exploration; (2) success rate; (3) response time; (4) the participants' fixation, dwell, and transition of the AOIs during their task-solving stages; (5) the feedback from the participants. The eye-tracking data is published online (https://github.com/Map-based-Dashboard/eye-tracking-experiment).

4.1. Fixation in Free Exploration

The fixation distribution of the participants reflects the visual attraction of different sections of the visual interface. We show the fixation distribution in the first 90 s at the free exploration stage on heatmaps (shown in Figure 6). Considering our sample size, we chose 10 s as the interval to include enough fixations in forming clusters in each interval. The heatmaps exhibit different patterns of the fixation distribution in 0–20 s (Figure 6a,b) and 20–90 s (Figure 6c–i). In the first 10 s (Figure 6a), we can see that the fixations were mostly on the title, AOI Task, AOI Enterprise, and AOI GDP. Between 10 to 20 s (Figure 6b), the fixations were more on AOI Enterprise, AOI GDP, AOI History, and less on AOI Task. From 20 to 90 s (Figure 6c–h), the attention of the participants was located more evenly on each dashboard panel. In general, panels located in the center attracted more attention than other panels at the free exploration stage, for example, AOI Enterprise and AOI GDP were focused on at the beginning of the exploration, while AOI Ranking only received a small amount of attention. The visualizations with a high information density also attracted more attention from the participants, because there were more fixations on the maps than on the bar charts. The dynamic visualizations attracted a significant amount of attention. For example, the AOI History drew a large amount of attention, as the chart in the panel changed when there was a mouse hover or click. In contrast, AOI Ranking did not involve much interaction and it received less attention. Additionally, a large amount of attention went to AOI Task in the beginning because the participants needed to click and read the task items. We inferred that bright colors also play an important role in attracting users' attention. Last but not least, anomalous patterns, such as incomplete data and outliers, also attracted the participants' attention. In the 0–30 s, more fixations were at the AOI Logistic than AOI Population, where a large gray area indicated unavailable data.

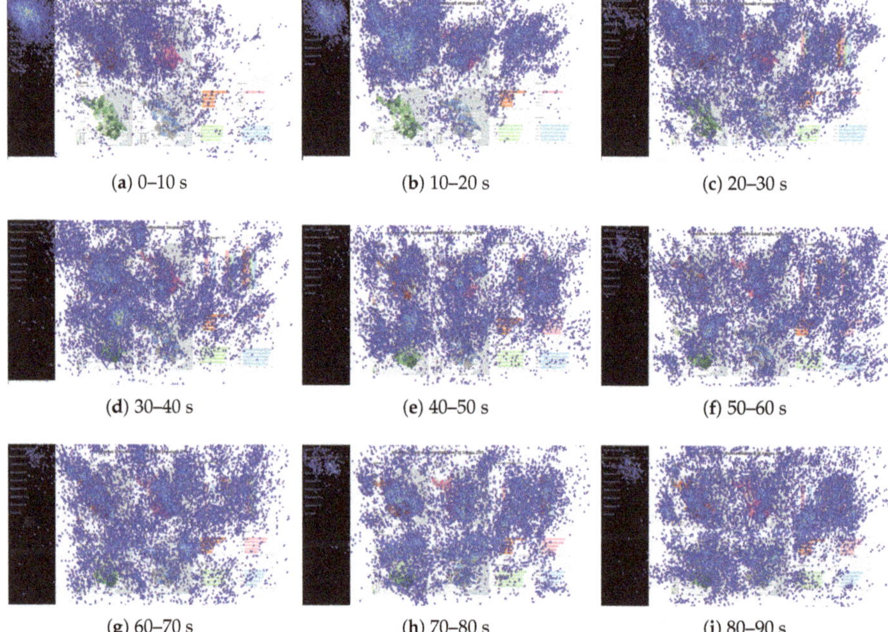

(**a**) 0–10 s (**b**) 10–20 s (**c**) 20–30 s

(**d**) 30–40 s (**e**) 40–50 s (**f**) 50–60 s

(**g**) 60–70 s (**h**) 70–80 s (**i**) 80–90 s

Figure 6. The heatmaps of all the participants' fixations during at free exploration stage. The blue dots represents fixations.

4.2. Success Rate

The success rate reflects the effectiveness of the dashboard design. In this section, we first compared the success rates according to the perception difficulties of the tasks and the familiarities of the proposed dashboard of the participants. Specifically, we compared the success rates among the proposed benchmark tasks (Section 3.3). The levels of difficulty in perception raised from the tasks T1 to T6. The familiarity levels of the dashboard increased from the statements S1 to S6 on the experiment tool (Section 3.4). Recall that the experiment tool shows each participant the tasks in random order. We assume that the participants were becoming familiar with the dashboard while they were carrying out the tasks.

Figure 7 shows the success rates according to the order of increasing difficulty (from T1 to T6) (Figure 7a) and the order of increasing familiarity (from S1 to S6) (Figure 7b). From Figure 7a, we can see that in general the success rates are very high for all the tasks. Among the identification tasks from T1 to T3, we can see that T1 and T3 have higher success rates than T2. Compared to T2, T1 requires a lower cognitive effort than T2 and more dashboard interactions, while T3 requires less dashboard interaction and more complex cognitive operation. For the reasoning tasks from T4 to T6, the success rates decrease significantly. Figure 7b shows in general the success rates did not increase with the participants' familiarity with the dashboard.

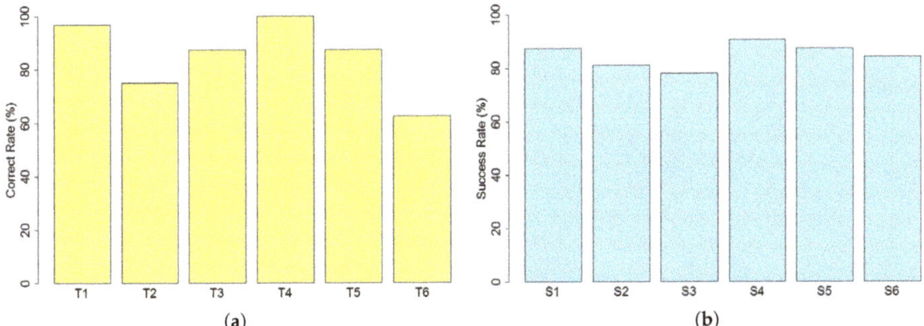

Figure 7. The bar charts of the success rates. (**a**) The success rates of the tasks in increasing difficulties of perception. (**b**) The success rates across their position in the task sequence.

Figure 8 further shows the success rates of each task across the position of the task sequence. The success rates of T3–T6 did not increase along with the increasing of the familiarity of the dashboard. However, the success rates of T1 and T2 had an upward trend. It might indicate that the participants were performing better in location searching when they became more familiar with the dashboard.

Finally, we compared the success rates among different groups of participants including aspects such as gender, educational background, and usage experience. Table 4 shows the success rates of each task of different groups. Most of the tasks were correctly conducted in every group. The average success rate of all the participants is 84.9%. The success rates did not show great differences among these groups.

4.3. Response Time

To analyze the efficiency of the dashboard, we examined the response time in completing the tasks. Similar to the success rate analysis, we compared the median response time first among the increasing cognitive complexities, second among the increasing familiarities of the dashboard, and third among the different participant groups. Additionally, we compared the response time between the successful and failed tasks.

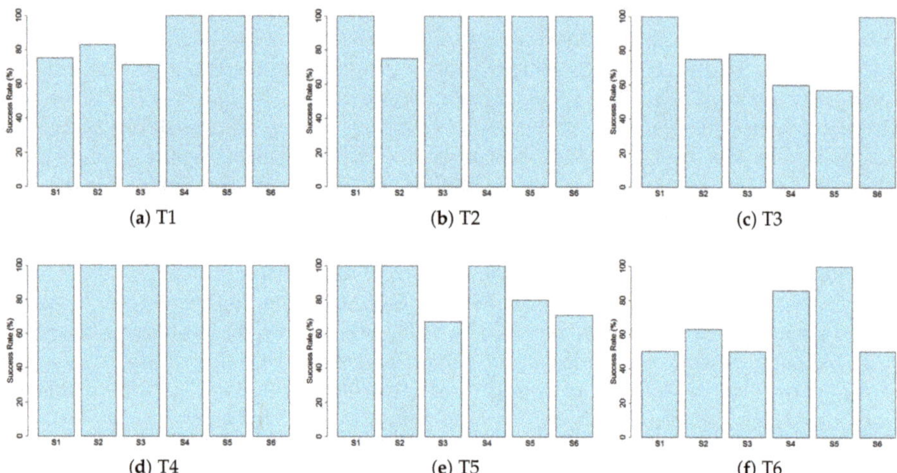

(**a**) T1 (**b**) T2 (**c**) T3

(**d**) T4 (**e**) T5 (**f**) T6

Figure 8. The bar charts of the success rates of each task across the position in the task sequence.

Table 4. The task success rate (%) of each task in different groups.

Category	Group	Number of Participants	T1	T2	T3	T4	T5	T6	Overall
Overall	Overall	32	96.9	75.0	87.5	100.0	87.5	62.5	84.9
Gender	Female	17	94.1	64.7	94.1	100.0	94.1	58.8	84.3
	Male	15	100.0	86.7	80.0	100.0	80.0	66.7	85.6
Education	Master	14	92.9	64.3	85.7	100.0	85.7	78.6	84.5
	Bachelor	17	100.0	82.4	94.1	100.0	88.2	52.9	86.3
	High School	1	100.0	100.0	0.0	100.0	100.0	0.0	66.7
Previous Dashboard Usage Experience	Used > 5 times	12	100.0	83.3	91.7	100.0	83.3	75.0	88.9
	Used $\leq$ 5 times	5	100.0	60.0	80.0	100.0	60.0	40.0	73.3
	Heard only	8	100.0	87.5	87.5	100.0	100.0	50.0	87.5
	Never heard	7	85.7	57.1	85.7	100.0	100.0	71.4	83.3

Figure 9 uses boxplots to show the response time of each task. The median response time of all the tasks was 51.5 s. In Figure 9a, we can see that all the identification tasks (T1–T3) took relatively less time, while the respond time of reasoning tasks (T4–T6) varied a lot. The response time of T5 and T6 was much longer and more dispersed than T4. Figure 9b shows a general downward trend of the response time along the increasement of the familiarity. This could indicate that the more the participants interact with the dashboard, the shorter the time they need to solve a task. The median response time did not continue to decrease after the completion of four tasks. It might indicate that the participants were familiar enough with the dashboard after carrying out the first four tasks, and the median duration was 3.9 min. The median response time of the last two completed tasks was at a relatively low level, but the increasing variation might suggest that some of the participants were tired towards the end of the experiment.

Figure 10 shows in detail the response time of each task in the carried out sequence. We cannot see an obvious downward trend from each task, which could be caused by the small sample size. However, we can see that T5 and T6 had a relative longer response time when they were carried out in the first position.

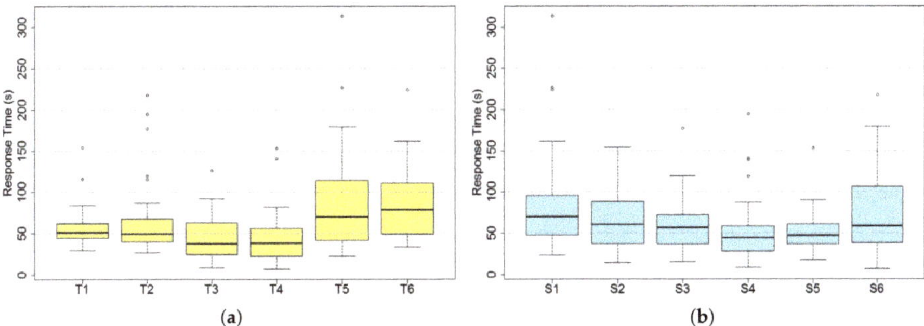

(**a**) (**b**)

Figure 9. The boxplots of the response time. (**a**) The response time of the tasks in increasing difficulties of perception. (**b**) The response time across their position in the task sequence.

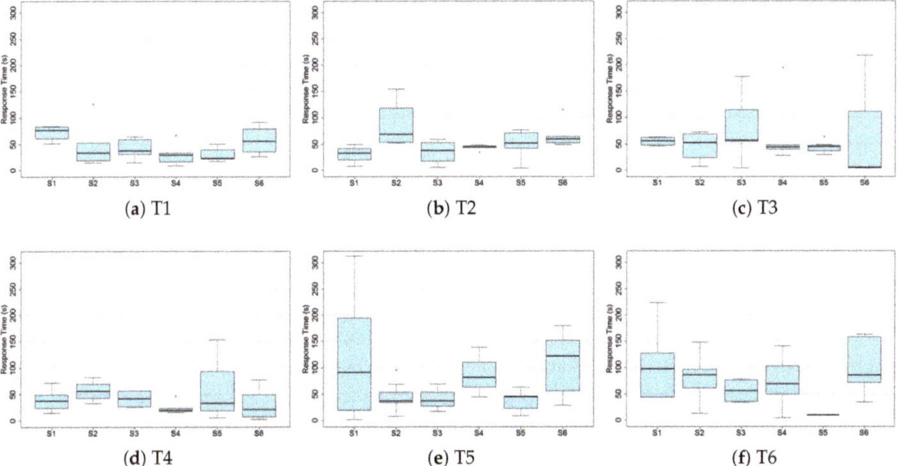

(a) T1 (b) T2 (c) T3

(d) T4 (e) T5 (f) T6

Figure 10. The boxplots of the response time of each task across the position in the task sequence.

Table 5 gives an overview of the median response time of each task in different groups. We found that the response time does not exhibit many differences among groups with different demographic attributes.

Table 5. The median response time (s) of tasks in different groups.

Category	Group	Number of Participants	T1	T2	T3	T4	T5	T6	Overall
Overall	Overall	32	37.7	51.2	49.8	38.3	70.0	78.8	357.6
Gender	Female	17	40.1	52.6	56.9	32.3	53.8	87.3	348.4
	Male	15	36.7	49.4	47.9	43.1	81.9	74.8	377.4
Education	Master	14	37.7	49.1	48.8	48.7	89.2	88.7	386.0
	Bechalor	17	36.7	52.6	50.3	25.1	53.8	75.6	339.7
	High School	1	64.3	41.7	217.8	32.9	91.4	48.7	496.8
Previous Dashboard Usage Experience	Used > 5 times	12	32.2	52.8	59.5	38.0	86.2	74.2	388.3
	Used ≤ 5 times	5	40.1	45.2	43.9	56.0	96.5	97.9	445.9
	Heard only	8	31.6	50.7	44.0	33.3	67.3	94.3	315.1
	Never heard	7	53.0	51.4	63.2	46.7	57.8	75.6	348.4

The differences in response time between the successful and failed tasks is shown in Figure 11. The figure shows that except for T6, the median response time of the failed tasks is longer than the successful ones. This could indicate that the failed tasks were caused by the wrong search strategy of the participants.

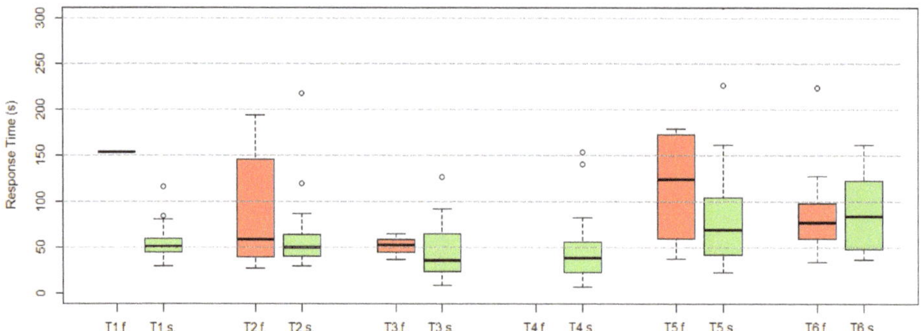

Figure 11. The response time of the tasks. Failed tasks (**red**), and the successful tasks (**green**) in the increasing difficulty levels of perception.

4.4. *Dwell and Transition during Tasks*

To find more differences in the task-solving strategies between the successful and failed tasks, we compared the fixation sequences, as well as the dwell and transitions in performing the tasks. For each task, we illustrated the fixation sequence of the participants in sequence charts, using the same color scheme of the AOIs in Figure 5. The time range of the sequence charts was set as 0–250 seconds as it fits well to all the tasks, so that we can better compare the fixation among the tasks. From the sequence charts, we could easily interpret the fixation time and the viewing orders of AOIs of each individual participant. We also visualized the average dwell time and transition probabilities among AOIs for each task. In all the sequence charts and the dwell and transition charts, the successful and failed groups are visualized separately. For each task, we described the most effective solution and compared the participants' solutions with it. In this section, we examine in detail the strategies of the participants in solving the tasks with the aim of identifying clues on how to improve further the effectiveness and efficiency of the dashboard.

For T1 "In 2015, the Tertiary Industry value of Qidong is 85 billion CNY. (Wrong)", the participants needed to find a GDP-related value of a county. To solve it, the most effective solution was to (1) locate County Qidong with the search function, (2) switch the layer to Layer Tertiary Industry in AOI GDP, and (3) move the mouse onto AOI GDP or AOI History to read the value in the pop-up window. As shown in Figure 12, the participants spent most of the time in checking AOI GDP and AOI History. The failed user was spending a lot of time on AOI Ranking, which did not provide the needed information. We found that the viewing in the combination of AOI GDP and AOI Task often happened shortly before some participants (P2, 4, 14, 17, 19, 31, 36) finished the task. This indicates that when the participants knew where to find the ranking information, they can quickly finish the task.

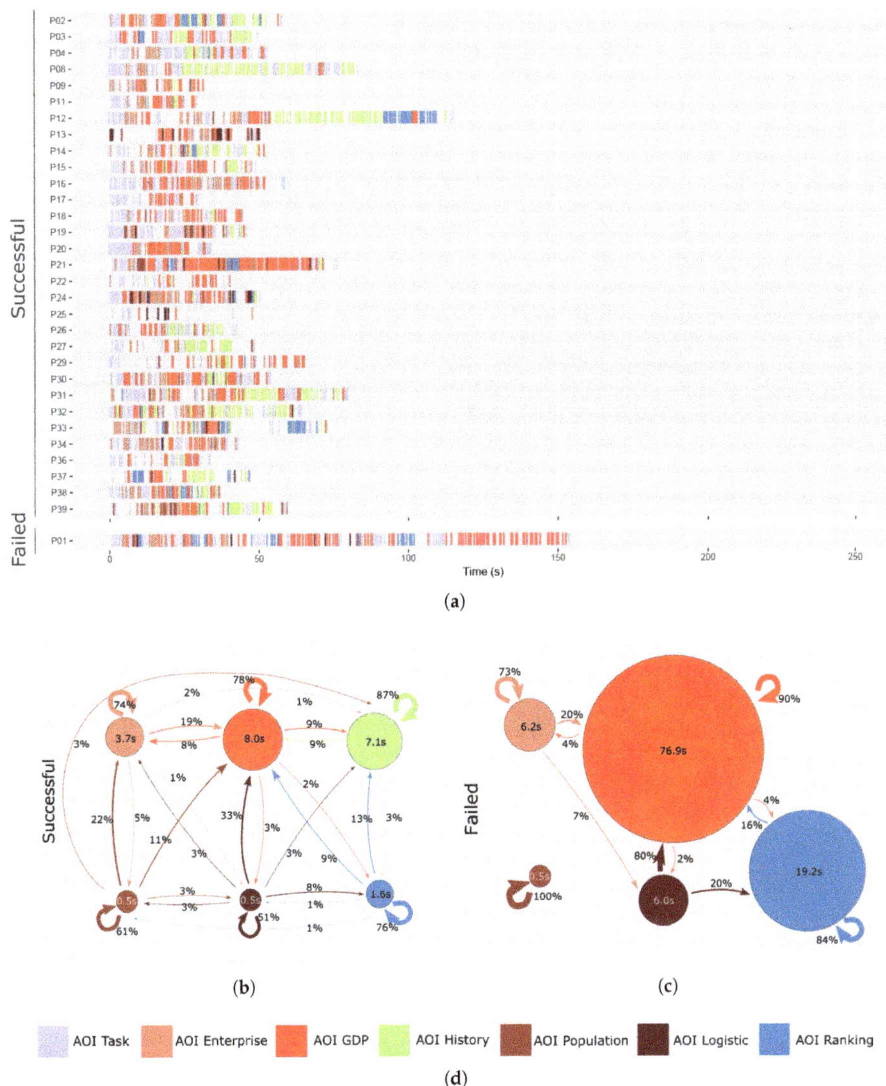

Figure 12. The visualization of search strategies of T1: (**a**) The fixation sequence chart. (**b**) The dwell and transition chart of the successful group. (**c**) The dwell and transition chart of the failed group. (**d**) The legend of the AOIs.

For T2 "The number of enterprises in Jintan increases from 2013 to 2015. (Unknown)", it required the participants to find the temporal trend of an enterprise-related factor of a county. To solve T2, the best solution was to (1) locate Jintan with the search function, (2) make sure the active layer is Enterprise in the Enterprise category, and (3) read the temporal trend of number of enterprises in the bar chart on AOI History. From Figure 13, we can see that the participants spent the most time on viewing AOI Enterprise and *AOI Ranking*. This is because they were using the search bar to locate the place in the task. Some participants (P1, 3, 22) spent relatively longer time on AOI Enterprise and AOI History, and most of the transitions are from AOI GDP. It suggests that they were having difficulties in solving step (1) and (3) of this task.

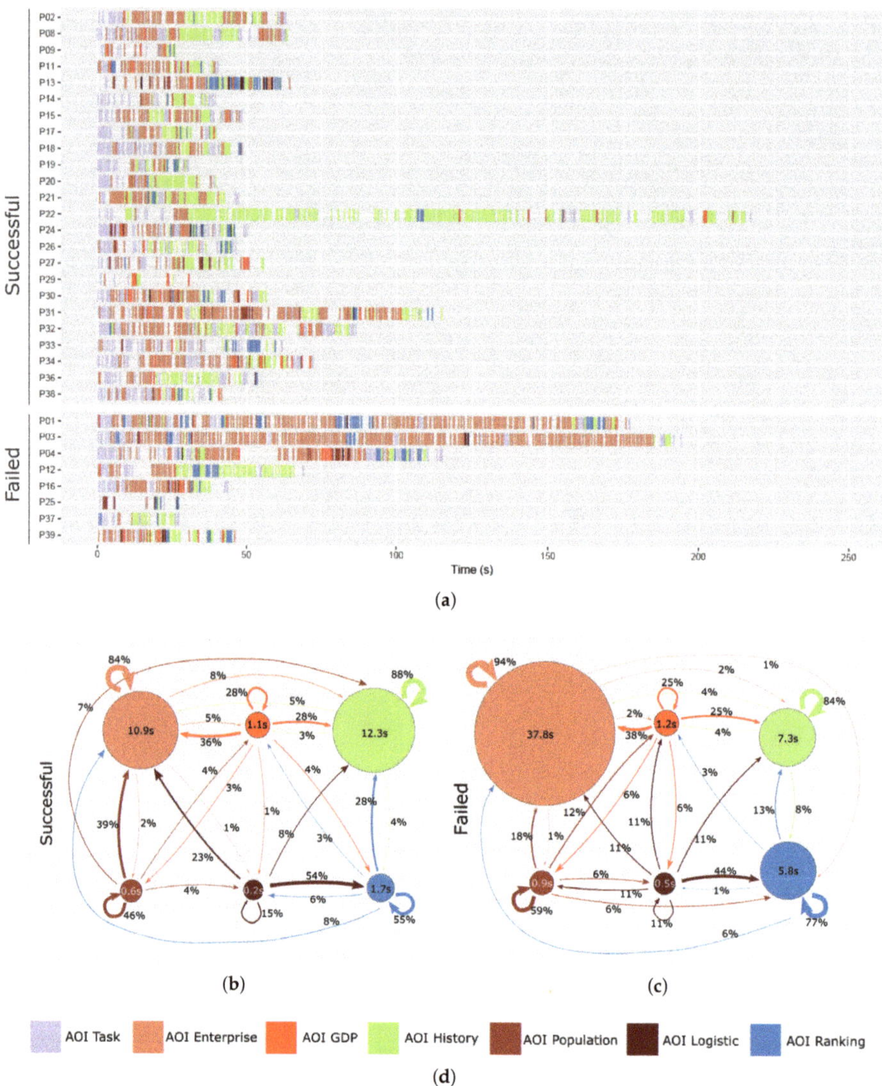

Figure 13. The visualization of search strategies of T2: (**a**) The fixation sequence chart. (**b**) The dwell and transition chart of the successful group. (**c**) The dwell and transition chart of the failed group. (**d**) The legend of the AOIs.

T3 "In 2015, among all the counties in Jiangsu, Kunshan has the largest number of enterprises. (Correct)" required the participants to find the county with the highest value of an enterprise-related factor. The most effective solution was (1) make sure the active layer was the Layer Enterprise in AOI Enterprise, and (2) read the name of the top county on AOI Ranking. Figure 14 shows that the participants were mainly paying attention to view the AOI Enterprise. They used the search bar in the AOI Enterprise to locate the place, and checked the AOI Ranking and the AOI History. The AOI Ranking took on average only 3 seconds in the successful group. This may indicate that the view was efficient. However, two of the four failed participants (P22, 25) completely missed the AOI Ranking. This suggests that the two participants were not familiar with the dashboard interface.

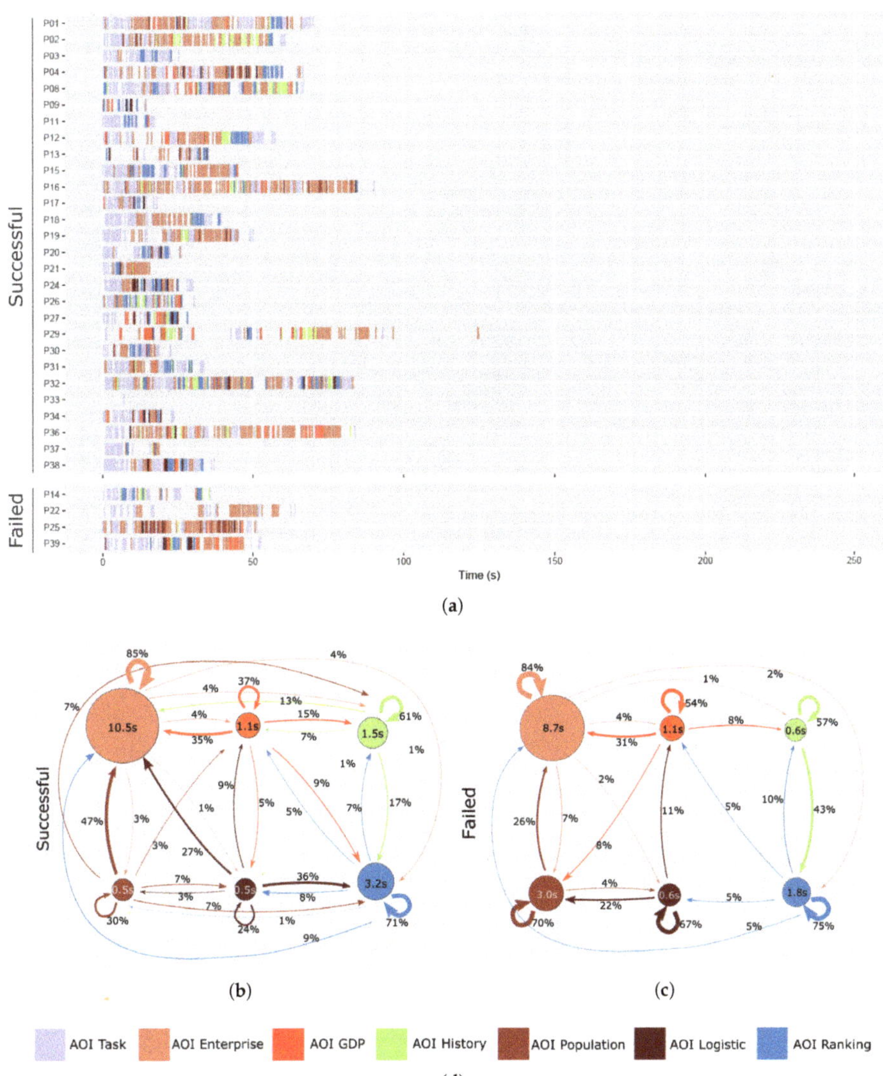

Figure 14. The visualization of search strategies of T3: (**a**) The fixation sequence chart. (**b**) The dwell and transition chart of the successful group. (**c**) The dwell and transition chart of the failed group. (**d**) The legend of the AOIs.

T4 "The south part of Jiangsu is economically stronger than the north part. (Correct)" required the participants to find a GDP-related factor's spatial distribution. Its effective solution was to check the spatial distribution of the layers in AOI GDP. From Figure 15, we can see that all the participants finished the task quickly and successfully. Most of the time, the participants were viewing AOI GDP. From the transitions chart, we infer that the participants knew where to find the spatial distribution information. Two participants (P3 and P30) finished T4 within 10 seconds with the search strategy of first reading the statement, then viewing the related information in AOI GDP, and then confirming their answers by rereading the statement again.

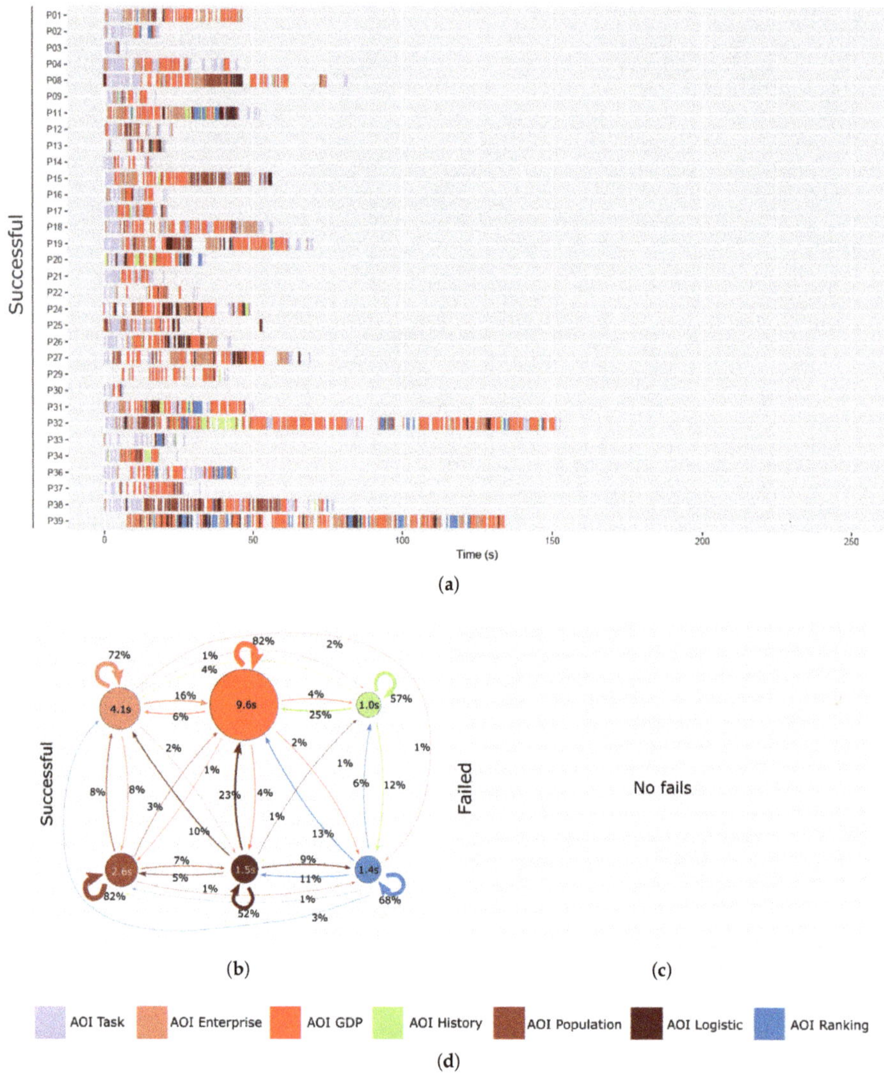

Figure 15. The visualization of search strategies of T4: (**a**) The fixation sequence chart. (**b**) The dwell and transition chart of the successful group. (**c**) The dwell and transition chart of the failed group. (**d**) The legend of the AOIs.

T5 "In Jiangsu, the more employees in a county, the higher the citizens' disposable income is. (Wrong)" required the participants to compare the spatial correlation between two population-related factors. The effective solution was to compare the spatial distributions between the Layer Employee and the Layer Citizen disposable income in the AOI Population. As seen from Figure 16, both successful and failing groups focused on the AOI Population to check the spatial correlation. A few participants (P29, 32, and 33) checked AOI Ranking for some time. Some participants (P16, 31, and 32) viewed AOI History for some time. We infer that the participants knew where to find the answers on the dashboard, but the high perception difficulty caused a longer viewing time.

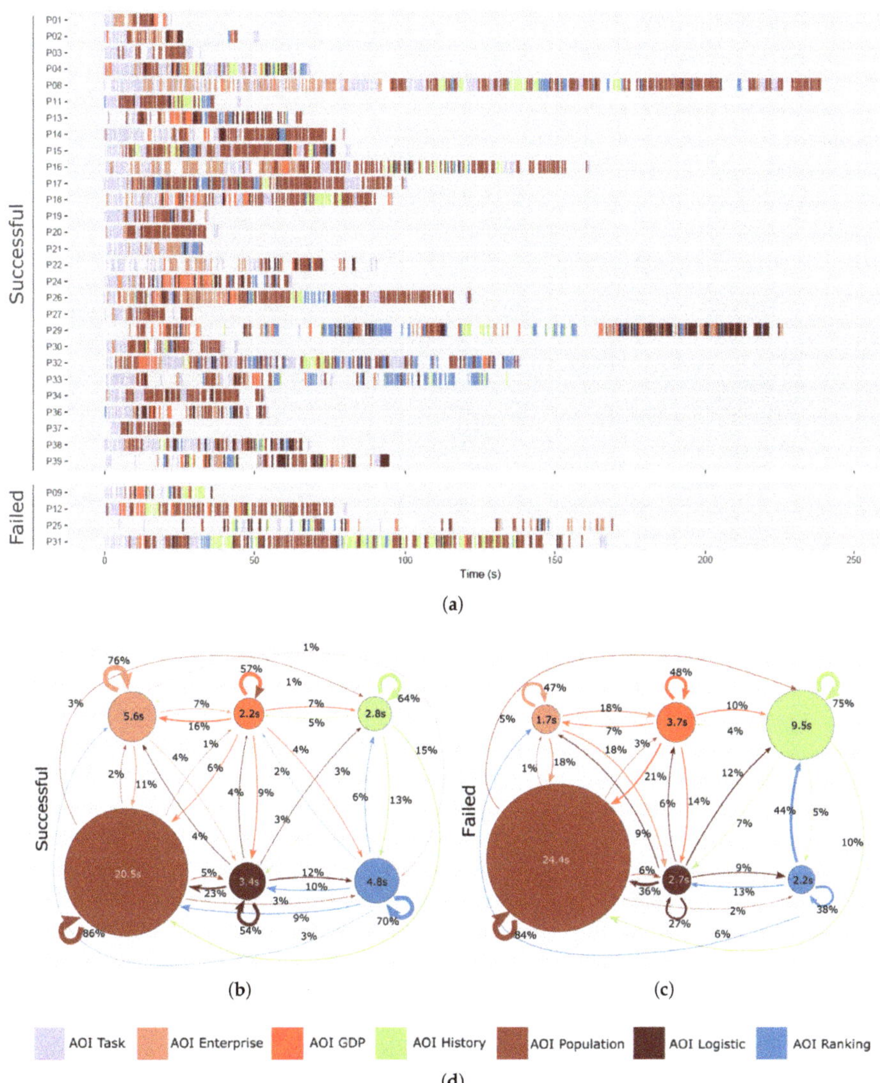

Figure 16. The visualization of search strategies of T5: (**a**) The fixation sequence chart. (**b**) The dwell and transition chart of the successful group. (**c**) The dwell and transition chart of the failed group. (**d**) The legend of the AOIs.

T6 "In Jiangsu, the longer the total length of the road of a county, the higher the industrialized level is. (Wrong)" required the participants to find the correlation between a logistics-related factor and a GDP-related factor, and the data of the logistics-related factor were incomplete. The most effective solution for T6 was to compare the spatial distribution between the Layer Road length in the AOI Logistic and the Layer Secondary industry in the AOI GDP. From Figure 17, we can see that the participants were comparing AOI Logistic and AOI GDP. Some participants were trying to find the correlation by comparing the factors in AOI Ranking, which was a wrong strategy. The unavailable data required the participants to summarize the correlation with less data than T5, which was more challenging and thus caused more failed cases.

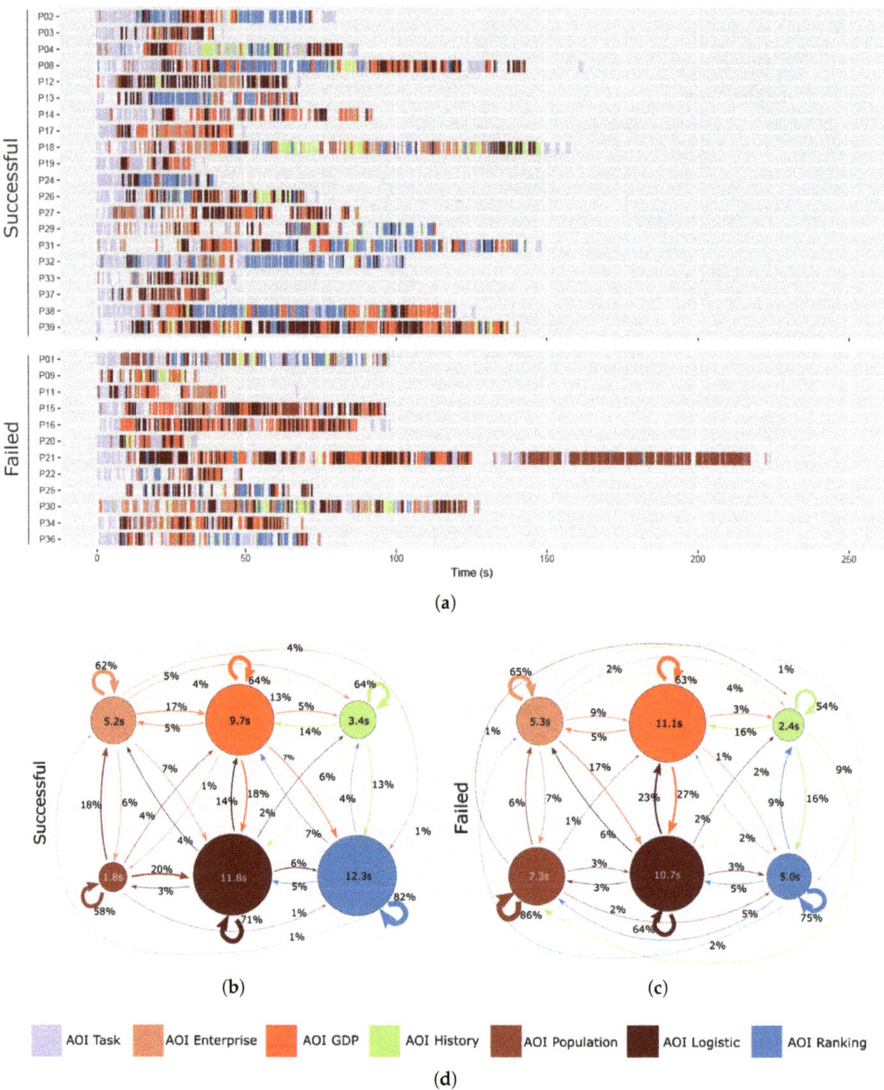

Figure 17. The visualization of search strategies of T6: (**a**) The fixation sequence chart. (**b**) The dwell and transition chart of the successful group. (**c**) The dwell and transition chart of the failed group. (**d**) The legend of the AOIs.

In summary, the reading sequence during the task solving is mainly driven by two reasons. One reason is the task requirement. The participants viewed the dashboard focused on searching for the most relevant information required by the tasks. Therefore, labeling of the panels is very important in navigating the users' attention to the needed information. Another reason is the layout of the panels. The participants tended to read the adjacent panel after reading the current panel. Thus the layout of the panels plays an important role in dashboard reading.

4.5. Feedback

In this section, we evaluated the usability of the dashboard by analyzing the feedback from the interview. First, we discussed the rating of the participants on the usability of the dashboard. Second, we classified the keywords in the interview protocol in groups of dashboard design elements.

The participants rated on average 8.25 (1–10 from less to more confident) on the confidence of their answers, and 7.89 (1–10 from very hard to very easy to use) on the general usability of the dashboard. The overall results were very positive. Moreover, some of the participants reported they were less confident with the answers of T5 and T6. There were two reasons for this: one was that the unavailable data made them unsure of their answers, and the other one was because of the lack of economic knowledge background.

We asked the participants to list the design items that helped them during the task-solving procedure without predefined options. The feedback is grouped into four categories: panel, layout, interaction, and others. Table 6 summarizes the positive feedback with the associated frequency in each group. In the panel group, the most helpful panel was the spatial panel (nine mentions). In the layout group, the color scheme was very helpful in supporting the participants in organizing the information. In the interaction group, the search function was identified as the most helpful item.

Table 6. Grouped positive items mentioned by the participants in the interview.

Group	Item	Frequency
Views	The spatial panel is helpful	9
	The temporal panel is helpful	6
	The ranking panel is helpful	2
Layout	The color scheme helped in organizing information	13
	The grouped layers helped in factor finding	4
	The juxtaposition benefits for comparison	3
	The structured design gives a good overview	2
Interaction	The search function is useful in finding places	15
	The interactions of the temporal panel helped them find data quickly	6
	Mouse hovering and clicking are helpful	5
	Layer switching is efficient	2
Other	Natural to use	1

Similarly, we have grouped the negative design items named by the participants in the interview. Table 7 shows the items in detail. Compared to the positive items, the negative ones are more related to specific issues. The most frequently mentioned items are the top margin of the bars in the temporal panel is sometimes too narrow, the shifting of the maps is disturbing, and the font size is too small. It is important to note that three participants thought the temporal panel is too informative, and one participant pointed out the temporal panel did not follow the four category structure as other panels did. Two participants also commented that they did not know where to look for the information that they needed.

In summary, map-based dashboards are very useful media to show the spatiotemporal information. Maps are the main element to bridge the knowledge from different perspectives and their location information. A uniformed style (color scheme, interaction, layout) of all the panels can help the users in dashboard reading.

Table 7. Grouped negative items mentioned by the participants in the interview.

Group	Item	Frequency
Penal	The top margin of the the bars in the temporal panel is sometimes too narrow	5
	The maps shift when the mouse moves close to their boundaries	5
	The temporal panel is too informative	3
	The legend intervals are confusing	3
	The axises in the temporal panel change their ranges	1
	The axises in the temporal panel are not necessary	1
	The temporal panel should be split into four charts as other panels	1
	Mark the important places in the spatial panel	1
Layout	Hard to compare two layers in one map	3
	The color scheme is not good for color-blind people	3
	Only one map in the spatial panel is preferred	2
	The color hue should be increased in the temporal panel and ranking panel	2
	The color of the unavailable data should be lighter	1
Interaction	The search bar should be in each map / outside the spatial panel	5
	The ranking panel should be clickable	4
	The selected place should be highlighted on all the maps	2
	The map legends should be clickable	2
Other	The font size is too small	5
	The unavailable data increases the difficulty	3
	No idea where to look at on the dashboard	2
	The dashboard is too informative	2
	The listing of top five municipalities is not interested to the participant	1
	A learning time is required	1
	Lack of the economic background information	1

5. Discussion

In this section, we discuss the strengths and weaknesses of the proposed map-based dashboard for spatiotemporal knowledge acquisition and analysis, and the limitations of the experiment.

The map-based dashboard is designed to enable users to acquire and analyze spatiotemporal knowledge. The panels are designed to reveal different perspectives of georeferenced knowledge about what happens where and how it happens. The linked panels provide users with the real-time response of the subsets of the data. Each panel is placed on a fixed position and outlined with an enclosure, giving users a necessary anchor to quickly navigate in the data space. The juxtaposition reduces the difficulties in comparing multiple factors and facilitates the correlation analysis. The arrangement of the panels is very important. We placed the spatial panel in the middle area and in a large size to guide the users' attention to it. Last but not least, we applied a uniformed design style (color, layout, interaction) to each panel with the aim to foster the habit formation for an efficient perception and interaction.

However, the current design can be improved in some aspects. The annotations in dashboards have not received sufficient attention in previous studies. We realized that the font plays a very important role in dashboard understanding. The font size should be big enough to read at a glance. To better guide users to visually explore the data, the labeling of the panels should clearly express what types of information it conveys. Therefore, the panels in our dashboard should be labeled as the spatial panel, temporal panel, and ranking panel. Furthermore, the arrangement of panels should follow a logical order. The panels with similar content should be placed adjacently. When visualizing multi-granularity or multi-temporal data in several panels, the sequence of the panels should follow a common reading pattern, such as from large to small, or from old to new. Moreover, the function of each panel should remain simple. In this dashboard, we should integrate the search and layer switch functions to the toolbar. In addition, the design style of each panel should be kept uniform. The temporal panel should be split into four charts, as in the spatial and ranking panel. The clicking and hovering function should also be designed in the ranking panel. Moreover, we should prepare a second color scheme for the color-blind users.

The evaluation experiment of the dashboard has led to some new insights. Compared to similar experiments with only interviews in [18] or only an eye-tracking experiment in [35], our eye movement data and the interview provide complementary results reflecting the user experience and the usability of the map-based dashboard. With regard to the sample size of participants, 40 participants divided into smaller groups may seem to be small, but it is still an acceptable sample size. In similar eye-tracking experiments, the number of participants was usually not large, e.g., 21 and 17 participants in [25,35]. Our dashboard has been experimentally proved both effective and efficient for different groups. In future studies, more people with economics-related background or domain experts will be invited. Another insight is related to how feedback was collected. We adopted semi-structured interviews, and the participants were asked to list the positive and negative design items. Although the interviews were carried out immediately after the experiment, the participant tended to ignore some items or focus on the last issues they encountered prior to ending their participation. One alternative to this problem is to use a predefined questionnaire to measure each design item as suggested by Pezanowski et al. [17]. Think-aloud would be another method as a complement during the experiment [7]. It can allow participants to talk more about how they made their decisions in answering the benchmark tasks.

6. Conclusions

Map-based dashboards have opened up convenient opportunities for stakeholders to perceive and analyze complex spatiotemporal knowledge from multi-dimensional data with an at-a-glance overview and details on demand. By integrating the high-interactive and high intuitive features of visual analytics into the dashboard, we expanded the dashboard with more analytical functions. Moreover, we have contributed the design lessons of map-based dashboards.

In this study, we designed and developed a map-based dashboard displaying geo-economic environmental data targeting decision-makers in SMEs and citizens. To evaluate the effectiveness and efficiency of our map-based dashboard, we specially designed an experiment consisting of an eye-tracking study, benchmark tasks, and an interview. We analyzed the collected eye movement data in terms of fixation, success rate, response time, and dwell and transition metrics. Furthermore, we analyzed the feedback and summarized the positive and negative items on views, layout, interaction, and others. The analysis results from the eye-tracking study and the interviews have verified the map-based dashboard for spatiotemporal knowledge acquisition along with a number of findings related to the limitations of the current design of map-based dashboards and user studies.

Our future work involves three main tasks. First, the interface design will be improved with the focus on the study of how different layouts of the multiple views and their labeling influence the efficiency of the corresponding dashboards. Second, further user experience and usability experiments will be conducted. We will quantitatively study how the visualization, panel arrangement, color scheme, and user background influence users' attention and the spatiotemporal knowledge acquisition and analysis. Lastly, we will extend our dashboard design by adding more visual analytical methods. For instance, we will add correlation calculation and anomaly detection function and dashboard panels. We also plan to conduct more experiments on different datasets, e.g., social media data and volunteered geographic information.

Author Contributions: Conceptualization, Chenyu Zuo, Linfang Ding and Liqiu Meng; methodology, Chenyu Zuo; software, Chenyu Zuo; Formal analysis, Chenyu Zuo; investigation, Chenyu Zuo; resources, Chenyu Zuo; data curation, Chenyu Zuo; writing—original draft preparation, Chenyu Zuo; writing—review and editing, Chenyu Zuo, Linfang Ding and Liqiu Meng; visualization, Chenyu Zuo; supervision, Linfang Ding and Liqiu Meng; project administration, Chenyu Zuo and Linfang Ding. All authors have read and agreed to the published version of the manuscript.

Funding: This work is funded by the project "A Visual Computing Platform for the Industrial Innovation Environment in Yangtze River Delta" supported by the Jiangsu Industrial Technology Research Institute (JITRI) and Changshu Fengfan Power Equipment Co., Ltd.

Acknowledgments: The census and land use datasets are provided by Yangtze River Delta Science Data Center, National Earth System Science Data Sharing Infrastructure, National Science & Technology Infrastructure of China (http://nnu.geodata.cn:8008/). The authors would also like to thank the editor and anonymous reviewers for their efforts to review this paper.

Conflicts of Interest: The authors declare no conflict of interest.

Abbreviations

The following abbreviations are used in this manuscript:

AOI	area of interest
CEO	chief executive officer
SME	small and medium-sized enterprise
GDP	gross domestic product
CNY	Chinese Yuan

Appendix A

Table A1. The selected socioeconomic factors and the statistics of the data.

Category	Factor	Explanation
Enterprise	Enterprise	The total number of the enterprises in a county
	Enterprise above designated size	The number of enterprises with annual main business revenue of 20 million CNY or more
GDP	GDP per county	The total gross domestic product in one county
	GDP per capita	The average GDP per person
	Primary industry	The GDP value of the county from natural raw materials, such as mining, agriculture, or forestry
	Secondary industry	The GDP value of industry which converts the raw materials provided by primary, such as manufacturing industry
	Tertiary industry	The GDP value concerned with the provision of services
Population	Population per county	The total population of one county
	Population per km^2	The population density of one county
	Employee	The total number of employees in one county
	Citizen disposable income	The average citizen disposable income of one county
	Citizen consumption	The average citizen consumption of one county
	Engel coefficient	The proportion of income spent on food falls
Logistic	Road length	The total road length in one county
	Road passenger	The total transported passenger number in one county in one year
	Road cargo	The total weight of the transported cargo in one county in one year
	Car parc	The number of cars and other vehicles in a region or market

References

1. Jing, C.; Du, M.; Li, S.; Liu, S. Geospatial Dashboards for Monitoring Smart City Performance. *Sustainability* **2019**, *11*, 5648. [CrossRef]
2. Few, S. *Information Dashboard Design: The Effective Visual Communication of Data*; O'Reilly Media, Inc.: Newton, MA, USA, 2006.
3. Huijboom, N.; Van den Broek, T. Open data: An international comparison of strategies. *Eur. J. ePractice* **2011**, *12*, 4–16.
4. Kitchin, R.; Lauriault, T.P.; McArdle, G. Knowing and governing cities through urban indicators, city benchmarking and real-time dashboards. *Reg. Stud. Reg. Sci.* **2015**, *2*, 6–28. [CrossRef]
5. Liu, Z.; Jiang, B.; Heer, J. imMens: Real-time Visual Querying of Big Data. *Comput. Graph. Forum* **2013**, *32*, 421–430. [CrossRef]
6. Reda, K.; Johnson, A.E.; Papka, M.E.; Leigh, J. Effects of Display Size and Resolution on User Behavior and Insight Acquisition in Visual Exploration. In Proceedings of the 33rd Annual ACM Conference on Human Factors in Computing Systems (CHI '15), Seoul, Korea, 18–23 April 2015; Association for Computing Machinery: New York, NY, USA, 2015; pp. 2759–2768. [CrossRef]
7. Yalçın, M.A.; Elmqvist, N.; Bederson, B.B. Keshif: Rapid and Expressive Tabular Data Exploration for Novices. *IEEE Trans. Vis. Comput. Graph.* **2018**, *24*, 2339–2352. [CrossRef] [PubMed]
8. Vornhagen, H. Effective visualisation to enable sensemaking of complex systems. The case of governance dashboard. In Proceedings of the International Conference EGOV-CeDEM-ePart, Donau, Austria, 3–5 September 2018; pp. 313–321.
9. Gurstein, M.B. Open data: Empowering the empowered or effective data use for everyone? *First Monday* **2011**, *16*. [CrossRef]
10. Batty, M.; Axhausen, K.W.; Giannotti, F.; Pozdnoukhov, A.; Bazzani, A.; Wachowicz, M.; Ouzounis, G.; Portugali, Y. Smart cities of the future. *Eur. Phys. J. Spec. Top.* **2012**, *214*, 481–518. [CrossRef]
11. Batty, M. A perspective on city dashboards. *Reg. Stud. Reg. Sci.* **2015**, *2*, 29–32. [CrossRef]
12. Census Mappong Module: Dublin City. Available online: http://airo.maynoothuniversity.ie/Instant_Atlas/Updated%20Modules/dd/Dublin%20City/atlas.html (accessed on 25 June 2020).
13. Smart City 2-Boston Smartcity. Available online: https://boston.opendatasoft.com/page/smart-city-2/ (accessed on 25 June 2020).
14. Visualising Enterprise, Innovation & Employment in Galway City and County. Available online: http://galwaydashboard.ie/enterprise (accessed on 25 June 2020).
15. Keim, D.; Andrienko, G.; Fekete, J.D.; Görg, C.; Kohlhammer, J.; Melançon, G. Visual Analytics: Definition, Process, and Challenges. In *Information Visualization: Human-Centered Issues and Perspectives*; Kerren, A., Stasko, J.T., Fekete, J.D., North, C., Eds.; Springer: Berlin/Heidelberg, Germany, 2008; pp. 154–175. [CrossRef]
16. Robinson, A.C.; Peuquet, D.J.; Pezanowski, S.; Hardisty, F.A.; Swedberg, B. Design and evaluation of a geovisual analytics system for uncovering patterns in spatio-temporal event data. *Cartogr. Geogr. Inf. Sci.* **2017**, *44*, 216–228. [CrossRef]
17. Pezanowski, S.; MacEachren, A.M.; Savelyev, A.; Robinson, A.C. SensePlace3: A geovisual framework to analyze place–time–attribute information in social media. *Cartogr. Geogr. Inf. Sci.* **2018**, *45*, 420–437. [CrossRef]
18. Li, J.; Chen, S.; Zhang, K.; Andrienko, G.; Andrienko, N. COPE: Interactive Exploration of Co-Occurrence Patterns in Spatial Time Series. *IEEE Trans. Vis. Comput. Graph.* **2019**, *25*, 2554–2567. [CrossRef] [PubMed]
19. Nazemi, K.; Burkhardt, D. Visual analytical dashboards for comparative analytical tasks—A case study on mobility and transportation. *Procedia Comput. Sci.* **2019**, *149*, 138–150. [CrossRef]
20. Roth, R.E. Interactive maps: What we know and what we need to know. *J. Spat. Inf. Sci.* **2013**, *2013*, 59–115. [CrossRef]
21. Roth, R.E.; Çöltekin, A.; Delazari, L.; Filho, H.F.; Griffin, A.; Hall, A.; Korpi, J.; Lokka, I.; Mendonça, A.; Ooms, K.; et al. User studies in cartography: Opportunities for empirical research on interactive maps and visualizations. *Int. J. Cartogr.* **2017**, *3*, 61–89. [CrossRef]
22. Chang, R.; Ziemkiewicz, C.; Green, T.M.; Ribarsky, W. Defining Insight for Visual Analytics. *IEEE Comput. Graph. Appl.* **2009**, *29*, 14–17. [CrossRef]

23. Roth, R.E.; Ross, K.S.; MacEachren, A.M. User-Centered Design for Interactive Maps: A Case Study in Crime Analysis. *ISPRS Int. J. Geo-Inf.* **2015**, *4*, 262–301. [CrossRef]
24. Andrienko, N.; Andrienko, G.; Gatalsky, P. Exploratory spatio-temporal visualization: An analytical review. *J. Vis. Lang. Comput.* **2003**, *14*, 503–541. [CrossRef]
25. Bogucka, E.; Jahnke, M. Feasibility of the Space–Time Cube in Temporal Cultural Landscape Visualization. *ISPRS Int. J. Geo-Inf.* **2018**, *7*, 209. [CrossRef]
26. Brinck, T.; Gergle, D.; Wood, S.D. *Usability for the Web: Designing Web Sites That Work*; Morgan Kaufmann Publishers: Burlington, MA, USA, 2001.
27. Brooke, J. SUS: A "quick and dirty'usability. In *Usability Evaluation in Industry*; Routledge: Abingdon, UK, 1996; p. 189.
28. Seebacher, D.; Häuäler, J.; Hundt, M.; Stein, M.; Müller, H.; Engelke, U.; Keim, D. Visual Analysis of Spatio-Temporal Event Predictions: Investigating the Spread Dynamics of Invasive Species. *IEEE Trans. Big Data* **2018**. [CrossRef]
29. Cao, N.; Lin, C.; Zhu, Q.; Lin, Y.; Teng, X.; Wen, X. Voila: Visual Anomaly Detection and Monitoring with Streaming Spatiotemporal Data. *IEEE Trans. Vis. Comput.* **2018**, *24*, 23–33. [CrossRef]
30. Liu, D.; Xu, P.; Ren, L. TPFlow: Progressive Partition and Multidimensional Pattern Extraction for Large-Scale Spatio-Temporal Data Analysis. *IEEE Trans. Vis. Comput. Graph.* **2019**, *25*, 1–11. [CrossRef] [PubMed]
31. Shi, L.; Huang, C.; Liu, M.; Yan, J.; Jiang, T.; Tan, Z.; Hu, Y.; Chen, W.; Zhang, X. UrbanMotion: Visual Analysis of Metropolitan-Scale Sparse Trajectories. *IEEE Trans. Vis. Comput. Graph.* **2020**. [CrossRef]
32. McKenna, S.; Staheli, D.; Fulcher, C.; Meyer, M. BubbleNet: A Cyber Security Dashboard for Visualizing Patterns. *Comput. Graph. Forum* **2016**, *35*, 281–290. [CrossRef]
33. Hegarty, M.; Smallman, H.S.; Stull, A.T. Choosing and using geospatial displays: Effects of design on performance and metacognition. *J. Exp. Psychol. Appl.* **2012**, *18*, 1–17. [CrossRef] [PubMed]
34. Opach, T.; Gołębiowska, I.; Fabrikant, S.I. How Do People View Multi-Component Animated Maps? *Cartogr. J.* **2014**, *51*, 330–342. [CrossRef]
35. Popelka, S.; Herman, L.; Řezník, T.; Pařilová, M.; Jedlička, K.; Bouchal, J.; Kepka, M.; Charvát, K. User Evaluation of Map-Based Visual Analytic Tools. *ISPRS Int. J. Geo-Inf.* **2019**, *8*, 363. [CrossRef]
36. Mckenna, S.; Staheli, D.; Meyer, M. Unlocking user-centered design methods for building cyber security visualizations. In Proceedings of the 2015 IEEE Symposium on Visualization for Cyber Security (VizSec), Chicago, IL, USA, 25 October 2015; pp. 1–8.
37. Howson, C. *Successful Business Intelligence: Unlock the Value of BI & Big Data*; McGraw-Hill Education Group: New York, NY, USA, 2013.
38. Zuo, C.; Liu, B.; Ding, L.; Bogucka, E.; Meng, L. Usability Test of Map-based Interactive Dashboards Using Eye Movement Data. In Proceedings of the 15th International Conference on Location Based Services, Vienna, Austria, 11–13 November 2019; Vienna University of Technology: Vienna, Austria, 2019. [CrossRef]
39. Holmqvist, K.; Nyström, M.; Andersson, R.; Dewhurst, R.; Halszka, J.; van de Weijer, J. *Eye Tracking: A Comprehensive Guide to Methods and Measures*; Oxford University Press: Oxford, UK, 2011.
40. Tran, V.T.; Fuhr, N. Using Eye-Tracking with Dynamic Areas of Interest for Analyzing Interactive Information Retrieval. In Proceedings of the 35th International ACM SIGIR Conference on Research and Development in Information Retrieval (SIGIR '12), Portland, OR, USA, 12–16 August 2012; Association for Computing Machinery: New York, NY, USA, 2012; pp. 1165–1166. [CrossRef]

Publisher's Note: MDPI stays neutral with regard to jurisdictional claims in published maps and institutional affiliations.

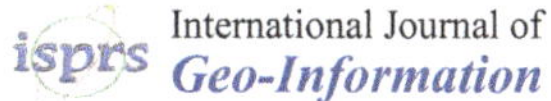
International Journal of
Geo-Information

Article

Time, Spatial, and Descriptive Features of Pedestrian Tracks on Set of Visualizations

Łukasz Wielebski [1,*], Beata Medyńska-Gulij [1], Łukasz Halik [1] and Frank Dickmann [2]

1 Research Division Cartography and Geomatics, Adam Mickiewicz University, 61-680 Poznań, Poland; bmg@amu.edu.pl (B.M.-G.); lhalik@amu.edu.pl (Ł.H.)
2 Geography Department, Cartography, Ruhr University Bochum, 44801 Bochum, Germany; frank.dickmann@rub.de
* Correspondence: lukwiel@amu.edu.pl

Received: 27 April 2020; Accepted: 20 May 2020; Published: 26 May 2020

Abstract: The aim of the paper is to elaborate on and evaluate a multiperspective cartographic visualization of the spatial behavior of pedestrians in urban space. The detailed objective is to indicate the level of usefulness of the proposed visualization methods for analyzing and interpreting the following features: track shape (trajectory geometry), topographical truth, track length, track visibility, walking time, motivation for getting to the finish point, walking speed, stops, spatial context (spatial surroundings, street names, and so on), and trajectory similarity. Each of the elaborated visualization presents spatial data from a different perspective and visually strengthens other aspects of the behavior of participants of the experiment. Recording the movement of participants by means of global positioning system (GPS) receivers was the first method used in the research, with the other one being a questionnaire that made it possible to determine what kind of motivation pedestrians had when selecting a track leading to the finish point. The results demonstrate different levels of usefulness of the six presented visualizations for reading selected features of the spatial behavior of pedestrians.

Keywords: multiperspective cartographic visualization; usefulness of visualizations; pedestrian tracks; complementary visualizations; time–spatial behavior; space–time paths; trajectories; set of visualizations

1. Introduction

The urban and communication systems are important determinants influencing spatial behavior. However, field obstacles are not the only factor that determine the kinds of tracks selected by pedestrians. According to Hamid [1] (p. 265), what decides on the way that pedestrians move in space is "the complexity of the space and how the pedestrian understands the space". The way pedestrians perceive space is significantly connected with the so-called 'cognitive map', a concept of the track that everybody creates in their mind on the basis of previous experiences [2]. These experiences are not only linked to movement in space, but they also result from interpreting space on the basis of maps, a skill developed throughout the entire life [3]. Empirical studies have shown that "spatial information from analogue maps takes the user to a more detailed mental model [than maps for navigation devices], which indicates a different level of efficiency of the two forms of cartographic representation" [4] (p. 67). The cognitive map may be also interpreted as "a cartographic representation of an individual's estimates of distance relations among points in his spatial environment" [5] (p. 19). According to Hamid [1] (p. 265), "pedestrians' behavior depends on the individual's preferences" and the choice of the track may be motivated by its visual aspects (attractivity), habits of pedestrians, knowledge of buildings, or the shortest distance. The impact of numerous external factors and decisions of pedestrians conditioned by individual preferences makes the analysis (of the behavior of pedestrians) complex and multifaceted. Particularly, drawing conclusions about many people's behavior in the same geographic space requires the analysis to be supported with the evidence in the cartographic, and generally graphic, form. To process and visualize behavior patterns, most of the studies make use of spatial analysis such like kernel density [6–8], heat maps [9,10], or density maps [11,12]. The visualization and analysis of movement is used in many different disciplines like geography, sociology, informatics, ecology [13], and many others.

The main aim of our research is to elaborate on and evaluate a multiperspective cartographic visualization of the spatial behavior of pedestrians in urban space, taking into consideration individual attitudes towards multiple different tracks leading from the same starting point to the same finish point. The detailed objective is to indicate the level of usefulness of the proposed visualization methods for analyzing and interpreting the following features: track shape (trajectory geometry), topographical truth, track length, track visibility, time, motivation for getting to the finish point, speed, stops, spatial context (spatial surroundings, street names, and so on), and trajectory similarity. In this study, we want to present answers to the following questions. To what extent can static cartographic visualizations facilitate the analysis of the behavior of pedestrians in urban space and can this analysis be more complete thanks to complementary visualizations? What is the usefulness of proposed visualizations for the analysis of pedestrian behavior? What research methods can be applied to correlate the information from those visualizations?

Multiperspective cartographic visualization of the aforementioned features can be treated as the new approach and the development of other studies, for example, aimed at finding the optimal representation for visualizing pedestrians. Such studies were conducted by, among others, Biadgilgn et al. [14], who analyzed four different mapping techniques in terms of their suitability for the presentation of trajectory characteristics, such as "speed change, returns, stops and path of movement" (p. 80).

In turn, the use of numerous both complementary and integral methods of graphic and cartographic presentation of geographical space in this research casts new light on the cartographic research method in the context of alternative ways of visualization in their interdisciplinary depiction [15]. Static visualizations, that is, static two-dimensional, surface three-dimensional, and interactive, treated as creating a complementary visualization included in the process of geovisualization, have been conceived so far [16]. The advantages and disadvantages of adding new methods of presentation are another problem taken into consideration, as indicated in the research by Medyńska-Gulij and Cybulski [17]. This research constitutes a part of the trend toward seeking alternative and interdisciplinary ways of visualizing information.

2. Materials and Methods

2.1. Literature Review

Studies of the analysis of the movement of pedestrians are most frequently conducted with the use of global navigation satellite systems (GNSSs), such as global positioning system (GPS) [18]. They may be carried out by means of GPS receivers, but recently also smartphones with the built-in GPS modules or the more precise localization system for outdoor pedestrians with smartphones called APT (accurate outdoor pedestrian tracking) ([19]). The increasingly expanded monitoring network with public webcams also enables research related to pedestrian traffic and congestion in the streets [6,20] or pedestrian movement simulation and visualization [21] based on video recordings. The usage of GPS technology releases researchers from the necessity to continuously monitor and disturb the 'freedom' of action of participants, as the execution of individual activities takes place more naturally, thus improving the quality of data obtained. Direct observation may affect the behavior of pedestrians [1], which results from "the nature of human behavior when being observed and tendencies to react differently" ([1], p. 276 after [22]). GPS receivers record the location of pedestrians in space and time in a quasi-continuous way. Track points are recorded by receivers automatically in accordance with the specified time interval. The higher the sampling rate, the better the documentation of the movement of pedestrians. However, a larger amount of data is also connected with a longer processing time and more difficult analysis. Movement tracks are the result of recording the tracks of pedestrians. According to Parent et al. [23] (p. 3), "Segments of the object's movement track that are of interest for a given application" are called trajectories. Trajectory of movement "is a sequence of positions in a two-dimensional (2D) geographic environment with time stamps" that allows one to represent the path of a moving object [24] (p. 86). Apart from advantages, GPS also has limitations connected with variable quality of the received signal, dependent on external conditions, affecting quality of localization record and translating into the picture on visualizations.

Raw data obtained from GPS constitute only a record of the geometry of the track movement. To gain some information about behavior patterns, measures need to be applied that allow one to analyze the phenomenon visually. A map, understood in this research as a form of visualization, is the key element on which one can present spatial behavior. It is the most basic, fundamental graphic solution that enables one to mark the trajectory of the movement of pedestrians in a broader spatial context. However, that solution has its limitations and sometimes it fails to provide exhaustive answers to questions connected with analyzed features of the behavior of pedestrians, hence the need to apply other methods of cartographic visualization that make it possible to follow various aspects of the behavior of research participants and analyze them from various perspectives. Mapping techniques are of considerable importance for the conveyance of information about the dynamics of behavior in urban space and the considerable possibilities of presenting the penetration of geographical space by pedestrians at a specific point and time.

The overview of literature leads to the conclusion that both interactive and classic static cartographic visualizations or animated maps may be helpful in the analysis of the behavior of pedestrians [25,26]. Inasmuch as interactive solutions prove highly useful during exploratory analysis of large databases, static visualizations turn out to be extremely helpful when communicating the results of such an analysis to the wide audience, a fact resulting from the nature of both types of visualizations ('MacEachren's cube' [27]). If both solutions consist of sets of visualizations, it is important to design visualizations in such a way that they can be linked to one another. In interactive solutions this link is literal, that is, highlighting an element on one visualization leads to the respective element on another visualization being marked. In static solutions, a graphic link between pieces of information presented on various visualizations becomes of great significance, providing their mutual complementarity [16]. The very concept of complementary visualizations derives "from a professional practice of architects and urban planners, who use complementary visualizations to envision, think, innovate, communicate, disseminate and document complex knowledge" [28] (p. 137). Burkhard [28] (p. 137) defined complementary visualizations as "the use of at least two visual representations that complement each other to augment knowledge-intense processes". Complementary visualizations can also be used purposely to reduce complexity [29].

The opportunity to use complementary visualizations enables one to look at the same problem from multiple perspectives [30]. In our study, the multiperspective cartographic visualization is taken to mean a set of separate (but mutually linked) cartographic presentations, each of which made use of a different mapping technique and different means of graphical expression. Each one of these visualizations may be analyzed separately, however, the comprehensive cognition thereof, according to various perspectives of perceiving human behavior in the geographical-temporal context, may significantly increase the effectiveness of the analysis of spatial behavior.

In our publication, we want to demonstrate that cartographic visualizations can simplify the perception of the behavior of pedestrians in the city and the use of complementary visualizations makes it easier to analyze specificity of the behavior. The set of static mapping methods proposed by us constitutes a sui generis documental (because it is both complete and permanent) record of behavioral-geographical information. The most important objectives when elaborating visualizations include capturing the dependence between the intentions of pedestrians and the distance actually covered thereby, and a comparison of the geometry of routes and times for numerous pedestrians, who have the same starting and finish point. Methods of cartographic visualization used in this study to present the spatial behavior of pedestrians include the following: map with non-symbolized trajectories, understood by us as a cartographic visualization saved in a digital landscape model (DLM), topographic map with trajectories, schematic map, route graph, route miniatures, and space–time cube with trajectories (space–time paths). The methods of cartographic visualization used to deal with the problem of pedestrian trajectories can have an analytical or synthesizing nature. Most studies of that kind examine many observations (hundreds or thousands of samples). In such cases, it is important to reach for advanced methods like V-Analytics by Andrienko & Andrienko (http://geoanalytics.net/V-Analytics/) supporting data analysis. V-Analytics enables to produce maps by aggregation of input lines, which is important and can be indispensable by big samples of trajectories. Flow maps give one the opportunity to have a more synthesizing image of the most frequented pedestrian routes. In our study, we focused on a relatively small group of people, with the ability to compare their trajectories, but still being possible to analyze them separately (with the only one exception in the form of map with non-symbolized trajectories). It was the key to choosing the set of presentation methods. This is the reason we used the schematic map with parallel presentation of people trajectories instead of the more common flow map.

In order to achieve the objective set forward above, an experiment concerning the spatial behavior of pedestrians, whose assumptions and course will be presented first and foremost, was conducted. Then, we will discuss the transformation of the source data from GPS and the way they were subsequently processed and visualized. We will demonstrate the proposed methods of cartographic visualization and discuss their advantages and disadvantages. In the summary of the research, we will analyze to what extent specific methods of visualization answer questions about examined features and which of them prove the most effective when displaying those features. We will also present a path that allows one to link pieces of information from six methods of visualization with one another.

2.2. Purpose, Basic Assumptions, Participants, and Place of Experiment

To collect spatio-temporal data documenting the behavior of pedestrians in urban space, indispensable to create a set of complementary visualizations helpful in the analysis of that behavior, we conducted a field experiment in the city center of Poznań, Poland. A group of 30 people was invited to take part in the experiment. The participants comprised a homogenous group, consisting of geography students with an average age of 22 years—16 women (53%) and 14 men (47%). All participants were master students, who confirmed that they have lived in Poznań for at least three years (after bachelor study level) and know the city center well. Confirmation of knowledge of the city center was a prerequisite for participating in research. Participants were not remunerated for taking part in the research and participated in it voluntarily and without coercion. The experiment was accompanied by a short survey, before which it was explained to the respondents how their data would be used and their privacy would be protected, and all respondents gave their consent. The aim of the experiment was also explained to the participants beforehand.

The aim of the task for experiment participants was to move from point A to point B, taking a freely chosen route, whose course depended entirely on participants and their choices. Both points are characteristic and all participants knew their location, as they constitute places significant to every city. The main railway station (an important communication point where many visitors begin their tour around the city) was the starting point, while the town hall building (located on the Old Market Square, the heart of the city, a frequent meeting spot for locals, and one of the must-see sights for tourists) served as the finish point. The straight-line distance between those two points is 1.5 km, whereas the shortest distance indicated by Google Maps at that time is 2 km, and it is possible to cover it on foot in 25 min.

Each participant attended the experiment separately, covering the distance on their own, and received the following task: Please, walk from the main train station to the town hall, using a freely selected route. Time intervals between consecutive participants were arranged in such a way as to prevent them from meeting on the route. The course of the route of each participant was recorded by means of the tourist GPS receiver (GPS60csx Garmin) with position accuracy of approximately 3 m. Aside from signal reception conditions and parameters of the receiver that may significantly affect position accuracy in the recording, time intervals between automatic recording of pace, track points, and changes in walking speed were important parameters for data accuracy (sampling rate). In our research, the GPS receiver recorded current position of the participant in urban space every 15 s (which was the default value set in many receivers of that type). Participants were not permitted to use GPS receivers for navigation, however, they were allowed to use mobile navigation apps in their smartphones. They made use of their smartphones, for they were able to operate them intuitively, which was one of the assumptions of the experiment.

2.3. *The Course of the Experiment*

The field experiment was coordinated by two research assistants. The first research assistant was standing next to the starting point nearby the way out of the main railway station. His responsibility was to explain the task in detail to the participant and to prepare the device for recording the route. The assistant was also supposed to properly attach the GPS receiver to the participant's arm by means of a special band, which kept the antenna of the receiver at shoulder height. The aim of that was to set the receiver so that its position towards the participant was fixed, as that boosted the quality of the signal sent to the receiver from the satellite. During the research conducted, temporal signal loss occurred most frequently in underpasses and inside the buildings, such as shopping centers, which often make it possible to move from one street to another, thus becoming a part of the city communication corridors.

The second research assistant was waiting at the finish point at the town hall for successively arriving participants. His task was to save in the memory of the device an active track recorded automatically by the GPS receiver. He also interviewed the participant about the route, for example, asking about the motivation of the participant when selecting the route. At the finish point, the following question was asked: What influenced you when selecting a route leading to the finish point? Out of five possible answers, select the one that describes the most important reason for you: interesting objects/situations, route length, habits, traffic lights, and the map in your smartphone.

2.4. *Transformation and Categorization of the Data Obtained*

GPS tracks obtained during field work were downloaded to a computer and imported into geoinformation software. Such raw data, that is, "data as captured from the device [without any processing], may be used as such for further analysis or be transformed into other kinds of representation of movement" [23] (p.2), as we did in the research. GPX (the GPS Exchange Format) files components were converted to shapefiles and, in the "semantic enrichment process" [24], every route received annotations, that is, attribute data (first name of the participant, numeric code of pedestrian, motivation for selecting a specific route, length of the track, and walking time). Changing the coordinate system from the default WGS 84, where GPS data were recorded, to projected coordinate system was a significant measure. It made it possible for the data to be presented in the system without any distance distortions and position could be determined on the basis of coordinates in the more friendly and intuitive metric system instead of decimal degrees. It was vital for the opportunity to determine the distance on the basis of some of the prepared visualizations.

The data required also the so-called trajectory data cleaning by removing GPS errors, which is a very important stage in the development of correct trajectories [23]. First and foremost, it was necessary to delete parts of the recording from the beginning and the end of the route, on which the assistant was talking to the participant and the receiver was handed over when already/still recording. Then, random GPS errors, connected with signal loss or obvious position misrecord, resulting from the weak signal sent by satellites, had to be removed.

The Web Map Service along with some fragments of an orthophotomap and a topographic map, displayed in a geoinformation app as a base map and Google Maps with the Street View service, were extremely helpful in the process of analysis and recreation of routes [31]. After trajectory data cleaning, track length and track time were calculated for each track on the basis of feature geometry and time stamps. Next, each participant was assigned a code from 01 to 30. That order corresponded to the time taken to cover the distance—from the shortest (21 min and 30 s) to the longest (39 min), which made a difference of 17 min and 30 s. Discrepancies in the trajectory length of the movement of pedestrians were smaller and reached 450 m.

2.5. Assumptions Adopted in the Process of Designing Complementary, Multiperspective Cartographic Visualizations

The data obtained during the field experiment were used for creating a set of complementary visualizations. Maintaining cohesion in graphic and cartographic link between numerous methods of visualization was the key element of that process. A simple scale bar with a meter unit for cartometric visualizations or a scale bar described as 'not scalable' became a cartographic and mathematical element. Color and text markings, assigned to the routes of specific research participants, constituted graphic and descriptive elements that made it possible to correlate all the visualizations. Color is often the first distinguishing feature used for identification and may override differences in shape [32], hence that visual variable was chosen to distinguish between participants. The large number of participants, much bigger than the number of colors that can be easily distinguished by a person without color blindness, was a fundamental technical problem. The problem of distinguishable colors is important for public transportation maps, where each line must be shown with a clearly different color [32]. Rougeux [33] made an interactive application "Global Subway Spectrum. An exploration of colors used for lines in every rapid transit system" (https://www.c82.net/spectrum/). The application makes it possible to explore the colors used on subway maps in cities all over the world. Having analyzed the colors most frequently used on subway maps in various cities, Trubetskoy [34,35] made a "List of 20 Simple Distinct Colors" (https://sashat.me/2017/01/11/list-of-20-simple-distinct-colors/), which he used in his Roman Roads project—"A subway-style diagram of the major Roman roads, based on the Empire of ca. 125 AD" (https://sashat.me/2017/06/03/roman-roads-index/).

In the search for 26 colors necessary for the research connected with coding the alphabet with colors, Green-Armytage [32] took into consideration, among others, Kelly's study [36] with "the set of 22 colors of maximum contrast", but also the set of colors used in practice on the maps of tram lines, subways and suburban railways in such cities as Gothenburg, Tokyo, Paris, or London and the color zones system connected with universal color language (UCL), which was the base for selected colors in that study. In the discussions and conclusions from the conducted research, Green-Arymtage [32] state that the answer to the question, how many colors can be successfully used, may not be definitive owing to the number of "different contexts in which color coding is used" (p. 22), but he indicates that, "for practical purposes, the 26 colors of the alphabet can be regarded as a provisional limit" (p. 22) and that "the system would surely break down much beyond that number" (p. 20).

In our study, to make pedestrians trajectories distinguishable, we adopted a set of colors, visible in Table 1, presented by Green-Armytage (code PG-A) in the study connected with color coding alphabet (8 of 26 colors) and by Trubetskoy (code ST) in Roman Road project (19 of 20 colors), as well as four individually selected colors (code "own") for marking 30 routes.

Similarity of routes constituted a basis for assigning colors to research participants. The main objective was to make colors of similar characteristic denote routes that had common stretches of a significant distance, considering first navigation decisions of pedestrians. The choice of, for instance, an underpass that enabled one to get into the other side of the street was such a decision. To make a clear-cut identification of the routes possible, each color received a numeric code consistent with the number of the participant (numbers from 01 to 30, Table 1). The same solution was applied on geological maps, on which a great number of geological features of rocks was distinguished by means of colors, as well as letter and numeric codes [32]. To suggestively present the starting and finish point, common for all participants, yellow circles with a black arrow or square were used, known for START and STOP buttons.

Table 1. Selected set of colors for representing pedestrians in the study according to the source (PG-A—Paul Green-Armytage; ST—Sasha Trubetskoy; own).

Code	Original Name by Source	Source	RGB	Color
01	iron/grey	PG-A/ST	128-128-128	
02	black	ST	0-0-0	
03	pistachio	own	215-255-190	
04	brown	ST	170-110-40	
05	olive	ST	128-128-0	
06	azure	own	204-255-255	
07	yellow	own	255-255-0	
08	magenta	ST	240-50-230	
09	lime	ST	210-245-60	
10	red	PG-A	255-0-16	
11	maroon	ST	128-0-0	
12	pink	ST	250-190-190	
13	beige	ST	255-250-200	
14	orange	ST	245-130-48	
15	blue	ST	0-130-200	
16	damson	PG-A	76-0-92	
17	mint	ST	170-255-195	
18	uranium	PG-A	224-255-102	
19	green	ST	60-180-75	
20	lime	PG-A	157-204-0	
21	purple	ST	145-30-180	
22	coral	ST	255-215-180	
23	quagmire	PG-A	66-102-0	
24	forest	PG-A	0-92-49	
25	lavender	ST	230-190-255	
26	navy	ST	0-0-128	
27	zinnia	PG-A	255-80-5	
28	teal	ST	0-128-128	
29	grey	own	204-204-204	
30	cyan	ST	70-240-240	

Colors were also employed in the research for distinguishing the motivation that participants had for selecting a specific route. The following colors and letters indicated the motivation of individual participants for the method of movement used to reach the finish point: red—shortest distance/route (D), blue—habits (H); yellow—interesting objects/situations (I); purple—street signaling/traffic lights (T); and green—the map in the smartphone (M).

The basis for creating visualizations is the appropriate obtainment of data and the classification of spatial and attribute data [37]. Each set of geographical data may be presented in many different ways, using various techniques [38]. To render visualizations complementary and to make it possible to examine chosen features from various perspectives, it was necessary to select pieces of information and graphic elements in such a way so that each feature could be demonstrated on at least two visualizations. In the following paragraphs, we will present and discuss individual visualizations, focusing specifically on the features they demonstrate, as well as their potential and limitations.

2.6. Set of Elaborated Cartographic Visualizations

A map with non-symbolized trajectories (Figure 1) is the only visualization on which color was not employed to distinguish 30 specific tracks, hence it is not possible to analyze individual tracks on the basis of this visualization. Raw movement routes are visible as black lines presented in the spatial context, that is, on the background of the network of squares with a side length 100 m and streets with visible names. Although, thanks to including street names, it is possible to determine streets that were frequented more often, it is not possible to provide a specific number of people owing to overlapping and crossing trajectories. This method of presentation refers to Golledge's [39] work and is considered in the present article as a simple, 'raw' view, without cartographic editing, which forms the basis for creating successive visualizations of spatial behavior.

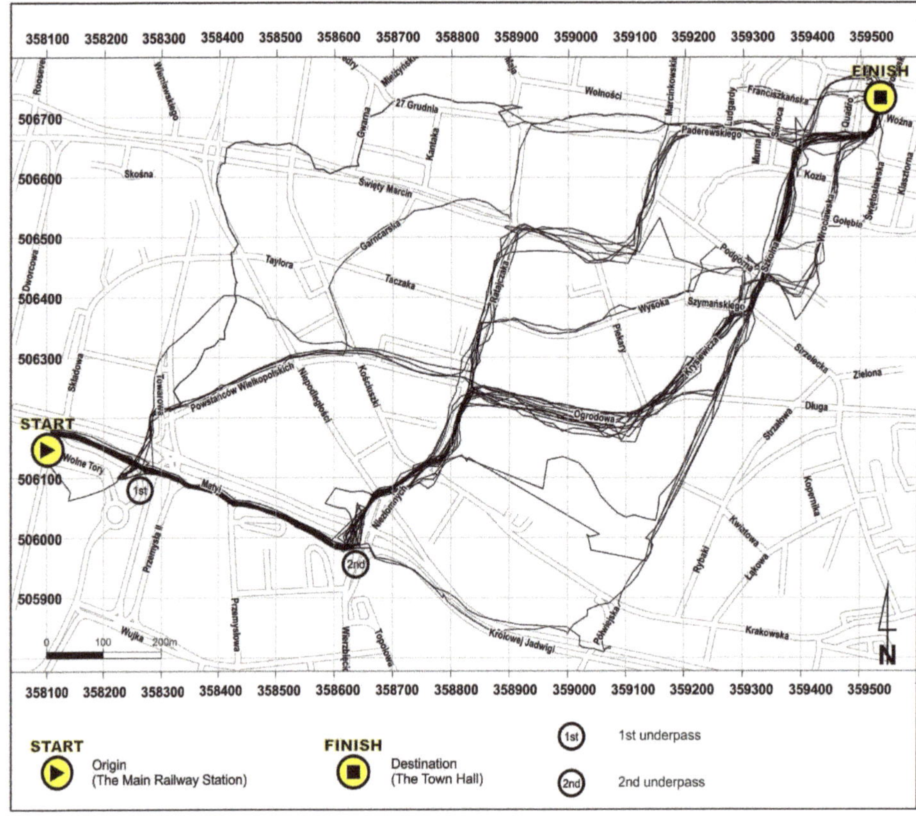

Figure 1. Map with non-symbolized trajectories—'raw' presentation of 30 tracks.

A map on Figure 2 enables one to follow the trajectory of the movement of pedestrians on a traditional, topographical map. Streets, buildings, green areas, and water reservoirs create spatial context and make it possible to simply recreate the route. Cartometric properties of a topographic map with trajectories also facilitate estimating the distance. Presented trajectories of movement faithfully reflect the routes recorded by GPS. The fact that tracks overlap in street sections selected by many pedestrians is the greatest flaw of this form, as that significantly hampers identifying and following individual routes, despite their being marked with separate colors. Variable power of the signal reaching the GPS receiver and the level of generalization of base map elements make specific routes not fully geometrically consistent with base content. Paradoxically, weaker signal and location errors, although they affect the geometry of the routes recorded, contribute to better route visibility owing to their larger dispersion. The great number of colors, necessary because of the number of routes demonstrated, makes some of them merge with base content, despite highly bright colors used in there. The more tracks recorded, the less effective that presentation method becomes. Additionally, a black dashed line is used to mark the Euclidean distance on the map as a graphic element of the visualization's complementarity with the next method of route miniatures.

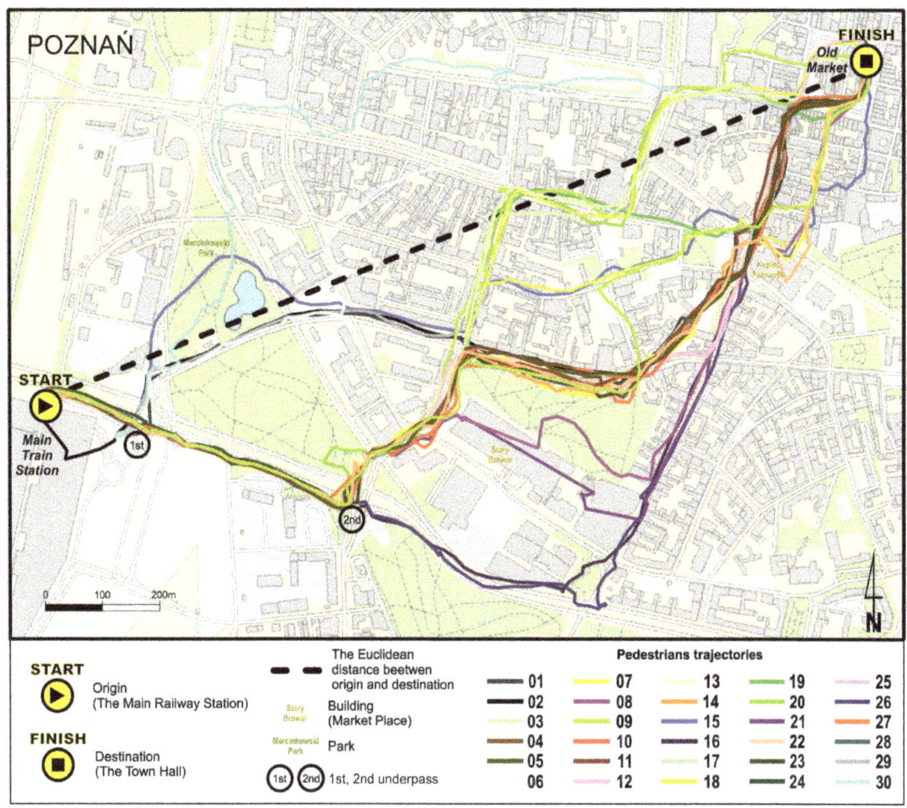

Figure 2. A topographic map with trajectories.

Route miniatures (Figure 3) demonstrate the actual geometry of routes best of all the visualization methods presented in this article. Each trajectory is presented in the separate rectangular box, all of them being maintained in the same scale, owing to which their shapes are comparable. Each miniature consists of three elements: a color-marked route with the participant's code, the line of the Euclidean distance between the starting and the finish point, and the surface area lying between them and marked in grey. The level of complexity of each route owing to the number and frequency of direction changes may be evaluated visually thanks to the Euclidean distance, which constitutes a point of reference for each route, making every deviation clearly visible (compare Figure 2). Moreover, thanks to that method, one can also, at a rough estimate, draw conclusions about track length. Complexity of tracks was also specified mathematically and determined numerically on the basis of the trajectory shape factor. It is the quotient of the circumference of an area defined by the route line and the Euclidean distance connecting the starting and the finish point, as well as the circumference of a circle of the surface area identical with the surface area of the defined area, that allow researchers to rank 30 routes according to their complexity, assigning them values ranging between 3.92 to 1.52. Values of the coefficient are situated in the right bottom corner of each box on Figure 3. The shortest possible route, calculated on the basis of Google Maps, is presented in the 'Google' box. Arranging routes according to their courses and similarity of shapes is also important. The first column consists of routes on which pedestrians used the first underpass nearby the railway station. The second, third, and fourth columns present routes of participants that used the second underpass.

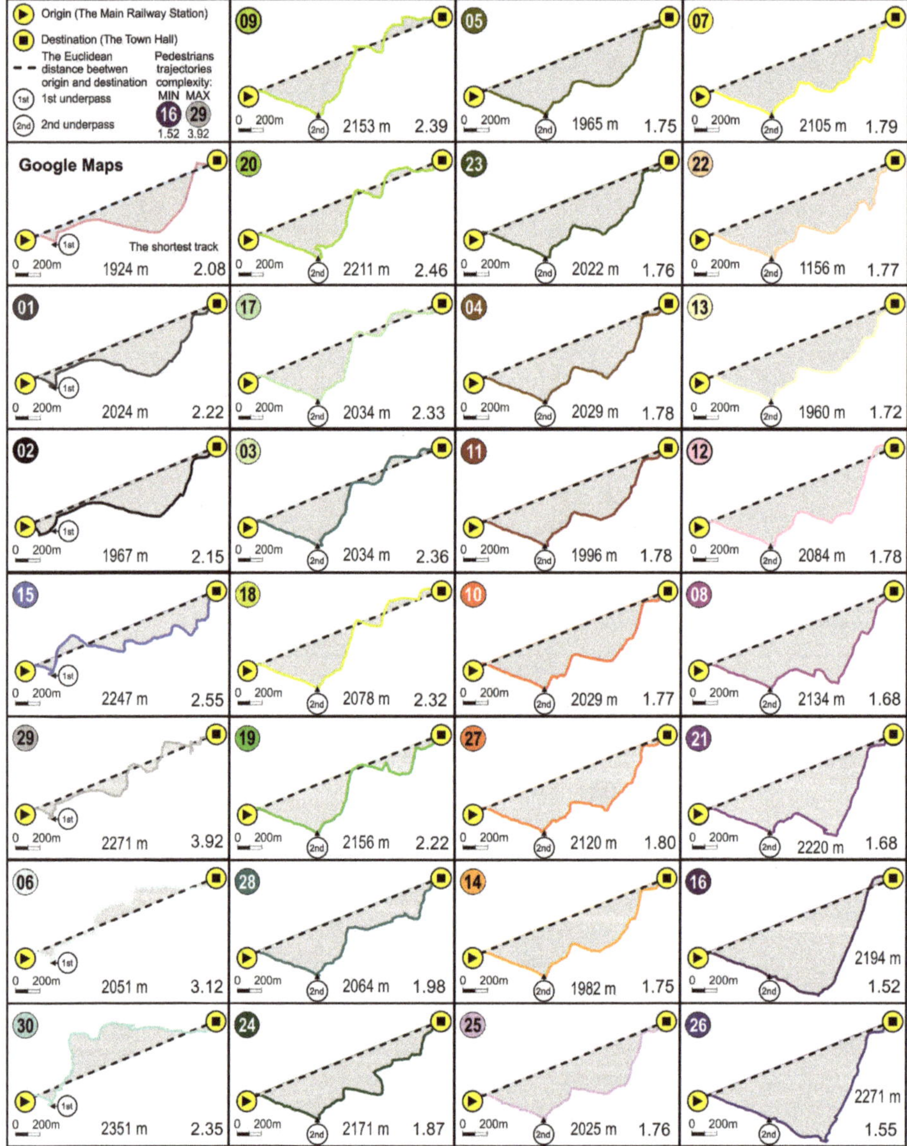

Figure 3. Miniatures of 30 routes of the research participants and the shortest route indicated by Google Maps along with the value of trajectory shape factor.

Small size of boxes that may limit visibility of miniatures is a disadvantage of that solution [25], especially when presenting more routes.

The analysis of Figure 4 makes it possible to draw a conclusion that simplified geometry is predominant, with the pedestrian maintaining one direction to the finish point, with 3–5 changes in each direction, coinciding with the course of the shortest route marked out in the Google Maps application. Two routes (01 and 02) had a course similar to the shortest Google route, with route 01 being the most similar. The routes of all pedestrians who used the first underpass are demonstrated in the first column. Other miniatures present the routes of those participants who went further and used the second underpass. The third column consists of routes of identical course, with the only exception being the route in the last line, which differs slightly, similarly to four routes from the last column.

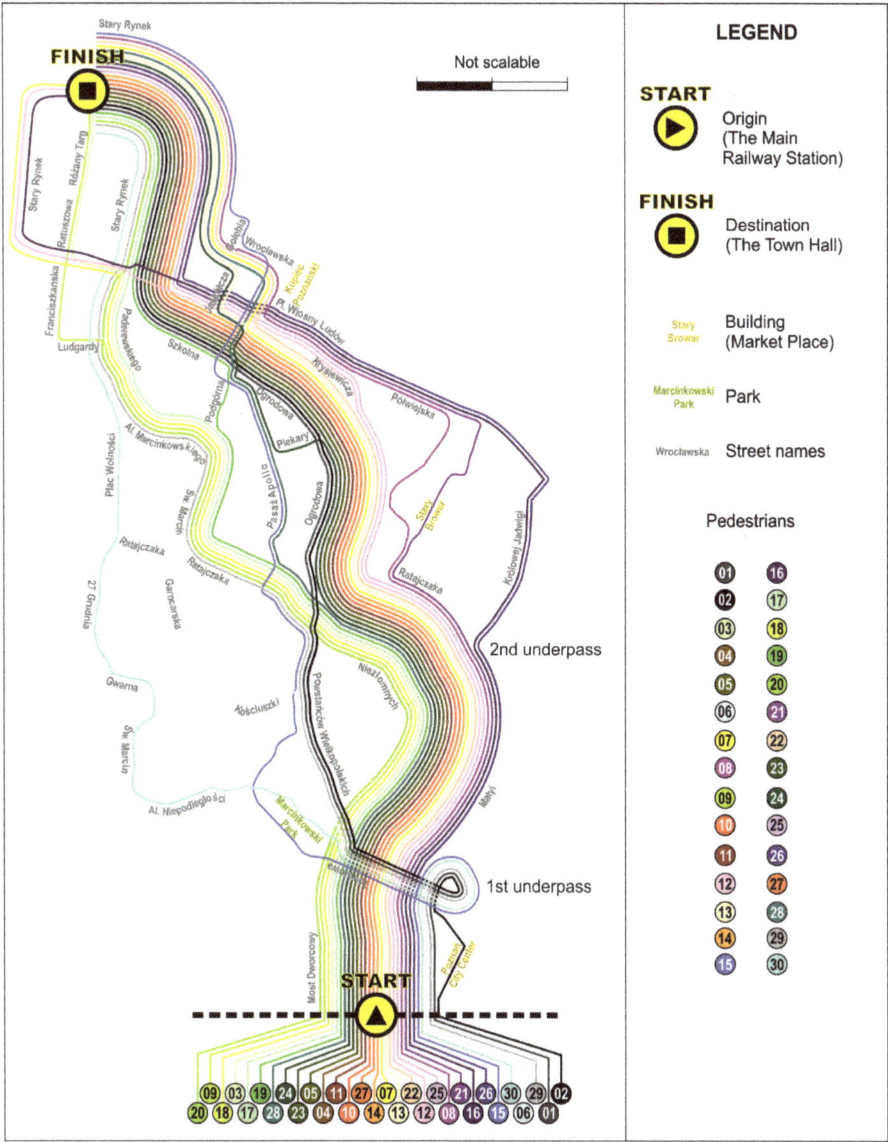

Figure 4. A schematic, non-scalable map demonstrating similarities and differences in the course of 30 routes.

A schematic map of routes (Figure 4) is based upon a design idea similar to a subway map, while this solution imitates the topographic course of routes more faithfully than visualizations of that type. The geometry of routes is significantly generalized and distorted, presented in a simplified and schematic way. What results from that fact is that it does not maintain actual distances, hence the 'not scalable' note next to the scale. Presenting sections of streets on which routes overlap is the most significant objective of that visualization. Routes are traced out as parallel lines, thus making all the routes well visible, and the number of crossings is reduced to the minimum. Lines in Figure 4 are adjusted according to colors and the route order, taking the similarity of their courses into consideration. It is precisely the arrangement of routes in that visualization that decides upon colors assigned to individual participants on all the complementary visualizations presented in this publication. Thanks to this visualization, it is possible to easily determine what street sections were frequented most often and by which participants, just by counting colorful lines. The orientation of this visualization towards cardinal points is purposely altered from the regular northern to the one oriented to the final destination that participants were supposed to arrive. Thus, routes are presented from the point of view of pedestrians arriving at their destination in such a way that, at the bottom of the visualization, there is a starting point and the finish point is situated at the top of it.

The fact that one can follow the routes of all pedestrians and analyze them both individually (elementary questions [40]) and in the context of the entire group (synoptic questions [40]) is a great strength of this presentation, compared with the topographic map with trajectories. It can be easily established what street sections are frequented most often and at what stage routes become common for many participants, with the opportunity to identify each one of them. The lack of cartometric property, manifesting itself in the distortion of proportions and making it impossible to estimate the distance, is a flaw of this method. However, the spatial context is maintained thanks to the topology of routes and names of streets used to describe groups of lines. Considering the manner in which the route is reflected, one may draw a conclusion that this visualization resembles a cognitive map the most, that is, it fails to present inaccurate distances that we tend to roughly estimate, instead showing places of more significant alterations and changes of the movement direction that we remember and are capable of recreating much more precisely.

A route graph (Figure 5) is an example of a graphic solution in which trajectories are synchronized to facilitate the comparative analysis. The idea of a time graph or a 'route graph' was presented and described in the article by Andrienko & Andrienko [25] (p. 9) as a way to solve the problem of "visual clutter and overlapping of lines" in the case of "representing multiple trajectories". In this method, each trajectory is represented by a horizontal (or vertical) bar. In the case of our visualization, the horizontal position and length of the bar correspond to the starting point and the length of the bar to the length of the trajectory. Hence, each bar is one 'straightened' trajectory with the starting point on a common left base. The lengths of bars allow us to observe the coincident value of the length of covered distances, for all of them fall in the range of 1960–2351 m. As Andrienko & Andrienko [25] (p. 9) hold, "the vertical dimension of the display is used to arrange the bars, which can be sorted based on one or more attributes of the trajectories". In this case, trajectories are sorted out according to time. The time required to pass from the starting point to the finish point is more diverse, especially if we take extreme values into consideration. On this visualization, one can clearly see that the distance covered does not directly impact the walking time. Thanks to the distribution of track points, we may indicate stopping places, and the variability or regularity of walking speed (the closer dots are to one other, the slower the speed). Owing to the size of dots in individual rows, it is possible to place track points every one minute, which allows us to point out only longer stops during the walk. However, "trajectories can be gapped by missing or bad GPS-signal" [41] (p. 26); therefore, places where the loss of signal occurred do not have track points. This method included an important feature, namely the motivation that participants had when selecting a route. It was marked with colorful squares with black letters. The research on decisive factors that influence motivation when selecting a route proved that distance is the most important thing, while factors like "safety, visual attractions or the level of congestion"

(p. 4) are less important ([42] after [43]). The results of our research confirm the significance of distance when choosing a route. The shortest distance was the dominant factor, indicated by 18 participants, habits ranked second (5), and attractiveness of the route ranked third (interesting things to see on the way were mentioned by four participants). Only two people claimed that favorable traffic lights on pedestrian crossings were the most relevant factor and one person was motivated by a map on a smartphone.

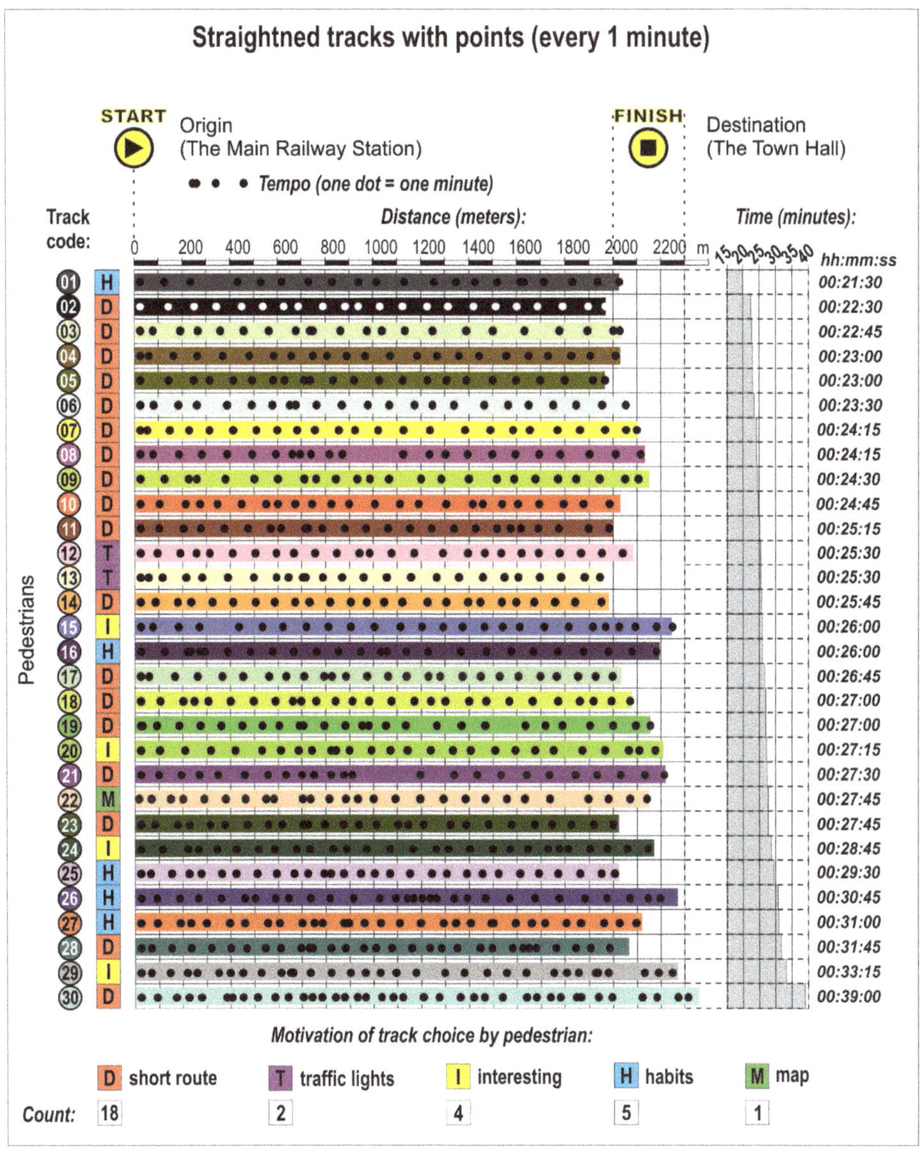

Figure 5. Horizontal graph in a linear composition with 30 classified routes.

A space–time cube (STC) with trajectories (Figure 6) derives from time geography created by Hägerstrand [30], "who introduced a space-time model which included features such as a

Space-Time-Path, and a Space-Time-Prism" [44] (p. 189). In this method, trajectories of movement are presented in the form of a perspective view. The 2.5-dimensional space (a pseudo 3D view) enables one to present the position of the pedestrian in space and time, but the opportunity to read this piece of information accurately in the case of static visualizations is significantly limited [26,30]. The reading of spatial behavior from the STC is very effective when using a monitor, when one can manipulate the angle and height of observation, as well as rotate and zoom in. Among the disadvantages, we should mention the necessity to analyze the visualization from multiple views at different angles. At the same time, using a perspective view greatly hampers the estimation of the distance.

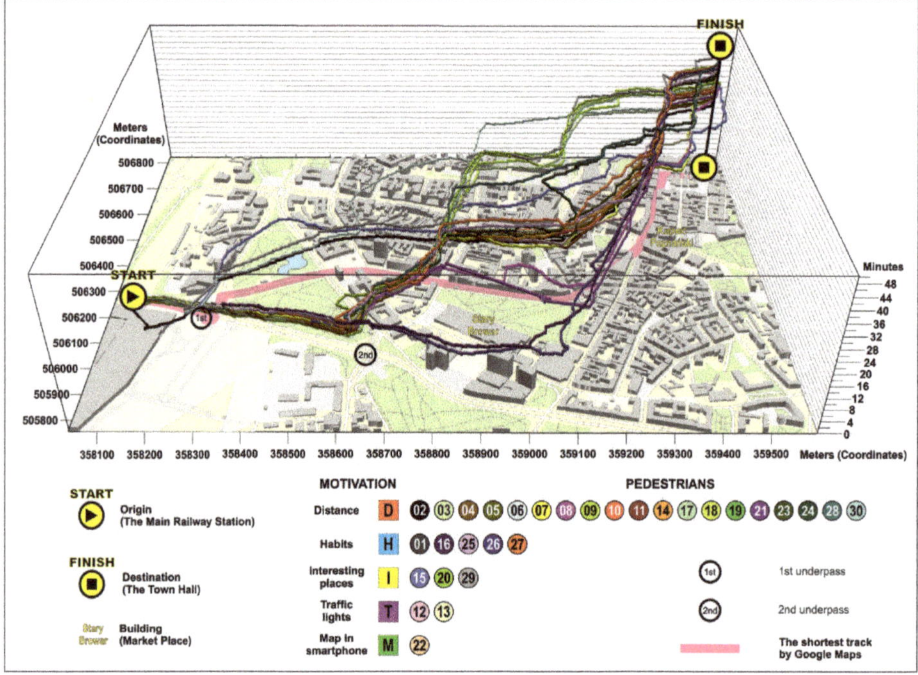

Figure 6. Space–time cube with trajectories (space–time paths).

At the bottom of the STC, some elements of a topographic map can be seen. Buildings are presented in the form of solid figures (City GML Level of Detail 1). Their shape and variable height, created on the basis of the number of floors, reflect actual space dominants, which are a frequent orientation element when selecting a route. This method enables the user to have a bird's eye view (or convergent perspective) on space visualization. The orientation of STC is northern as a reference to the topographic map with trajectories. Figure 6 contains one selected image, which best corresponds with the map from Figure 2, for it retains a northern orientation. The visually attractive image requires the appropriate method of interpreting features for one route and the entire set of routes, as routes 'rise' upwards with each minute.

3. Results

Presented methods of spatial visualizations of the behavior of pedestrians constitute a part of a set of complementary visualizations designed in a way that enables one to correlate pieces of information. In the presented research, each of the analyzed features may be assessed in at least two visualizations. The analysis may look as follows: correlating the schematic map with the map with non-symbolized trajectories on the basis of street names, we gain information about the location and actual distance relations contained in the form of accurate geometry of the track. In turn, a comparison with the topographic map with trajectories enables one to precisely determine the course of tracks and their location towards buildings. The visualization with track miniatures provides knowledge of an accurate course of tracks and similarities between them, whereas the schematic map informs about which street sections were common for various tracks. This one also makes it possible to identify individual pedestrians, which cannot be done by means of the map with non-symbolized trajectories, with only the one that lacks the marking allowing that. The comparison of the space–time cube with the topographic map with trajectories adds the information about the height of buildings, as a spatial dominant important for orientation, to the analysis. Moreover, it provides such data as Cartesian coordinates and route time. Visualizations have been evaluated by their authors (university lecturers of cartography) and the results of that evaluation have been presented in the form of a table, frequently utilized in textbooks (e.g., [45,46]).

The tabular evaluation according to four levels of usefulness of the visualization methods for the analysis of selected features of spatial behavior of pedestrians is presented in Table 2. The table, which is a breakdown of the proposed methods of visualization and analyzed features, presents authors' evaluation (by four academic cartographers) of the usefulness level of each visualization for the demonstration of specific features of spatial behavior. Colors of table cells denote the level of usefulness for presenting specific features (dark grey—very good, grey—good, light grey—possible, and white with crossing—inapplicable/no usefulness). Naturally, the readability and usefulness of those methods will be also influenced by many graphical details connected with graphical design, like map style, scale, quality colors, size of symbols, and so on.

Table 2. Usefulness of the proposed visualization methods for analyzing and interpreting selected features of spatial behavior and pedestrian trajectories (dark grey: very good; grey: good; light grey: possible, white with strikethrough: inapplicable).

	Map with Non-Symbolized Trajectories	**Topographic Map with Trajectories**	**Route Miniatures**	**Schematic Map**	**Route Graph**	**Space-Time Cube**
Track geometry (topographic accuracy)				X	X	
Trajectories similarity (shape)					X	
The ease of differentiation of pedestrian tracks	X					
Track length						
Walking time	X	X	X	X		
Motivation of route choice	X	X	X	X		
Spatial context: location in geographic space					X	
Pace/stops/tempo	X	X	X	X		

To estimate how strong the analyzed feature is represented in the set of mapping techniques, the following questions connected with the rows of the Table 2 were formulated:

A. Is the route shown in a topographically faithful way, free from distortions other than those resulting from GPS?

B. Is it possible to assess the similarity of the route and to what extent?

C. Is it possible to distinguish individual pedestrians' routes and to what extent?

D. Is it possible to estimate the length of the route and with what accuracy?

E. Is it possible to estimate travel time and with what accuracy?

F. Is there information about the most important motivation prompting the pedestrian to choose a route?

G. Is the spatial context known (e.g., route surrounding: buildings, parks, the arrangement of streets and their names, topographic coordinates, reference line: Euclidean distance between start and finish point)?

H. Is it possible to evaluate how quickly the pedestrians moved on to the next stages of the route?

The largest number of features is possible to read on the basis of a route graph. However, these methods do not expose the spatial context, which is very important. The next presentation method is route miniatures, where this context appears partially. Less features show the topographic map with trajectories and the schematic map, while the worst result was obtained by the map with non-symbolized trajectories.

The most exposed feature in the visualization set is the spatial context, which, on visualizations, is manifested in many ways. A strong emphasis is also placed on the ability to distinguish individual pedestrians' routes. The degree of similarity of the routes traveled is a bit harder to analyze, while the time and pace of the route are the least represented.

4. Discussion

This research constitutes a part of the trend of graphically aided spatial behavior analysis, the interest in which results from the general increase in mobility, as well as from constantly developing technical opportunities to collect information on that subject. All potential benefits from the knowledge obtained through such analyses matter greatly in that context.

Similar to other studies of that kind, our research also indicates problems connected with visualizations of pedestrians routes, such as overlapping of graphic elements. In the most of research connected with this subject, the sample size is much bigger than in ours. This generates problems with visibility and makes data aggregation indispensable. The number of 30 participants could seem to be very small, but if we want to analyze each pedestrian's trajectory separately on static visualization and try to compare them, this number seems to be very close to the limit of unique lines, which can be shown. The problem is connected with graphical restrictions, which other publications cited in this article point to, such as, for example, distinguishable colors. The selection of sufficiently distinct colors to distinguish such a number of pedestrians is difficult. Another solution than color can be usage of different line styles (continuous, dashed) or widths, which, however, may suggest another feature of object (e.g., importance). Trajectories with different line styles and the same color can be visually merged when overlapping. The problem with line overlapping is also strongly connected with map scale.

Creating visualizations, the large number of participants constituted a challenge, as it made it difficult to present tracks in the graphic form in a transparent, clear, and distinguishable way. Finding a set of colors that would be distinguishable enough and would enable one to graphically mark participants was a fundamental problem. At the same time, readability of some visualizations and the evaluation of their usefulness for the analysis of specific features could be significantly better, if the number of participants had been smaller and tracks had been more dispersed. The number of participants and the complexity of track geometry cause difficulties and may be a serious limitation on the process of creating static visualizations for the analysis of the movement of pedestrians in urban space. The space–time cube is an example of a visualization that proves more effective in the interactive form, with this fact confirmed by previous studies, such as that by Biadgilgn et al. [14], who indicated that "space-time cube requires higher expertise than other visualizations" (p. 101) and "perhaps needs the real interactive interface" (p. 101). Another form of presentation that could facilitate the interpretation of prepared visualizations, such as route graph, is a cartographic animation, which, in special way, enables one to present spatial and temporal changes simultaneously [47,48].

This research, along with previous studies on similar subjects, shows the limitations of GPS, which, when employed in a heavily developed urban space, affect the shape of visualizations and opportunities for the analysis. The research is yet another example supporting the thesis that pedestrians find distance to be the most significant factor that determines their choice of the route. Adding Google Maps data (the shortest trajectory) to visualizations was an interesting idea, as the interpretation of static visualization methods might thus immediately relate to a popular navigation service. All the research participants were aware that the research area was located in the center of the city and, even though they all had smartphones, just one participant used a map on the smartphone. The final and practical confirmation of the results obtained in the study and the level of usability of the whole set of visualizations considered as a system of related cartographic presentations could be the use of the eye tracking method and test exercises connected with the visualization-based analysis of the behavior of pedestrians, which remains the main task for the future.

5. Conclusions

To examine and comprehend the behavior of pedestrians in urban space, one cannot limit itself to just one analyzed feature. It is also unfeasible to present all examined features on a single visualization. Hence, when examining the problem of spatial behavior, the need arises to plan a set of visualizations that makes it possible to adopt multiple various perspectives. This problem can be solved by means of complementary visualizations, both interactive and static, just as those presented in this publication. In both cases, particularly when dealing with static visualizations, graphic cohesion is of key importance to the potential and effectiveness of proposed solutions. Analyzing the literature, it seems that interactive solutions dominate in this field. However, the authors of this publication still value the advantages of static visualizations, particularly when the previously processed information is supposed to be conveyed to the wider audience. Greater control of the author over transparency and finiteness of the form constitute significant strengths of static visualizations. In static visualizations, it is much more challenging to maintain transparency of various elements and features, as their scale cannot be altered. Lack of interactivity demands that correlations between visualizations presenting the same problem from multiple perspectives are thus coherent and consistent. When the large number of tracks is presented, static visualizations may easily lose their readability. They are also limited when it comes to demonstrating features and are strongly influenced by map scale. However, their finished form that is presented to the audience, adjusted by their author to highlight a given feature graphically, by means of visual variables or by the appropriate data hierarchization, is their great advantage.

Static methods do not allow including all the pieces of information on one visualization, as that would significantly affect its graphic readability and the audience's understanding. For that reason, visualizations need to be correlated in a logical way by retaining common graphic elements. That enables one to analyze connections between various features. Including many visualization methods in the analysis has both benefits and drawbacks. On the one hand, complementary visualizations require the researcher to simultaneously analyze multiple images, while on the other, it serves to strengthen synthetic inferences. The complementary approach may be necessary in such a complex process as pedestrian trajectory analysis, which requires consideration of many variables and factors. The system (set) of visualizations may be the key to know and understand all of them.

Static complementary visualizations boost the opportunity for analysis, as various features are emphasized, but a graphic correlation with other visualizations is maintained. Therefore, one feature can be analyzed from multiple different perspectives or in relation to another feature. It is hence possible to examine the behavior of pedestrians more insightfully. Each of the elaborated visualizations presents spatial data from a different perspective and visually strengthens other aspects of the behavior of research participants. That means that, although the same features may occur on a few visualizations, they are not highlighted to the same extent. Depending on the kind of the adopted graphic solution, various methods of visualization emphasize the same feature to different extents. A comparison of the elaborated visualizations leads to the conclusion that there is no universal method of presenting numerous features. It seems that a good strategy should allow creating visualizations in such a way that each of them clearly shows at least one feature or a correlation between two or more analyzed features. Apart from the choice of graphic solutions that highlight a specific feature, the arrangement of visualization elements according to the examined feature would significantly support the analysis. The proposed methods of presentation allow us to interpret selected features of the spatial behavior of pedestrians in the city, and their utilization may facilitate the search for dependences between these features. The set of complementary visualizations needs to be designed as a system consisting of elements that belong together. Hence, the six methods of cartographic presentation, utilizing different perspectives of spatial perception, should be considered as a single set of multiperspective visualizations of spatial behavior.

Author Contributions: Conceptualization, Beata Medyńska-Gulij; Data curation, Łukasz Halik; Investigation, Łukasz Wielebski, Beata Medyńska-Gulij, Łukasz Halik, and Frank Dickmann; Methodology, Beata Medyńska-Gulij; Project administration, Łukasz Wielebski; Resources, Beata Medyńska-Gulij and Łukasz Halik; Software, Łukasz Wielebski; Validation, Łukasz Halik and Frank Dickmann; Visualization, Łukasz Wielebski; Writing—original draft, Łukasz Wielebski and Beata Medyńska-Gulij. All authors have read and agreed to the published version of the manuscript.

Funding: This research received no external funding.

Acknowledgments: This paper is the result of research on visualization methods carried out within statutory research in the Research Division Cartography and Geomatics, Faculty of Geographical and Geological Sciences, Adam Mickiewicz University in Poznań, in Poland.

Conflicts of Interest: The authors declare no conflict of interest.

References

1. Hamid, S.A. Walking in the City of Signs: Tracking Pedestrians in Glasgow. *CUS* **2014**, *2*, 263–278. [CrossRef]
2. Lloyd, R. *Spatial Cognition. Geographic Environments*; Springer: Dordrecht, The Netherlands, 1997; ISBN 940173044X.
3. Lloyd, R. Cognitive Maps: Encoding and Decoding Information. *Ann. Assoc. Am. Geogr.* **1989**, *79*, 101–124. [CrossRef]
4. Dickmann, F. City Maps Versus Map-Based Navigation Systems – An Empirical Approach to Building Mental Representations. *Cartogr. J.* **2013**, *49*, 62–69. [CrossRef]
5. MacKay, D.B.; Olshavsky, R.W.; Sentell, G. Cognitive Maps and Spatial Behavior Of Consumers. *Geogr. Anal.* **1975**, *7*, 19–33. [CrossRef]
6. Plaue, M.; Chen, M.; Bärwolff, G.; Schwandt, H. Trajectory Extraction and Density Analysis of Intersecting Pedestrian Flows from Video Recordings. In *Photogrammetric Image Analysis*; Stilla, U., Rottensteiner, F., Mayer, H., Jutzi, B., Butenuth, M., Eds.; Springer: Berlin, Heidelberg, 2011; pp. 285–296. ISBN 978-3-642-24393-6.
7. Willems, N.; van de Wetering, H.; van Wijk, J.J. Visualization of vessel movements. *Comput. Graph. Forum* **2009**, *28*, 959–966. [CrossRef]
8. Yuan, M.; Nara, A. Space-Time Analytics of Tracks for the Understanding of Patterns of Life. In *Space-Time Integration in Geography and GIScience*; Springer: Berlin/Heidelberg, Germany, 2015; pp. 373–398.
9. Asahara, A. Decomposition of pedestrian flow heatmap obtained with monitor-based tracking. In Proceedings of the 2017 International Conference on Indoor Positioning and Indoor Navigation (IPIN), Sapporo, Japan, 18–21 September 2017; pp. 1–8, ISBN 978-1-5090-6299-7.
10. Gödel, M.; Köster, G.; Lehmberg, D.; Gruber, M.; Kneidl, A.; Sesser, F. Can we learn where people go? *arXiv* **2018**, arXiv:1812.03719. [CrossRef]
11. Qi, F.; Du, F. Trajectory Data Analyses for Pedestrian Space-time Activity Study. *JoVE* **2013**. [CrossRef]
12. Shibasaki, R.; Zhao, H.; Nakamura, K. A Laser-Scanner System for Acquiring Walking-Trajectory Data and Its Possible Application to Behavioral Science. In *GIS-Based Studies in the Humanities and Social Sciences, First Issued in Paperback*; Okabe, A., Ed.; CRC Press: Boca Raton, FL, USA, 2019; pp. 55–70. ISBN 978-0-8493-2713-1.
13. Demšar, U.; Buchin, K.; Cagnacci, F.; Safi, K.; Speckmann, B.; van de Weghe, N.; Weiskopf, D.; Weibel, R. Analysis and visualisation of movement: An interdisciplinary review. *Mov. Ecol.* **2015**, *3*, 5. [CrossRef]
14. Biadgilgn, D.M.; Blok, C.A.; Huisman, O. Assessing the Cartographic Visualization of Moving Objects. *Mejs* **2011**, *3*. [CrossRef]

15. Horbiński, T.; Cybulski, P.; Medyńska-Gulij, B. Graphic Design and Button Placement for Mobile Map Applications. *Cartogr. J.* **2020**, *6*, 1–13. [CrossRef]
16. Horbiński, T.; Medyńska-Gulij, B. Geovisualisation as a process of creating complementary visualisations. Static two-dimensional, surface three-dimensional, and interactive. *Geod. Cartogr.* **2017**, *66*, 45–58. [CrossRef]
17. Medyńska-Gulij, B.; Cybulski, P. Spatio-temporal dependencies between hospital beds, physicians and health expenditure using visual variables and data classification in statistical table. *Geod. Cartogr.* **2016**, *65*, 67–80. [CrossRef]
18. Ranacher, P.; Brunauer, R.; Trutschnig, W.; van der Spek, S.; Reich, S. Why GPS makes distances bigger than they are. *Int. J. Geogr. Inf. Sci.* **2016**, *30*, 316–333. [CrossRef]
19. Zhu, X.; Li, Q.; Chen, G. APT: Accurate outdoor pedestrian tracking with smartphones. In Proceedings of the 2013 Proceedings IEEE INFOCOM, IEEE Conference on Computer Communications, Turin, Italy, 14–19 April 2013; pp. 2508–2516, ISBN 978-1-4673-5946-7.
20. Petrasova, A.; Hipp, J.A.; Mitasova, H. Visualization of Pedestrian Density Dynamics Using Data Extracted from Public Webcams. *IJGI* **2019**, *8*, 559. [CrossRef]
21. Brandle, N.; Matyus, T.; Brunnhuber, M.; Hesina, G.; Neuschmied, H.; Rosner, M. Realistic Interactive Pedestrian Simulation and Visualization for Virtual 3D Environments. In Proceedings of the 2009 15th International Conference on Virtual Systems and Multimedia, Vienna, Austria, 9–12 September 2009; pp. 179–184.
22. Schwartz, M.S.; Schwartz, C.G. Problems in Participant Observation. *Am. J. Sociol.* **1955**, *60*, 343–353. [CrossRef]
23. Parent, C.; Pelekis, N.; Theodoridis, Y.; Yan, Z.; Spaccapietra, S.; Renso, C.; Andrienko, G.; Andrienko, N.; Bogorny, V.; Damiani, M.L.; et al. Semantic trajectories modeling and analysis. *ACM Comput. Surv.* **2013**, *45*, 1–32. [CrossRef]
24. McArdle, G.; Demšar, U.; van der Spek, S.; McLoone, S. Classifying pedestrian movement behaviour from GPS trajectories using visualization and clustering. *Ann. GIS* **2014**, *20*, 85–98. [CrossRef]
25. Andrienko, N.; Andrienko, G. Visual analytics of movement: An overview of methods, tools and procedures. *Inf. Vis.* **2013**, *12*, 3–24. [CrossRef]
26. Wang, Z.; Yuan, X. Urban trajectory timeline visualization. In Proceedings of the 2014 International Conference on Big Data and Smart Computing (BIGCOMP), Bangkok, Thailand, 15–17 January 2014; pp. 13–18, ISBN 978-1-4799-3919-0.
27. MacEachren, A.M. *Visualization in Modern Cartography*, 1st ed.; Elsevier: Amsterdam, The Netherlands, 1994; ISBN 9780080424163.
28. Burkhard, R.A. Towards a Framework and a Model for Knowledge Visualization: Synergies Between Information and Knowledge Visualization. In *Knowledge and Information Visualization*; Hutchison, D., Kanade, T., Kittler, J., Kleinberg, J.M., Mattern, F., Mitchell, J.C., Naor, M., Nierstrasz, O., Pandu Rangan, C., Steffen, B., et al., Eds.; Springer: Berlin/Heidelberg, Germany, 2005; pp. 238–255. ISBN 978-3-540-26921-2.
29. Nazemi, K.; Breyer, M.; Burkhardt, D.; Stab, C.; Kohlhammer, J. SemaVis: A New Approach for Visualizing Semantic Information. In *Towards the Internet of Services: The THESEUS Research Program*; Wahlster, W., Grallert, H.-J., Wess, S., Friedrich, H., Widenka, T., Eds.; Springer International Publishing: Cham, Switzerland, 2014; pp. 191–202. ISBN 978-3-319-06754-4.
30. Medyńska-Gulij, B.; Dickmann, F.; Halik, Ł.; Wielebski, Ł.W. Mehrperspektivische Visualisierung von Informationen zum räumlichen Freizeitverhalten. Ein Smartphone-gestützter Ansatz zur Kartographie von Tourismusrouten. Multiperspective visualisation of spatial spare time activities. A smartphone-based approach to mapping tourist routes. *Kartogr. Nachr.* **2015**, *67*, 323–329. [CrossRef]
31. Medyńska-Gulij, B.; Wielebski, Ł.; Halik, Ł.; Smaczyński, M. Complexity Level of People Gathering Presentation on an Animated Map—Objective Effectiveness Versus Expert Opinion. *IJGI* **2020**, *9*, 117. [CrossRef]
32. Green-Armytage, P. A Colour Alphabet and the Limits of Colour Coding. *Colour: Des. Creat.* **2010**, *5*, 1–23.
33. Rougeux, N. Global Subway Spectrum—An exploration of colors used for lines in every rapid transit system. Available online: https://www.c82.net/spectrum/ (accessed on 15 April 2020).
34. Trubetskoy, S. Roman Roads. Available online: https://sashat.me/2017/06/03/roman-roads-index/ (accessed on 3 January 2020).

35. Trubetskoy, S. List of 20 Simple, Distinct Colors. Available online: https://sashamaps.net/docs/tools/20-colors/ (accessed on 3 January 2020).
36. Kelly, K.L. Twenty-two colors of maximum contrast. *Color Eng.* **1965**, *3*, 26–27.
37. Dent, B.D.; Hodler, T.W. *Cartography. Thematic Map Design*, 6th ed.; McGraw-Hill Higher Education: New York, NY, USA, 2009; ISBN 9780072943825.
38. Wielebski, Ł.; Medyńska-Gulij, B. Graphically supported evaluation of mapping techniques used in presenting spatial accessibility. *Cartogr. Geogr. Inf. Sci.* **2019**, *46*, 311–333. [CrossRef]
39. Golledge, R.G. Wayfinding Behavior. In *Cognitive Mapping and Other Spatial Processes*; Johns Hopkins University Press: Baltimore, MA, USA, 1999; ISBN 080185993X.
40. Andrienko, N.; Andrienko, G.; Pelekis, N.; Spaccapietra, S. Basic Concepts of Movement Data. In *Mobility, Data Mining and Privacy: Geographic Knowledge Discovery*; Giannotti, F., Pedreschi, D., Eds.; Springer: Berlin/Heidelberg, Germany, 2008; pp. 15–38. ISBN 978-3-540-75177-9.
41. Blanke, U.; Guldener, R.; Feese, S.; Troester, G. Crowdsourced pedestrian map construction for short-term city-scale events. *Proc. First Int. Conf. IoT Urban Sp. ICST* **2014**.
42. Bierlaire, M.; Robin, T. Pedestrians Choices. In *Pedestrian Behavior*; Timmermans, H., Ed.; Emerald Group Publishing Limited: Bingley, UK, 2009; pp. 1–26. ISBN 978-1-848-55750-5.
43. Seneviratne, P.N.; Morrall, J.F. Analysis of factors affecting the choice of route of pedestrians. *Transp. Plan. Technol.* **2007**, *10*, 147–159. [CrossRef]
44. Kraak, M.-J.; Koussoulakou, A. A Visualization Environment for the Space-Time-Cube. In *Developments in Spatial Data Handling: 11th International Symposium on Spatial Data Handling*; Fisher, P.F., Ed.; Springer: Berlin/Heidelberg, Germany, 2006; pp. 189–200. ISBN 3-540-22610-9.
45. Slocum, T.A.; McMaster, R.B.; Kessler, F.C.; Howard, H.H. *Thematic Cartography and Geovisualization*, 3rd ed.; Pearson Prentice Hall: Upper Saddle River, NJ, USA, 2010; ISBN 0138010064.
46. Kraak, M.J.; Ormeling, F. Cartography. In *Visualization of Geospatial Data*, 3rd ed.; Prentice Hall: Upper Saddle River, NJ, USA, 2010; ISBN 0273722794.
47. Cybulski, P.; Medyńska-Gulij, B. Cartographic Redundancy in Reducing Change Blindness in Detecting Extreme Values in Spatio-Temporal Maps. *IJGI* **2018**, *7*, 8. [CrossRef]
48. Cybulski, P.; Wielebski, Ł. Effectiveness of Dynamic Point Symbols in Quantitative Mapping. *Cartogr. J.* **2019**, *56*, 146–160. [CrossRef]

Article

Panoramic Mapping with Information Technologies for Supporting Engineering Education: A Preliminary Exploration

Jhe-Syuan Lai *, Yu-Chi Peng, Min-Jhen Chang and Jun-Yi Huang

Department of Civil Engineering, Feng Chia University, Taichung 40724, Taiwan; M0702975@mail.fcu.edu.tw (Y.-C.P.); d0676303@mail.fcu.edu.tw (M.-J.C.); M0806485@mail.fcu.edu.tw (J.-Y.H.)

* Correspondence: jslai@fcu.edu.tw; Tel.: +886-4-2451725 (ext. 3118)

Received: 14 October 2020; Accepted: 18 November 2020; Published: 19 November 2020

Abstract: The present researchers took multistation-based panoramic images and imported the processed images into a virtual tour platform to create webpages and a virtual reality environment. The integrated multimedia platform aims to assist students in a surveying practice course. A questionnaire survey was conducted to evaluate the platform's usefulness to students, and its design was modified according to respondents' feedback. Panoramic photos were taken using a full-frame digital single-lens reflex camera with an ultra-wide-angle zoom lens mounted on a panoramic instrument. The camera took photos at various angles, generating a visual field with horizontal and vertical viewing angles close to 360°. Multiple overlapping images were stitched to form a complete panoramic image for each capturing station. Image stitching entails extracting feature points to verify the correspondence between the same feature point in different images (i.e., tie points). By calculating the root mean square error of a stitched image, we determined the stitching quality and modified the tie point location when necessary. The root mean square errors of nearly all panoramas were lower than 5 pixels, meeting the recommended stitching standard. Additionally, 92% of the respondents (n = 62) considered the platform helpful for their surveying practice course. We also discussed and provided suggestions for the improvement of panoramic image quality, camera parameter settings, and panoramic image processing.

Keywords: engineering education; image stitching; information technology; multimedia; panorama

---

1. Introduction

Panoramic images are being made available on an increasing number of online media platforms, such as Google Maps. Virtual reality (VR) technology is also becoming more common in modern life (e.g., video games and street view maps), providing immersive and interactive experiences for users. Various industries have incorporated this technology into their businesses; for example, companies in the leisure industry, such as the Garinko Ice-Breaker Cruise 360 Experience in Hokkaido, include VR images on their official websites to attract tourists [1]. Similarly, several real estate companies are showcasing furnished interior spaces to potential buyers by using panoramic images; this helps customers visualize the actual setting of the houses they are interested in [2]. These multimedia approaches integrating image, video, audio, and animations can further obtain better presentation and communication methods [3–7], such as visualization, map, graphical user interfaces, interactive recommendation system, etc.

Surveying practice is a fundamental and essential subject for civil engineering students. However, university students, with little civil engineering experience, mostly do not know how to accomplish a surveying task because of (a) unfamiliarity with surveying points, (b) inability to connect knowledge acquired in class with actual practices, (c) inability related to or unfamiliarity with establishing a record table, and (d) inability to operate or unfamiliarity with instrument operations. Lu [8] studied computer-assisted instruction in engineering surveying practice. According to the questionnaire survey results, 91% of the students either strongly agreed or agreed that virtual equipment helped increase their learning motivation. Additionally, 66% of the respondents correctly answered a question concerning azimuth, a foundational concept in civil engineering. Lu [8] indicated that the introduction of digital instruction is more likely to spark learning interest and motivation than conventional training would.

By integrating panoramic images into a virtual tour platform, this study adopted VR technology to create a webpage that facilitates the instruction of surveying practice. The present researchers selected panoramic images because of the low cost of image construction and ability to create realistic and immersive visual effects. The designed assistance platform included (a) surveying tips, (b) various surveying routes, (c) corresponding measurement principles, and (d) instructional videos. Therefore, students had access to the supplementary materials on this platform before or during a lesson, thereby increasing learning efficiency and helping students acquire independent learning skills. This study explored student acceptance of technology-aided instruction and the practicality of such an instruction method by using a questionnaire survey. Subsequently, the original web design was modified per the students' feedback. In this paper, the authors also discussed and proposed suggestions for the improvement of panoramic image quality, camera parameter settings, and panoramic image processing.

2. Related Works

2.1. Panorama

The word "panorama" originates from the Greek pan ("all") and horama ("view"). In 1857, M. Garrela patented a camera in England that could rotate around its own axis and take a horizontal 360° photo; it was the first camera for panoramic photos that employed mainspring control. According to the field of capture, panorama can be divided into three patterns, as presented in Table 1. In this study, the shooting targets were all objects above the ground, and the angle of coverage was mainly landscape; thus, each shot did not fully cover the landscape in the vertical direction. According to the range and angle settings, the photos taken in this study are considered 360° panoramas.

Table 1. Classification of Panorama [9].

Field of Capturing	Panorama	360° Panorama	Spherical Panorama
Horizontal	<180°	360°	360°
Vertical	<180°	<180°	180°

Image stitching is a crucial step in panorama generation. Zheng et al. [10] described and explored shooting strategies and image stitching methods in detail. Regarding research on the stitching process, Chen and Tseng [11], by identifying the corresponding feature point in different images (i.e., tie point), determined the tie point quality of panoramic images. By using panoramic photography and photogrammetry, Teo and Chang [12] and Laliberte et al. [13] generated three-dimensional (3D) image-based point clouds and orthophotos. Studies have also applied image stitching in numerous areas, including landscape identification, indoor positioning and navigation, 3D city models, virtual tours, rock art digital enhancement, and campus virtual tours [14–19].

2.2. Virtual Reality (VR)

VR is a type of computer-based simulation. Dennis and Kansky [20] stated that VR can generate simulated scenes that enable users to experience, learn, and freely observe objects in a 3D space in real time. When users move, the computer immediately performs complex calculations and returns precise 3D images to create a sense of presence. VR integrates the latest techniques in computer graphics, artificial intelligence, sensing technology, display technology, and Internet parallel computing. Burdea [21] suggested defining VR according to its functions and proposed the concept of three Is (i.e., interaction, immersion, and imagination), and suggested that VR must have said three characteristics. Furthermore, Gupta et al. [22] connected VR with soundscape data.

VR devices can be either computer-based (high resolution) or smartphone-based (portable). For example, the HTC VIVE-Pro is computer-based, whereas Google Cardboard is smartphone-based. We selected the portable VR device to conduct experiments because it is inexpensive and does not require a computer connection. To access the designed online assistance platform for teaching, the participating students only had to click on the webpage link or scan the quick response (QR) code using their smartphones.

2.3. Education with Information Technologies

As a result of technological advancement, e-learning has become prevalent. For example, information technologies, such as smartphones, multimedia, augmented reality, and internet of things, have been adopted to increase or explore learning outcomes [23–26]. Lee [27] maintained that through learners' active participation, computer-based simulation can effectively assist learners to understand abstract concepts, which in turn increases learning motivation and improves learning outcomes. VR can also help create a learning environment without time constraints; for example, Brenton et al. [28] incorporated VR into anatomy teaching to mitigate the major impediments to anatomy teaching, such as time constraints and limited availability of cadavers, by using 3D modeling as well as computer-assisted learning. The Archeoguide (Augmented Reality-based Cultural Heritage On-site Guide) proposed by Vlahakis et al. [29] demonstrated that VR is not bounded by spatial constraints. This on-site guide was used to provide a customized cultural heritage tour experience for tourists. ART EMPEROR [30] discovered that numerous prominent museums worldwide have established databases of their collections using high-resolution photography or 3D scanning and modeling. These databases enable users to explore art through the Internet.

Chao [31] surveyed and conducted in-depth interviews with VR users after they participated in VR-related scientific experiments; the results indicated that such experiments provide participants with the illusion that they are in a physical environment. Without spatial constraints, the VR-simulated environment offered the participants experiences that they could not have in real life. Therefore, VR helped the participants obtain information and knowledge of various fields in a practical manner. These experiences, compared with those obtained through videos and print books, left a stronger impression on students, prompting them to actively seek answers. Similarly, Chang [32] indicated that students receiving 3D panorama-based instruction significantly outperformed their counterparts who received conventional instruction. Liao [33] examined English learning outcomes and motivation among vocational high school students by using panorama and VR technology; the results revealed that the use of said technologies effectively improved learning outcomes, motivation, and satisfaction.

2.4. Summary

According to the aforementioned literature, panoramic photography and VR technology have advanced rapidly, have a wide range of applications, provide realistic 3D experiences, and enhance teaching effectiveness. However, few studies have applied said technologies to engineering education. The present study created panorama-based VR environments on a virtual tour platform and achieved a cost-effective display of on-site scenes. The research team hopes to help civil engineering students rapidly become familiar with the surveying practice elements in question and complement theories with practical knowledge and skills.

3. Materials and Methods

Figure 1 presents the study procedures. First, capturing stations were set up and images were collected (Section 3.1). Subsequently, image stitching was performed (Section 3.2) by combining multiple photos from a single station into a panoramic image. The combined panoramic images were then transformed into web format by using a virtual tour platform (Section 3.3); additional functions could be added to the web format. Next, a VR navigation environment was constructed, finalizing the development of a teaching assistance platform. The platform was assessed using a questionnaire survey (Section 3.4); the survey results and feedback from users were referenced to improve the platform design.

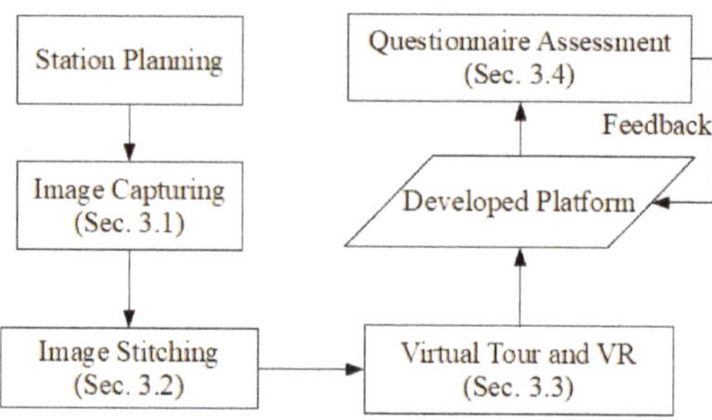

Figure 1. The conceptual procedure used in this study.

3.1. Image Capturing

The researchers mounted a full-frame digital single-lens reflex camera (CANON EOS 6D Mark II, Canon, Tokyo, Japan) equipped with an ultra-wide-angle zoom lens (Canon EF 16–35 mm F/4L IS USM, Canon, Tokyo, Japan) on a panoramic instrument (GigaPan EPIC Pro V, GigaPan, Portland, OR, USA; Figure 2) and tripod. To use the GigaPan, the horizontal and vertical coverage must be set, after which the machine automatically rotates and accurately divides the scene into several grid images; these images overlap, which facilitates the image stitching process. In addition, the camera took each photo by using bracketing and captured shots at three brightness levels (normal, darker, and brighter). These shots served as material for stitching and synthesis. Please refer to https://www.youtube.com/watch?v=JTkFZwhRuxQ for the actual shooting process employing the GigaPan.

Figure 2. Panoramic instrument of GigaPan EPIC Pro V.

3.2. Image Stitching

After we captured the images, Kolor Autopano was adopted in this study. The original images underwent feature point extraction, homography, warping, and blending to form the panoramic image of a station. The general process is detailed in Figure 3.

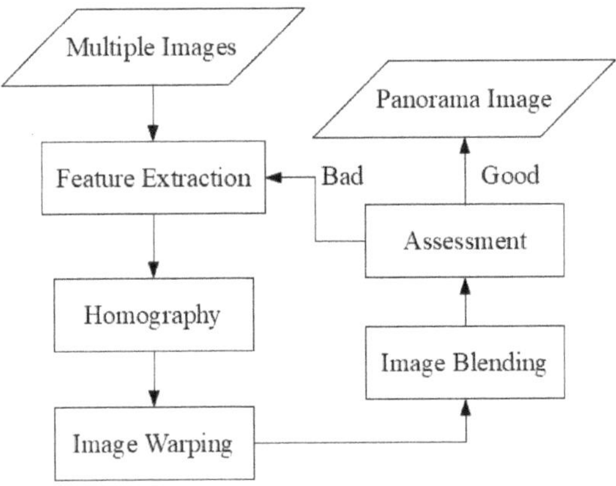

Figure 3. Procedure for image stitching.

3.2.1. Homography

When two images partially overlap, they share several corresponding feature points. These points can be connected using computations; this process is known as homography, and the corresponding feature points are called tie points.

Kolor Autopano extracts, matches, and transforms feature points into tie points by using the scale-invariant feature transform (SIFT) algorithm [10]. The features extracted using this technique are invariant to image rotation and scaling as well as changes in grayscale values. There are four major steps, including (a) scale-space extrema detection, (b) key-point localization, (c) orientation assignment, and (d) key-point descriptor. The results of detecting points of interest are termed as key-point candidates in the step of scale-space extrema detection. The Gaussian filter was used to convolve at different scales, and difference of Gaussian-blurred images were obtained. Key-points of the maximum and minimum Difference of Gaussians (DoG) are further extracted at multiple scales. A DoG image (D) is given by Equation (1), where L represents the convolution of the original image (x,y) with the Gaussian blur at scales of kσ; k and σ indicate a scale factor and standard deviation of the Gaussian blur, respectively. The second step is to localize the key-points. The scale-space extrema detection might produce too many unstable key-point candidates. This step is to fit the nearby data for accurate location in consideration of scale and ratio of principal curvatures. For assigning the orientation of key-points, the local image gradient directions in achieving invariance to rotation were determined by Equations (2) and (3), where θ and m represent orientation and gradient magnitude, respectively. Finally, the invariance to image location, scale, and rotation was checked by pixel neighborhoods and histogram-based statistics. Relevant details are provided in the research of Lowe [34].

$$\mathrm{D}(\mathrm{x},\mathrm{y},\sigma) = \mathrm{L}(\mathrm{x},\mathrm{y},\mathrm{k}\sigma) - \mathrm{L}(\mathrm{x},\mathrm{y},\sigma) \quad (1)$$

$$\theta(x,y) = \tan^{-1}\left(\frac{\partial \mathrm{L}}{\partial \mathrm{y}} / \frac{\partial \mathrm{L}}{\partial \mathrm{x}}\right) \quad (2)$$

$$\mathrm{m}(x,y) = \sqrt{\left(\frac{\partial \mathrm{L}}{\partial \mathrm{x}}\right)^2 + \left(\frac{\partial \mathrm{L}}{\partial \mathrm{y}}\right)^2} \quad (3)$$

After tie points were determined, homogeneous coordinates were used to generate a 3×3 matrix (H), which describes the spatial translation, scaling, and rotation of feature points in different images. If the feature point set of an image is (x,y,1) and that of another image (x′,y′,1), the mapping relationship between tie points of the two images is H. The relationship can be described using Equation (4), where H is to be calculated. At least four pairs of tie points are required to calculate H [35]. However, the degree of freedom must be zero and a unique solution. In practice, more than four pairs of tie points are often used, resulting in a degree of freedom > 0. Therefore, the match with minimum errors must be identified; this process is known as optimization. Kolor Autopano adopts the random sample consensus (RANSAC) algorithm, as shown in Equation (5), to minimize the errors between H and all tie points [10]. Specifically, random sample consensus randomly selects tie points as inliers to calculate matrix H and evaluate the errors between the matrix and other tie points. Subsequently, these tie points are divided into inliers and outliers before inliers are renewed. Said process is repeated until the matrix H with minimum errors is obtained, serving as the optimal solution. The number of iteration (N) in Equation (5) is chosen to ensure that the probability p (usually set to 0.99) and at least one of the sets of random samples exclude an outlier. Let u indicate the probability that the selected data point is an inlier, and v = 1 − u the probability of observing an outlier. N iterations of the minimum number of points show that m is required.

$$\begin{bmatrix} x \\ y \\ 1 \end{bmatrix} = \begin{bmatrix} h_{00} & h_{01} & h_{02} \\ h_{10} & h_{11} & h_{12} \\ h_{20} & h_{21} & h_{22} \end{bmatrix} \times \begin{bmatrix} x' \\ y' \\ 1 \end{bmatrix} \tag{4}$$

$$1 - p = (1 - u^m)^N \tag{5}$$

3.2.2. Image Warping and Blending

Image warping determines the distortion level of an image in a space by using the tie points identified in homography. One of the selected two images serves as a reference, whereas the other is projected onto the coordinate space according to the reference. After computation and distortion, the projected image is projected onto the reference image, thus achieving image warping. During image projection, the projected image is distorted to enable two images to be superposed. However, during image stitching, excessive distortion of the nonoverlapping area is likely. Cylindrical projection and spherical projection can mitigate such distortion; therefore, we confirmed that said projection methods are suitable for panoramic stitching and projected the images onto a cylinder or sphere.

Let the coordinates of an image be (x,y) and the projected coordinates on a sphere be (x,y,f). The spherical coordinate system is displayed as (r,θ,φ), where r denotes the distance between the sphere center and the target, θ is the included angle between r and the zenith (range = [0, π]), and φ represents the included angle between the r plane projection line and X-axis (range = [0, 2π]). The spherical coordinate system can also be converted into a Cartesian coordinate system. Accordingly, conversion between the image and spherical coordinate systems can be described using Equation (6). By converting the spherical coordinate system into a Cartesian one and implementing homography, we achieved image warping.

$$(r \sin\theta \cos\varphi, r \sin\theta \sin\varphi, r \cos\theta) \propto (x, y, f) \tag{6}$$

Blending, the last step of image stitching, involves synthesizing warped images by using color-balancing algorithms to create an image gradient on the overlapping area of two images. In this manner, chromatic aberration of the resulting image stitched from multiple images can be minimized. Common methods include feather blending, and multiband blending; please refer to [36] for further details.

3.2.3. Accuracy Assessment

Kolor Autopano was then used to calculate the root mean square errors (RMSEs) of the stitched panoramic images, as shown in Equation (7), where N denotes the number of tie points. The resulting RMSE value in this study represents the pixel distance (Diff) between tie points [37]. A value < 5 pixels indicates favorable stitching quality. Conversely, a value ≥ 5 indicates undesirable quality and the possibility of mismatch; in such cases, the tie points should be reviewed.

$$\text{RMSE} = \sqrt{\frac{\sum_{i=1}^{N} Diff_i}{N}} \tag{7}$$

3.3. *Virtual Tour with VR*

Panoramic images of several stations were obtained after image stitching. All the panoramic images in this study were imported into Kolor Panotour and displayed in a virtual tour. In addition to webpage construction, the Kolor Panotour software enables users to add data associated with a point of interest and attribute as well as insert images, videos, and hyperlinks. The software also facilitates the generation of VR navigation environments. Koehl and Brigand [16] provided an introduction to and outlined the application of Kolor Panotour.

3.4. *Designed Questionnaire*

The questionnaire employed in this study comprised four questions, which were rated on a 5-point Likert scale. A higher score indicates satisfaction with or interest in the designed platform. The questionnaire was designed using Google Forms with a quick response (QR) code attached. A total of 62 students completed the questionnaire, and some students were also interviewed. The survey questions are as follows:

Q1. After using the virtual tour webpages, compared with the scenario where only an introduction is provided by the instructor, can you more easily identify surveying targets on campus?
Q2. Are you satisfied with the overall webpage design?
Q3. Are you interested in research on surveying practice courses that employ panorama and VR technology?
Q4. Do you like courses that incorporate information technologies (e.g., e-learning)?

4. Results

4.1. *Camera Settings*

Appropriate exposure and focus are essential for taking a suitable picture [38]. Exposure is determined by the shutter speed, aperture size, and ISO value, whereas the image is in focus and becomes clear only once the lens–object distance is correctly adjusted.

A typical imaging device can only capture an extremely limited range within the complete dynamic range. Consequently, using general imaging devices can lead to a severe loss of scene information, particularly in highlights and shadows [39]. By using bracketing, we took normal, darker, and brighter photos and combined these photos, which have different exposure levels, to obtain a greater exposure dynamic range. After several tests, the suitable parameters for camera settings in the study case are shown in Table 2.

Table 2. Used parameters for camera settings in this study.

Shutter Speed (Sec.)	Aperture Size	ISO Value
1/500~1	F/11~F/16	100~400

4.2. Study Site and Routes

The study site was on the main campus (northwestern side) of Feng Chia University in Taichung City, Taiwan (Figure 4). The two routes (symbols 1 and 2), comprising surveying targets surrounding the Civil/Hydraulic Engineering Building and Science Building, were regarded as elevation-based measurement routes (i.e., the blue and green lines in Figure 4D). The site for angle-based measurement (symbol 3) was located on the lawn, which is represented by red lines in Figure 4D. According to the measurement tasks and targets, a total of 15 stations for panoramic photography were set up (Figure 5).

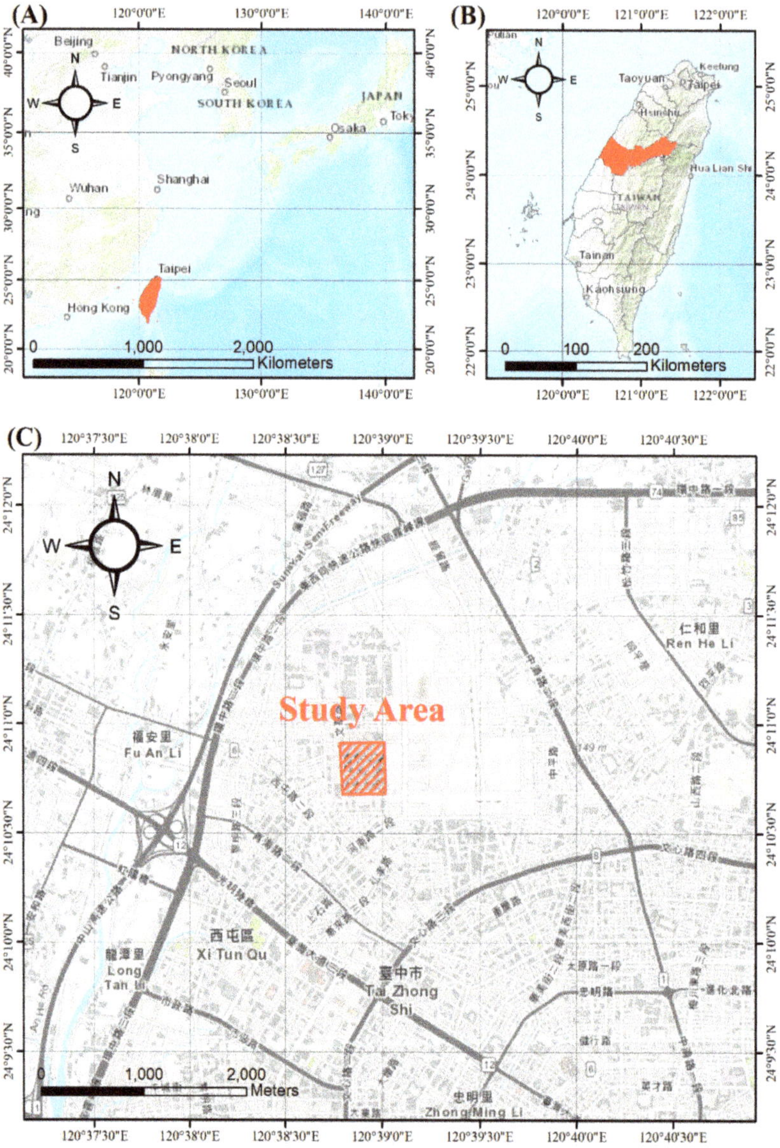

Figure 4. *Cont.*

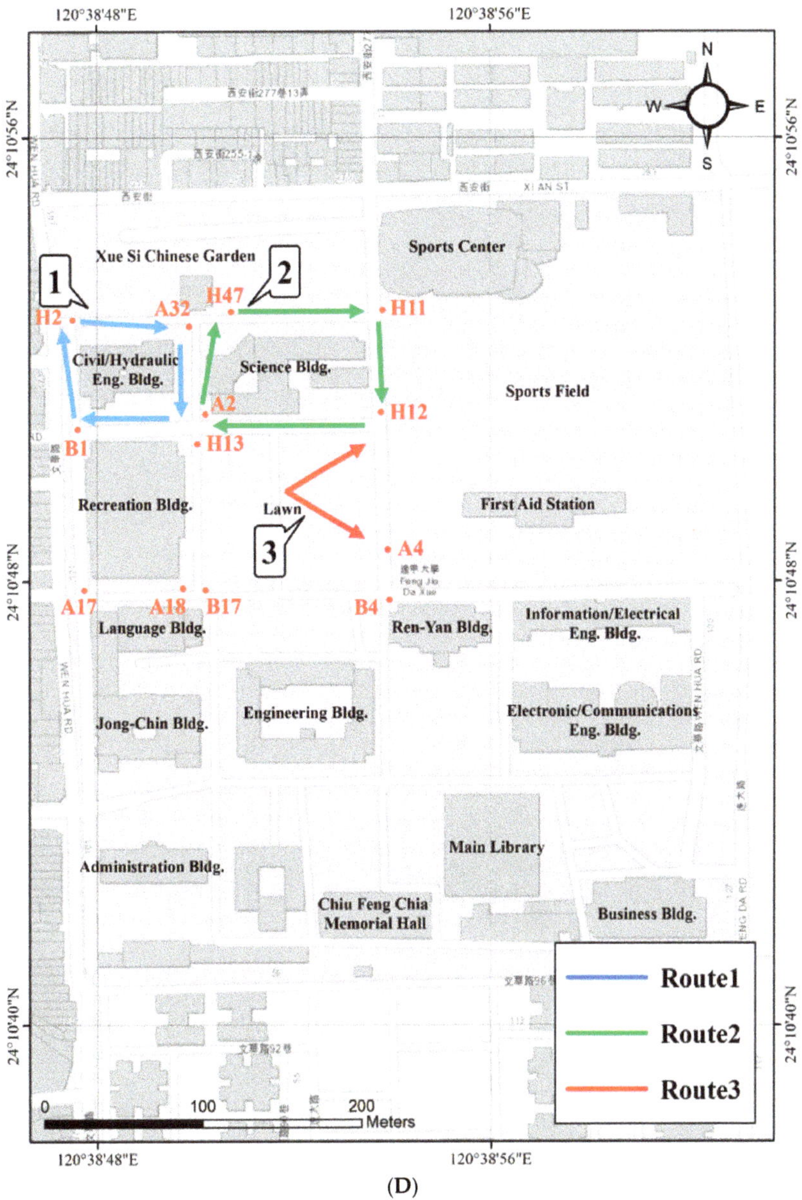

(D)

Figure 4. Study site (**A–C**), stations and routes (**D**) for the course of Surveying Practice in Feng Chia University (FCU), where blue and green colors indicate the routes for the elevation-based measurement, and red represents the station for the angle-based measurement.

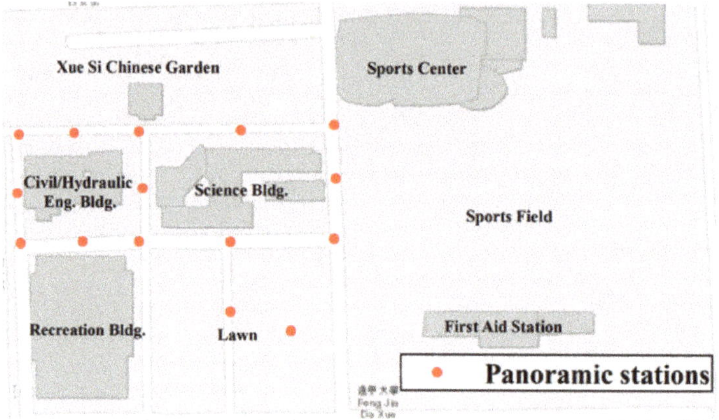

Figure 5. Panoramic stations (red points) for capturing images.

4.3. Developed Platform

4.3.1. Image Stitching

Figure 6 depicts the stitched image of a surveying station as an example; the green marks denote an RMSE < 5 pixels, whereas the red marks denote stitching errors with an RMSE ≥ 5. This figure demonstrates that most of the tie points meet the suggested standard. However, the stitching quality between trees and the sky was lower because when the camera took pictures, the target object moved, causing image stitching to fail. Therefore, manual adjustment of the tie points or postprocessing of the image was required. Figure 7 exhibits the stitching results of panoramas at four stations. These images were later imported into the virtual tour platform to enable panoramic navigation.

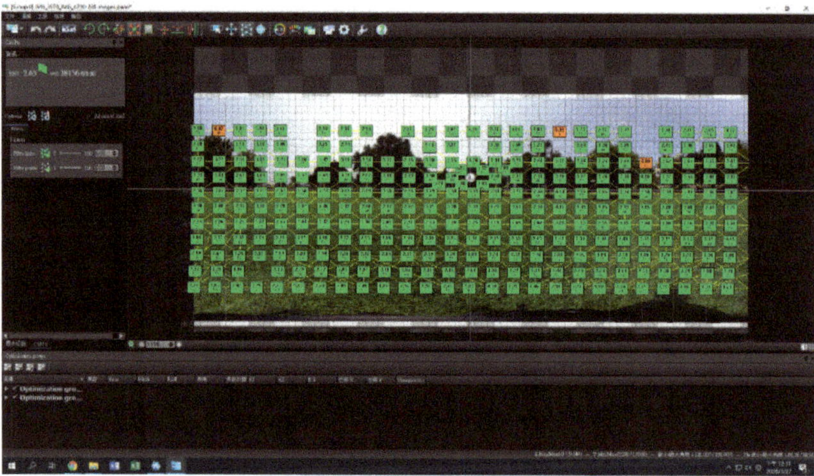

Figure 6. An example of image stitching with root mean square errors (RMSEs) where green color represents excellent results.

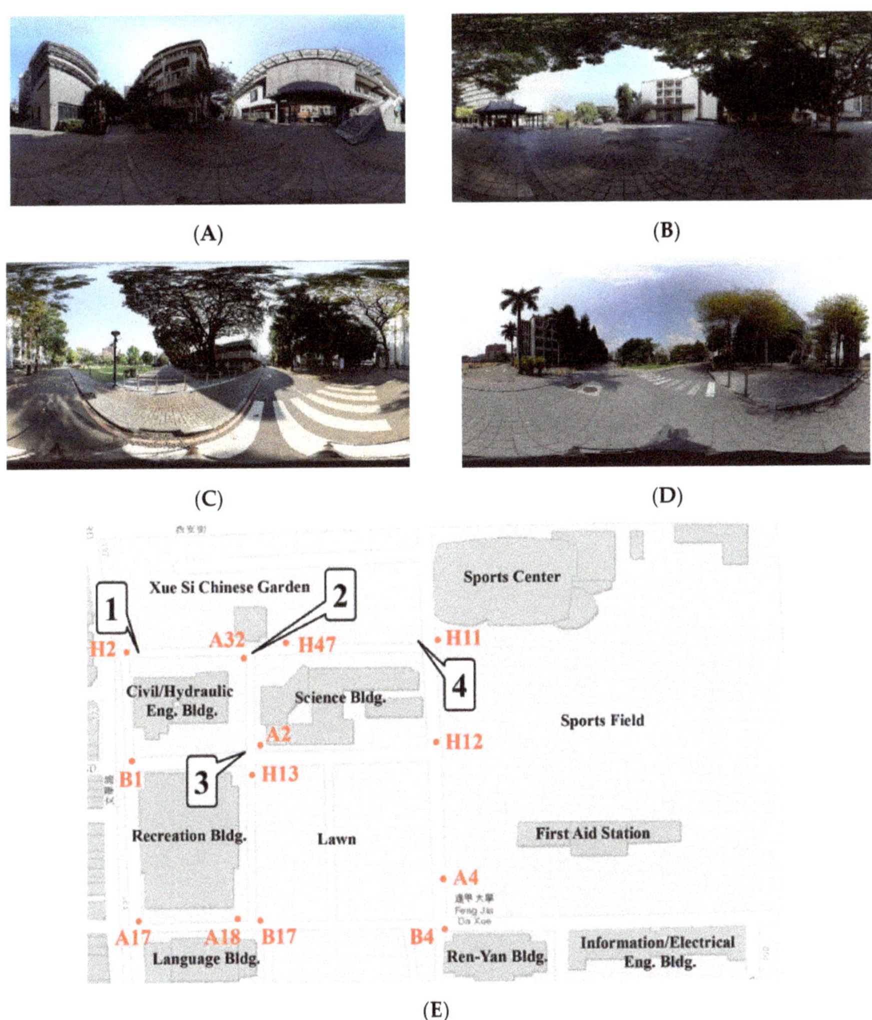

Figure 7. Examples of the stitched images, located (**A**) symbol 1, (**B**) symbol 2, (**C**) symbol 3, and (**D**) symbol 4 in (**E**).

4.3.2. Webpage and Virtual Tour

Figure 8 illustrates the designed webpage framework and panoramic stations, which have a spatial relationship that is consistent with that in Figure 5 from the top view. The homepage displays the panorama of Station 14. When users visit Station 15, they can read instructions on angle-based measurement and watch a tutorial on angular measuring device setup. Station 11-3 features instructions on elevation-based measurement and a tutorial on relevant instrument setup. This station also provides on-site views of the two routes available for surveying. Next, the research team connected the elements in Figure 8 to a virtual tour platform to establish a webpage, where various information can be added, including points of interest and attributes. Designers could also insert pictures, videos, and hyperlinks (Figure 9).

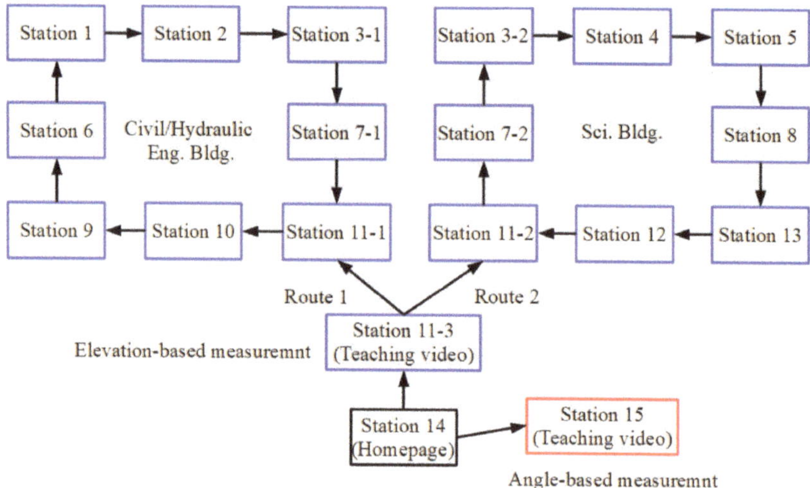

Figure 8. Webpage framework and panoramic stations, where blue and red colors represent elevation- and angle-based measurements.

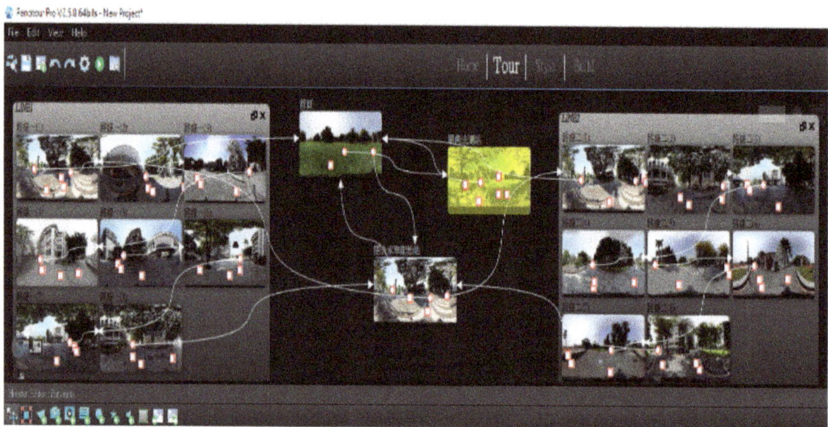

Figure 9. An example of setting a virtual tour.

4.3.3. Demonstration

The platform homepage is displayed in Figure 10. On the upper left (symbol 1) is a link to Feng Chia University's website; control bars (symbol 4) are located on the bottom left, enabling the user to zoom in, zoom out, or switch the image to VR mode. At the bottom middle (symbol 5) are relevant instructional documents, and on the bottom right (symbol 6) is a quick map showing the navigation direction. The homepage hyperlink is located in the upper right corner (symbol 3). By navigating around the homepage 360° panorama, the user can locate the entrances for tutorials on elevation-based and angle-based measurement (Figures 11A and 12A).

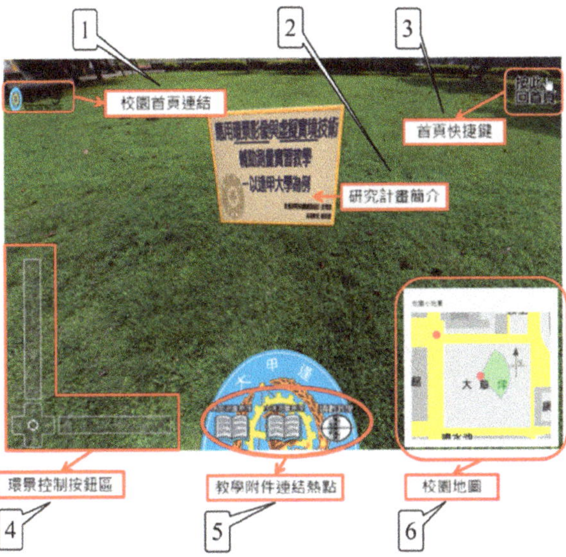

Figure 10. Homepage of the developed platform. Symbol 1: hyperlink to the FCU official webpage; symbol 2: information of the supporting project; symbol 3: back to the homepage control bar; symbol 5: hyperlink to the related documents; symbol 6: quick map.

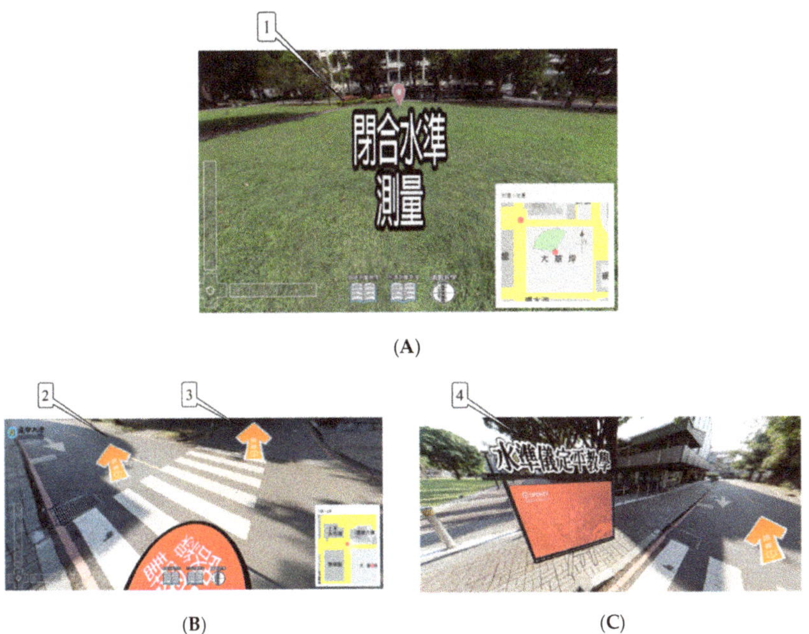

Figure 11. Webpage for teaching the elevation-based measurement. (**A**) Entrance for angle-based measurement (symbol 1, or called leveling measurement); (**B**) Route selection (symbols 2 and 3 represent routes 1 and 2, respectively); (**C**) Teaching video (symbol 4) for setting up the instrument.

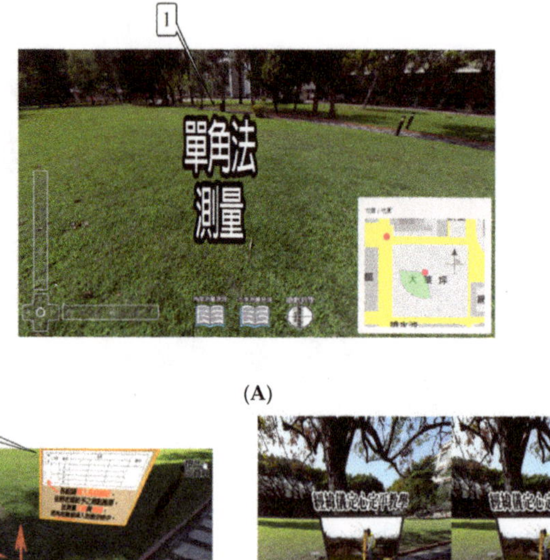

(**A**)

(**B**) (**C**)

Figure 12. Webpage for teaching the angle-based measurement. (**A**) Entrance for angle-based measurement (symbol 1); (**B**) Aimed direction from left for measurement (symbol 2) and how to record the observations (symbol 3); (**C**) Teaching video for setting up the instrument with VR mode.

After clicking on the entrance for elevation-based (or called leveling) measurement, users have two routes (symbols 2 and 3) to choose from in Figure 11B. Additionally, a tutorial video (symbol 4) on instrument setup for elevation-based measurement is available at Station 11-3 (Figure 11C). Figure 12b depicts the webpage shown when users click on the entrance for angle-based measurement instructions. This page not only instructs users on how to record surveying measurements (symbol 3), aim targets, and set the surveying direction (symbol 2), but also teaches them how to set up the measuring instrument in a tutorial video (Figure 12C). By clicking the VR icon on the control bars, the image switches to VR mode. Users can then place their smartphone inside a portable VR viewer, connect a joystick to the system, and enjoy the VR tour (Figure 12C).

4.4. Questionnaire-Based Results

We received 62 questionnaire responses from the participating students; the statistical results are listed in Table 3. A higher score indicates stronger approval of the use of the designed platform and information technologies in teaching. Q1 and Q2 aimed to investigate the effectiveness of the designed platform in teaching as well as its content display, and Q3 and Q4 attempted to determine whether the use of information technologies helps strengthen learning interest. Most of the ratings for these four questions were > 4 (92% of the total responses), signifying positive feedback from respondents. All the students used the web-based platform; only some of the students tested the head-mounted equipment because of limited devices. The major problem using a head-mounted instrument is the layout for display. The developed platform was revised based on feedback, such as adjusting the size of the quick map and adding other targets for measurement.

Table 3. Results of the questionnaire.

(%)	Score				
	5	4	3	2	1
Q1	53	44	3	-	-
Q2	48	44	8	-	-
Q3	37	60	3	-	-
Q4	45	55	-	-	-

5. Discussion

5.1. Capturing Modes

To capture a high-quality image, one should observe the surrounding light, estimate the area of the surrounding environment, and adjust camera parameters (e.g., shutter, aperture, ISO value, and bracketing level) accordingly. By increasing the rate of overlap in the panoramic instrument, we expanded the overlapping area of adjacent photos, increased the number of tie points, and reduced the RMSEs, thereby enhancing the success rate of image matching.

In case of undesirable matching results based on GigaPan capturing, the problem must be identified in the photos. If overexposure or underexposure is confirmed, the aforementioned bracketing and high-dynamic range mode can be selected to overcome difficulties in feature point extraction caused by exposure-related problems. If the scene in question has few feature points, we recommend taking a new photo to increase the rate of overlap. Therefore, a larger overlapping area can be obtained with corresponding increases in the numbers of feature and tie points.

For example, Figure 13 exhibits a stitched panorama. Table 4 lists the capturing time required and post stitching RMSEs at different rates of overlap. When the targets are far from the viewer's perspective in a scene, the rate of overlap contributes less to the stitching quality. However, when some of the targets are far from the viewer's perspective and others are nearer, the rate of overlap can positively influence the stitching quality. Furthermore, the rate of overlap is directly proportional to capturing time.

Figure 13. Cases for comparing different overlap ratios and RMSEs after image stitching. (**A**) Case 1; (**B**) Case 2.

Table 4. Statistics of Figure 13 for comparing different overlap ratios and root mean square errors (RMSEs).

Overlap	Capturing Time (Min)	RMSE (Pixel)	
		Case 1	Case 2
30%	~3	3.39	3.48
50%	~5	3.37	3.39
70%	~9	3.35	3.29

In outdoor photography, objects often move, causing ghosting effects or mismatches during image stitching (e.g., characters or clouds moving and grass moving in the wind) in the case of GigaPan capturing. To address this problem, we recommend on-site investigation and planning in advance; for example, pictures could be taken when atmospheric pressure differences are small or when fewer people are in the area. Alternatively, the effects of object movement can be mitigated by adjusting the shooting sequence of the panoramic instrument. For instance, letting the camera shoot vertical photos before moving to the left or right (Column-left/right) can reduce the ghosting effects of characters. The shooting process can also be paused and continue after people pass by the target site. Row-down shooting after the camera takes a 360° horizontal shot also helps avoid problems related to clouds and sunlight; the taken photos can later be adjusted using postproduction software.

5.2. Contributions, Comparison and Limitations

There are two approaches for capturing a panoramic image in general. One is to use a 360-degree spherical camera. Another is to adopt a platform with a camera for capturing and stitching the images, such as GigaPan. The former can easily and simply create a panoramic image, but the resolution is not better than the latter, and there is no chance to improve the image quality. On the other hand, the platform-based method requires stable conditions to capture images. This is a trade-off between these operations for producing a panoramic image. High resolution [40] is necessary in this study to display the targets for measurement in the field. Thus, the GigaPan-based approach was chosen in this study case.

This study contributes to knowledge by (a) discussing panoramic photography and image stitching quality as well as providing suggestions on camera parameters and GigaPan settings, (b) proposing strategies for the production of panoramic images that are more operable and have a higher resolution than Google Street View, (c) using information technologies (i.e., virtual tour tools and VR) to develop an assistance platform for teaching, (d) applying the designed platform to an engineering course, and (e) assessing teaching effectiveness through a questionnaire survey.

Ekpar [41] proposed a framework for the creation, management, and deployment of interactive virtual tours with panoramic images. Based on this concept, many cases for displaying campuses [42,43], cathedrals [44], and culture heritage sites [45] were explored in the previous literature. This study not only visualized reality-based scenes using panorama- and virtual tour-based technologies, but also connected related documents for engineering education. Furthermore, the teaching site was also emphasized in this study. E-learning is a trend of education; it can help teachers reduce the load of teaching and further concentrate on the professional issues in subjects. Furthermore, this study provided a useful platform to help students who cannot come to the classroom because of special circumstances (e.g., COVID-19).

In terms of limitations, before students could use the platform designed in this study, instructors had to explain relevant theories and provide detailed instructions for equipment operations. If students lack basic understanding of surveying, their learning outcomes might not meet expectations. Additionally, we did not recruit a control group and thus, could not compare the learning results between students who used the designed platform and those who did not. We endeavor to test and verify the effect of the designed platform on learning outcomes in future research. The questionnaires and assessments for improving engineering education could be designed to put more emphasis on user testing and the responses from students, for example,

1. Sampling design for statistical tests:
   - Testing differences of final grades between student groups with the IT (Information Technologies)-based method and without the IT-based method.
   - Grouping samples by the background of sampled students.
2. Asking more aspects for comprehensive assessment:
   - "Would you recommend this to your friends and colleagues?" followed by "What points do you recommend/not recommend?"
   - "How long did you take to complete the IT-based program?"
   - "Does the virtual tour seamlessly/comfortably guide you?"
   - "Does the virtual tour sufficiently represent the real world?"
3. Comparing and exploring the problems on IT-based and traditional learning.

6. Conclusions

To create a multimedia platform that assists students in a surveying practice course, we initially took multiple overlapping images and stitched them into panoramas; subsequently, we used information technologies including virtual tour tools and VR. A full-frame digital single-lens reflex camera with an ultra-wide-angle zoom lens was mounted on a GigaPan panoramic instrument to obtain a 360° horizontal field of vision. The effectiveness of said visualization and information technology application was verified through a questionnaire survey.

The research results indicated that the RMSEs of stitched images were mostly < 5 pixels, signifying favorable stitching quality. The designed platform also features elevation-based and angle-based measurement instructions as well as instrument setup tutorials and documents as supplementary materials. A total of 15 panorama stations were set up for students to navigate. Of the 62 survey respondents, more than 92% agreed that the information technology-based platform improved their engagement in the surveying practice course. Moreover, we discussed and explored the improvement of panoramic image quality (RMSEs), camera parameter settings, capturing modes, and panoramic image processing, as shown in Section 4.1 and Section 5.1. In the future, we plan to compare the learning outcomes in students who used the designed platform (experimental group) with those who did not (control group).

Author Contributions: Jhe-Syuan Lai conceived and designed this study; Jhe-Syuan Lai, Yu-Chi Peng, Min-Jhen Chang, and Jun-Yi Huang performed the experiments and analyzed the results; Jhe-Syuan Lai, Yu-Chi Peng, and Min-Jhen Chang wrote the paper. All authors have read and agreed to the published version of the manuscript.

Funding: This study was supported, in part, by the Ministry of Education in Taiwan under the Project B from June 2019 to February 2021 for New Engineering Education Method Experiment and Construction.

Acknowledgments: The authors would like to thank Fuan Tsai and Yu-Ching Liu in the Center for Space and Remote Sensing Research, National Central University, Taiwan, for supporting the software and panoramic instrument operation.

Conflicts of Interest: The authors declare that they have no conflict of interest.

References

1. AIR-360. Available online: https://www.youtube.com/watch?v=W3OWKEtVtUY (accessed on 30 September 2020).
2. Super 720. Available online: https://super720.com/?page_id=8603 (accessed on 30 September 2020).
3. Lorek, D.; Horbinski, T. Interactive web-map of the European freeway junction A1/A4 development with the use of archival cartographic source. *ISPRS Int. J. GeoInf.* **2020**, *9*, 438. [CrossRef]

4. Cybulski, P.; Horbinski, T. User experience in using graphical user interfaces of web maps. *ISPRS Int. J. GeoInf.* **2020**, *9*, 412. [CrossRef]
5. Kato, Y.; Yamamoto, K. A sightseeing spot recommendation system that takes into account the visiting frequency of users. *ISPRS Int. J. GeoInf.* **2020**, *9*, 411. [CrossRef]
6. Wielebski, L.; Medynska-Gulij, B.; Halik, L.; Dickmann, F. Time, spatial, and descriptive features of pedestrian tracks on set of visualizations. *ISPRS Int. J. GeoInf.* **2020**, *9*, 348. [CrossRef]
7. Medynska-Gulij, B.; Wielebski, L.; Halik, L.; Smaczynski, M. Complexity level of people gathering presentation on an animated map–objective effectiveness versus expert opinion. *ISPRS Int. J. GeoInf.* **2020**, *9*, 117. [CrossRef]
8. Lu, C.-C. A Study on Computer-Aided Instruction for Engineering Surveying Practice. Master's Thesis, National Taiwan University, Taipei, Taiwan, 2009.
9. Panos. Available online: http://www.panosensing.com.tw/2018/03/28/faq6/ (accessed on 30 September 2020).
10. Zheng, J.; Zhang, Z.; Tao, Q.; Shen, K.; Wang, Y. An accurate multi-row panorama generation using multi-point joint stitching. *IEEE Access* **2018**, *6*, 27827–27839. [CrossRef]
11. Chen, P.-Y.; Tseng, Y.-T. Automatic conjugate point searching of spherical panorama images. *J. Photogram. Remote Sens.* **2019**, *24*, 147–159.
12. Teo, T.-A.; Chang, C.-Y. The Generation of 3D point clouds from spherical and cylindrical panorama images. *J. Photogram. Remote Sens.* **2018**, *23*, 273–284.
13. Laliberte, A.S.; Winters, C.; Rango, A. A Procedure for orthrectification of sub-decimeter resolution image obtained with an unmanned aerial vehicle (UAV). In Proceedings of the ASPRS Annual Conference, Portland, OR, USA, 28 April–2 May 2008.
14. Wrozynski, R.; Pyszny, K.; Sojka, M. Quantitative landscape assessment using LiDAR and rendered 360° panoramic images. *Remote Sens.* **2020**, *12*, 386. [CrossRef]
15. Huang, T.C.; Tseng, Y.-H. Indoor Positioning and navigation based on control spherical panoramic images. *J. Photogram. Remote Sens.* **2017**, *22*, 105–115.
16. Koehl, M.; Brigand, N. Combination of virtual tours, 3D model and digital data in a 3D archaeological knowledge and information system. In Proceedings of the XXII ISPRS Congress, Melbourne, Australia, 25 August–1 September 2012.
17. Lee, I.-C.; Tsai, F. Applications of panoramic images: From 720° panorama to interior 3D models of augmented reality. In Proceedings of the Indoor-Outdoor Seamless Modelling, Mapping and Navigation, Tokyo, Japan, 21–22 May 2015.
18. Quesada, E.; Harman, J. A Step Further in rock art digital enhancement. DStretch on Gigapixel imaging. *Digit. Appl. Archaeol. Cult. Heritage* **2019**, *13*, e00098. [CrossRef]
19. Bakre, N.; Deshmukh, A.; Sapaliga, P.; Doulatramani, Y. Campus virtual tour. *IJARCET* **2017**, *6*, 444–448.
20. Dennis, J.R.; Kansky, R.J. Electronic slices of reality: The instructional role of computerized simulations. In *Instructional Computing: An Acting Guide for Educators*; Foresman: Glenview, IL, USA, 1984; 256p.
21. Burdea, G. Virtual reality systems and applications. In *Electro/93 International Conference Record*; Edison: Middlesex, NJ, USA, 1993; 504p.
22. Gupta, R.; Lam, B.; Hong, J.-Y.; Ong, Z.-T.; Gan, W.-S.; Chong, S.H.; Feng, J. 3D audio VR/AR capture and reproduction setup for Auralization of soundscapes. In Proceedings of the 24th International Congress on Sound and Vibration, London, UK, 23–27 July 2017.
23. Shambour, Q.; Fraihat, S.; Hourani, M. The implementation of mobile technologies in high education: A mobile application for university course advising. *J. Internet Technol.* **2018**, *19*, 1327–1337.
24. Wu, Z.-H.; Zhu, Z.-T.; Chang, M. Issues of designing educational multimedia metadata set based on educational features and needs. *J. Internet Technol.* **2011**, *12*, 685–698.
25. Hsieh, M.-C.; Chen, S.-H. Intelligence augmented reality tutoring system for Mathematics teaching and learning. *J. Internet Technol.* **2019**, *20*, 1673–1681.
26. Lin, C.-L.; Chiang, J.K. Using 6E model in STEAM teaching activities to improve university students' learning satisfaction: A case of development seniors IoT smart cane creative design. *J. Internet Technol.* **2019**, *20*, 2109–2116.
27. Lee, J. Effectiveness of computer-based instruction simulation: A meta-analysis. *Int. J. Instruct. Media* **1999**, *26*, 71–85.

28. Brenton, H.; Hernandez, J.; Bello, F.; Strutton, P.; Purkayastha, S.; Firth, T.; Darzi, A. Using multimedia and web3D to enhance anatomyteaching. *Comput. Educ.* **2007**, *49*, 32–53. [CrossRef]
29. Vlahakis, V.; Ioannidis, N.; Karigiannis, J.; Tsotros, M.; Gounaris, M.; Stricker, D.; Gleue, T.; Daehne, P.; Almeida, L. Archeoguide: An augmented reality guide for archaeological sites. *IEEE Comput. Graph. Appl.* **2002**, *22*, 52–60. [CrossRef]
30. ART EMPEROR. Available online: https://www.eettaiwan.com/news/article/20161226NT22-MIC-Top-10-technology-trends-for-2017 (accessed on 30 September 2020).
31. Chao, T.-F. The Study of Virtual Reality Application in Education Learning—An Example on VR Scientific Experiment. Master's Thesis, National Kaohsiung First University of Science and Technology, Kaohsiung, Taiwan, 2009.
32. Chang, S.-Y. Research on the Effects of English Teaching with 3D Panoramic Computer Skill Application. Master's Thesis, National University of Tainan, Tainan, Taiwan, 2018.
33. Liao, Y.-C. Research on Using Virtual Reality to Enhance Vocational Students' Learning Performance and Motivation. Master's Thesis, National Taiwan University of Science and Technology, Taipei, Taiwan, 2019.
34. Lowe, D.G. Object recognition from local scale-invariant features. In Proceedings of the 7th IEEE International Conference on Computer Vision, Kerkyra, Greece, 20–27 September 1999.
35. Wang, G.; Wu, Q.M.J.; Zhang, W. Kruppa equation based camera calibration from homography induced by remote plane. *Pattern Recogn. Lett.* **2008**, *29*, 2137–2144. [CrossRef]
36. Sakharkar, V.S.; Gupta, S.R. Image stitching techniques—An overview. *Int. J. Comput. Sci. Appl.* **2013**, *6*, 324–330.
37. Zeng, B.; Huang, Q.; Saddik, A.E.; Li, H.; Jiang, S.; Fan, X. Advances in multimedia information processing. In Proceedings of the 19th Pacific-Rim Conference on Multimedia, Hefei, China, 21–22 September 2018.
38. Busch, D.D. *Mastering Digital SLR Photography*; Course Technology: Boston, MA, USA, 2005; 254p.
39. Wang, Y.-C.; Shyu, M.J. A study of brightness correction method for different exposure in rebuilding scene luminance. *J. CAGST* **2015**, 271–284.
40. Li, D.; Xiao, B.J.; Xia, J.Y. High-resolution full frame photography of EAST to realize immersive panorama display. *Fusion Eng. Des.* **2020**, *155*, 111545. [CrossRef]
41. Ekpar, F.E. A framework for interactive virtual tours. *EJECE* **2019**, 3. [CrossRef]
42. Perdana, D.; Irawan, A.I.; Munadi, R. Implementation of web based campus virtual tour for introducing Telkom University building. *IJJSSST* **2019**, *20*, 1–6. [CrossRef]
43. Suwarno, N.P.M. Virtual campus tour (student perception of university virtual environment). *J. Crit. Rev.* **2020**, *7*, 4964–4969.
44. Walmsley, A.P.; Kersten, T.P. The imperial cathedral in Konigslutter (Germany) as an immersive experience in virtual reality with integrated 360° panoramic photography. *Appl. Sci.* **2020**, *10*, 1517. [CrossRef]
45. Mah, O.B.P.; Yan, Y.; Tan, J.S.Y.; Tan, Y.-X.; Tay, G.Q.Y.; Chiam, D.J.; Wang, Y.-C.; Dean, K.; Feng, C.-C. Generating a virtual tour for the preservation of the (in)tangible cultural heritage of Tampines Chinese Temple in Singapore. *J. Cult. Herit.* **2019**, *39*, 202–211. [CrossRef]

Publisher's Note: MDPI stays neutral with regard to jurisdictional claims in published maps and institutional affiliations.

International Journal of
Geo-Information

Article

Experts and Gamers on Immersion into Reconstructed Strongholds

Beata Medyńska-Gulij * and Krzysztof Zagata

Department of Cartography and Geomatics, Faculty of Geographical and Geological Sciences, Adam Mickiewicz University, 61-712 Poznań, Poland; krzysztof.zagata@amu.edu.pl

* Correspondence: bmg@amu.edu.pl

Received: 30 August 2020; Accepted: 23 October 2020; Published: 30 October 2020

Abstract: In this study, we have touched upon a problem in evaluating the method of immersion in specific historico-geographical virtual space constructed on the basis of traditional cartographic and graphic materials. We have obtained opinions from two groups of users on the perception of cultural objects reconstructed in a virtual reality previously unknown to them. To achieve our objective and answer the questions, we have adopted four main stages of research: to pinpoint concepts adopted by researchers by discussing two types of approach, to create a virtual reality application according to the scheme based on knowledge from analog sources and digital actions in several workspaces, to prepare and conduct a survey among experts and gamers, and to graphically juxtapose the results of the survey. The evaluation by experts in medieval strongholds and serious story game users of the specific ways of immersion in the VR of reconstructed buildings in the current area provides researchers with an extended view of its effectiveness and attractiveness as well as with suggestions for further design processes.

Keywords: evaluation of cartographic multimedia; expert opinion; gamer opinion; immersion way; graphical enrichment; medieval stronghold; virtual reality; design process; historico-geographical space; cultural heritage

1. Introduction

In this study, we have touched upon a problem in evaluating the method of immersion in specific historico-geographical virtual space constructed on the basis of traditional cartographic and graphic materials. Appropriate evaluation of a virtual reality application with comments from experts and gamers can help in determining its effectiveness and attractiveness as well as in the process of design. Designing spatial visualizations with reconstructions of historical objects in the virtual reality (VR) system on the basis of analog sources is another issue touched upon in this article [1,2]. The new media allow one to create an immersive virtual reality application for non-existing objects by integrating a virtual environment with traditional specialist graphics, frequently saved only as academic black-and-white sketches [3]. "Presence" in the context of VR is defined as a sense of being in the virtual world [4]. The user of story games is "subjectively" present in the virtual world thanks to the employment of the illusion of 3D movement inside a realistic 2D+ graphic [5]. In turn, "immersion" denotes an objective level of sensory fidelity provided by a VR system [6].

Historical objects, crucial for the country's history, have been frequently non-existent for centuries, and various sources need to be used to reconstruct them [7,8]. Schematic perspective drawings of buildings, horizontal and vertical maps, as well as situational maps made by scientists to present the hypotheses of the original medieval state constitute significant sources of information. On the other hand, there are digital 3D models of the current state of ruins and traces in relief obtained from TLS (Terrestrial Laser Scanning) and UAV (Unmanned Aerial Vehicle) photos in the form of raw data [9].

The process of integration of all the materials obtained in order to create a VR application can be ordered according to data types and formats, adjusting them to be then processed in suitable programming workspace, which is, again, carried out according to the established order of technological work in several workspaces.

The evaluation of the way space is presented and perceived by users constitutes a significant factor in establishing principles of good design in terms of spatial graphics, interactive maps and other products of multimedia cartography [10]. Creators of geographical space visualizations are searching for suitable ways of evaluation by different user types to be able to determine effectiveness, attractiveness and informativeness of the medium employed [11]. Testing products of multimedia cartography, such as atlases, by users is based on user-centered design principles as well as on principles of map design worked out by theorists and practitioners [12,13].

The so-called "public users" are usually invited to participate in the evaluation of cartographic visualizations [14]. These are frequently students as they constitute a group easily available to scientists and relatively homogenous, especially if they major in the same subject. The large number of students makes it possible to divide them into a few teams so that respondents from several teams can evaluate different versions of mapping techniques that present the same spatial data [15]. Surveys evaluating how specific fragments of urban space are perceived when on a tablet screen, e.g., with cartographic signs applied in the extended reality system, are carried out among random passers-by on the street much more seldomly [16]. The number of respondents in studies devoted to cartographic visualizations ranges between 10 and 30 for the evaluation of a single version, which frequently leads to over 200 respondents for several different versions [17].

The effectiveness of multimedia maps is tested through tasks performed by users on the computer or smartphone screen [18]. The objective effectiveness of visualizations is checked in online surveys by the user having to select the correct answer out of several possible options [16]. The attractiveness of presentations is verified by respondents providing their subjective opinions that not always coincide with objective effectiveness of the spatial information [19]. Surveys rarely include open questions or provide opportunities to comment on questions as that would complicate the entire analysis and make drawing clear-cut conclusions much more difficult.

The answers and marks provided by respondents are then placed in analytical tables, charts and diagrams to demonstrate specific features. To facilitate the process of drawing conclusions and to reveal the relations between users' evaluation, cartographic research employs forms of graphic enhancement, such as the use of color and other graphic variables, as well as graphically supported evaluation of mapping techniques in the form of advanced graphic forms [16,19,20].

The pragmatic usage of graphic elements resulted in the creation of a simple image that could be read intuitively, with high aesthetic appeal, and this has been described best as "Good design simply 'looks' right—it is simple (clear and uncomplicated)" [12]. The knowledge of cartographic design is passed on in the form of more or less precisely described rules, which apply to both simple and complex aspects of design process [21]. When designing immersive visualizations, it is important to consult various perspectives, both with regard to accuracy and artistic presentation of space as well as technical parameters of the application. It seems reasonable to look for a way of formulating and presenting principles and guidelines that can be adopted by designers in their own creative process.

2. Aim and Questions

Evaluating the way of becoming immersed in the VR of reconstructed buildings in present-day state of the area by both experts in medieval strongholds and serious story game users has become the main objective of the research. Apart from the goal set above, we have also raised a series of detailed questions about the evaluation of effectiveness and attractiveness of the designed VR application and the potential that traditional sources have for creating one:

- How to use traditional cartographic and graphic sources to create immersive virtual reality for cultural objects that currently have only their stone remnants and well-preserved traces in relief left?
- How to conduct a process of combined actions on both analog materials and digital spatial data to create a mobile VR application on game engine?
- How to carry out a study on respondents in two different user groups to determine features of the medium in terms of: effectiveness, attractiveness, and informativeness?
- Will graphic elements (i.e., colors, graphic symbols) in tables make it easier to reveal links and will they allow one to capture differences and similarities in comments?
- How to use opinions of two different user groups in designing a VR application?

3. Research Area and Research Objects

In order to conduct a questionnaire among experts and gamers, we have selected the residence of the Piast dynasty's first rulers on a holm on Lake Lednica (Ostrów Lednicki) with perfectly preserved 10th century ruins of the palatium and the chapel of Duke Mieszko I (ca. 960–992) and his son, the first King of Poland, Bolesław "the Valiant" [22]. What was relevant to our research was the fact that the ramparts of the stronghold were well-marked in the landscape of the island. The stronghold was developing in several phases, however, we focused on the second half of the 10th century [23]. This state of the stronghold included the large palatial-ecclesiastical structure and the small church together, which were accompanied by wooden constructions of varying size. These two stone buildings have become the objects of our virtual reconstruction.

The palatium was built from stones mortared together in the pre-Romanesque style, it served as one of the Piast ruler's palaces, and retained the court chapel, a baptistry with the two baptismal fonts [24]. This allows us to presume that it was here in 966 that Mieszko I, and symbolically the entire Poland, was baptized. The stronghold's church was yet another stone building erected in the second half of the 10th century [25]. At that time, the rampart was covered with wooden palisades and was approximately 12 m high [23]. Remnants of the stronghold ramparts, depicted in Figure 1, which are approximately 3 m high, are covered with grass. Current area of the stronghold is covered with grass and has hard-surfaced trails for tourists.

Figure 1. *Cont.*

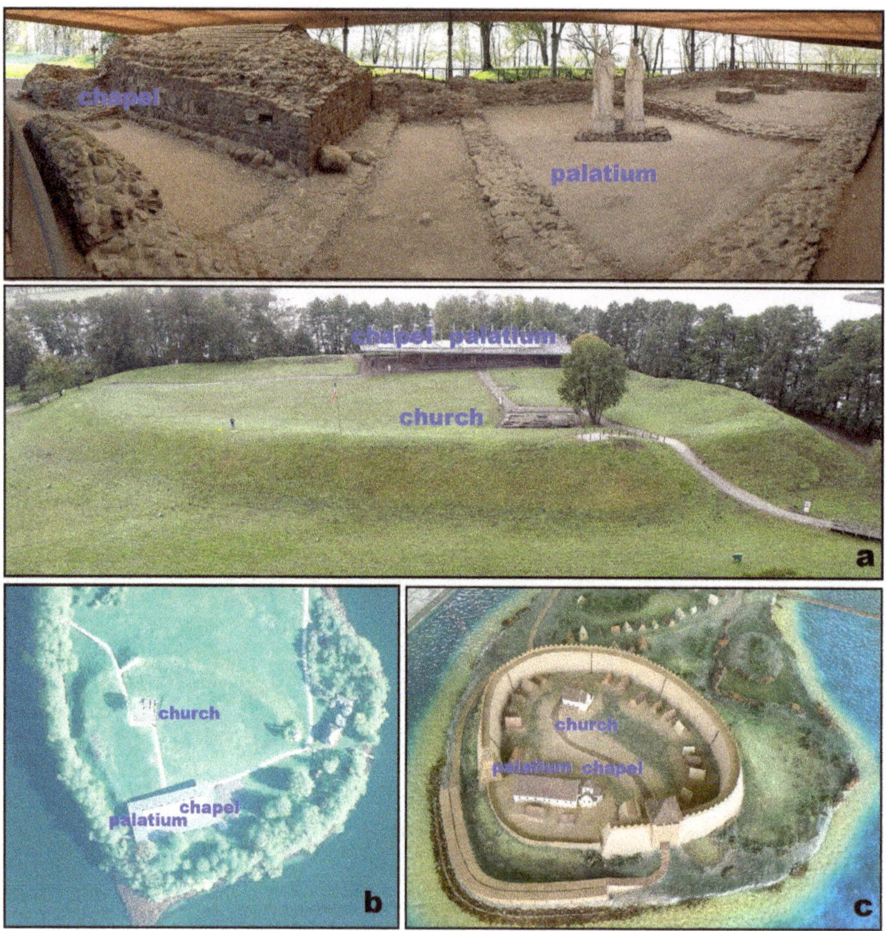

Figure 1. Stronghold in Ostrów Lednicki: (**a**,**b**) currently preserved remnants of the building's foundation and of the stronghold ramparts; (**c**) plaster model of the stronghold in the second half of the 10th century.

4. Methodology

To achieve our objective and answer the question, we adopted four main stages of research:

- to pinpoint concepts adopted by researchers by discussing two types of approach (Section 4.1.),
- to create a VR application according to the scheme based on knowledge from analog sources (Section 4.2, Figures 2–4),
- to prepare and conduct a survey among experts and gamers (Section 4.3, Figure 5),
- to graphically juxtapose the opinions and comments (Section 4.4, Figures 6–8).

4.1. Concept

Already, at the stage of concepts, two opposite approaches emerged. The first one was the approach of a cartographer-geographer, a frequenter of Ostrów Lednicki, that has been interested in medieval architecture for years, and the second one was the one of a cartographer-geomatics engineer interested in VR systems made on game engine and a gamer. Having discussed both approaches, we adopted a common concept, including the following initial assumptions:

- specificity of the multimedia presentation: virtual historico-geographical space connecting two distant moments in time: reconstructed palatium and stronghold church from the 10th century on current remnants of the rampart covered with grass,
- viewers of the virtual visualization: a possibly broad group of recipients, including both gamers and experts,
- medium: visual immersion in VR, supported by sound effects (adding artificial sounds: rustle of trees and the footsteps of the user),
- sources: various analog and digital materials, academic literature, LiDAR's official data, private UAV, and street view images,
- software and equipment: graphic and GIS programs; application worked out on the Unity game engine to fit VR goggles,
- technological process: data transformation and 3D modeling management in the geomatic process in several workspaces,
- respondents: experts on the stronghold in both practice and theory, as well as serious game users via a low-immersive screen-based (2D) presence who have never visited the stronghold in real life,
- geography of the virtual walk around the stronghold: possible exclusively along the designed trail with three viewing points,
- contents of the survey: the same issues/questions for two groups from the same viewing point with an opportunity to select one evaluation out of several versions and to make comments about all the issues,
- the way the survey was conducted: the survey was controlled; issues were read out without repeating; the reply without a time limit; during the survey the respondent was standing and wearing VR goggles; duration: around 20 min,
- presenting the results: the opinions were demonstrated with visual support, using graphical variables at the ordinal level and symbolic signs, and
- the results: formulating conclusions on effectiveness, attractiveness, and tips on designing similar VR applications.

4.2. Creating a VR Application

Adopting all the assumptions listed above, we started creating the application according to the following four main actions placed on scheme in Figure 2:

- obtaining materials and their initial classification,
- the analysis and selection of materials for 3D modeling and texture mapping,
- 2D texture designing, 3D building, and area modeling, and
- creating the VR application.

Obtaining materials and initial classification took place according to three data types: scientific analog stronghold reconstruction, visual documentation of the current state of the stronghold and official digital spatial data. The first types of sources included mainly perspective drawings and reconstruction plans from academic publications as well as a plaster model of the stronghold. Current visual documentation of the stronghold has been obtained by UAV and street view images. The employees of the museum have provided photographs of natural and artificial textures of walls and roofs. Having analyzed them, we have selected crucial materials for our research and considering their format, we have decided upon modeling in four workspaces.

Workspace is an area in the application (i.e., digital work environment dedicated to the specific application) that allows one to perform digital operations (framing, cut, rotation, adding points, georeferencing) for various data types (raster and vector) and formats (.png, .svg, .obj, .shp), operating on both desktop and mobile systems according to the interface and programming scripts. The first "3D Buildings Workspace" in our geomatic process developed manual 3D modeling in the architectural application SketchUp. In this operation, the medieval 10th century buildings of the palatium and the church were reconstructed with the use of the previously mentioned source materials (Figure 2). The reconstructed historical buildings were created as a result of the analysis of historical drawings, descriptive documentation and the plaster model on the one hand, and a digital model of the island on the other hand.

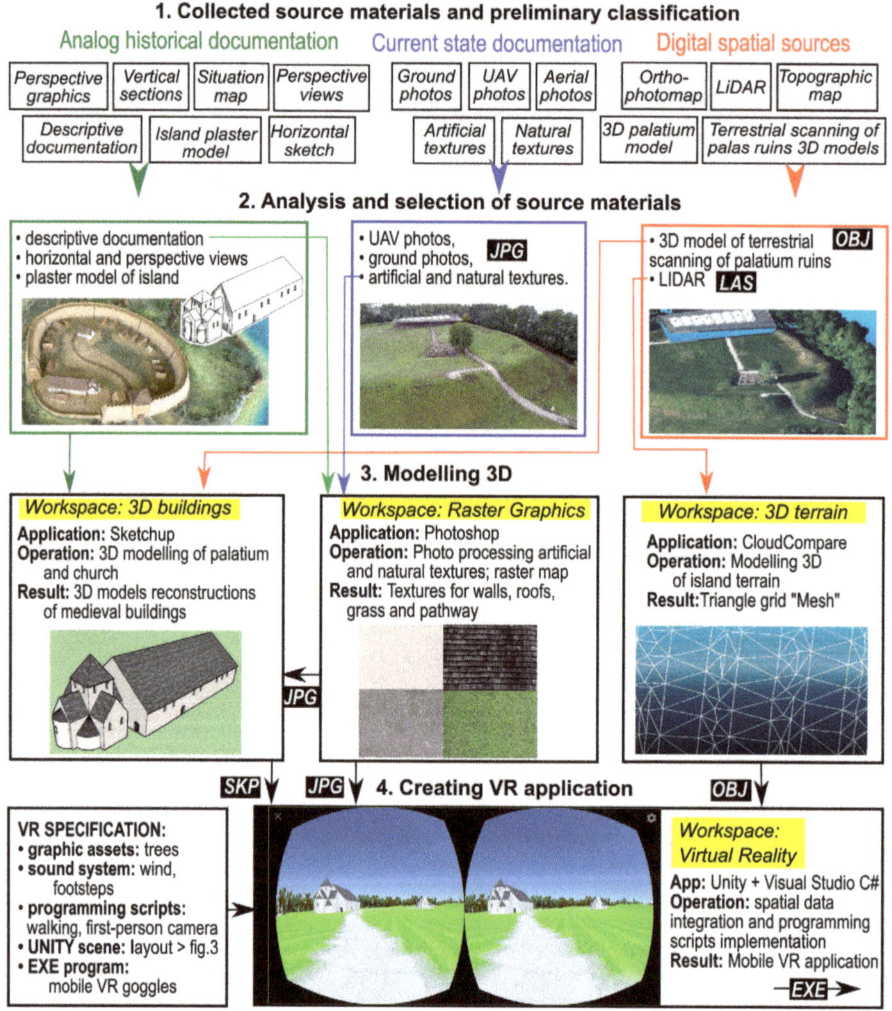

Figure 2. Scheme of creating the application according to four main actions and four workspaces.

In the second "Raster Graphics Workspace", photo processing of artificial and natural textures and raster maps took place. Adobe Photoshop was the application used for creating and preparing all textures. In this operation, some textures similar to the original medieval building materials, used for building the palatium with the chapel and the church, were designed. As a result, textures were developed for individual application elements, such as walls, roofs, grasses and pathway. In the third "3D Terrain Workspace", the 3D modeling of island terrain was the main operation. The point cloud from airborne laser scanning (LiDAR) constituted the basic source material. The triangle grid "Mesh" of Ostrów Lednicki, created with the CloudCompare application, was the final result.

In the last "Virtual Reality Workspace", spatial data integration and programming script implementation constituted the main operation (Figure 2). The Unity game engine was the application for this workspace. To develop the workspace, a separate specification of visualization parameters, such as the sound system (including the footsteps sound, the wind sound and the rustle of leaves), graphic assets (including trees), programming scripts (including the walking, first-person camera, VR controller, and light) and the Unity scene (Figure 3—layout of the geographical aspect of the virtual walk), was prepared. The EXE program, designed for mobile VR goggles (Figure 4), constitutes the final result of the VR application development process.

We assumed that each research participant will cover the same distance, stopping by the same viewing points, and will answer questions, standing by the last viewing point. Hence, before we prepared the survey, the designing cartographer, who had previously visited the stronghold, determined the starting point and the trails, as well as marked viewing points on the basis of the comparison of the 3D image seen in VR goggles with the real street view from the stronghold (Figure 3). We added rustle of trees and footsteps that are heard by the user during their virtual walk.

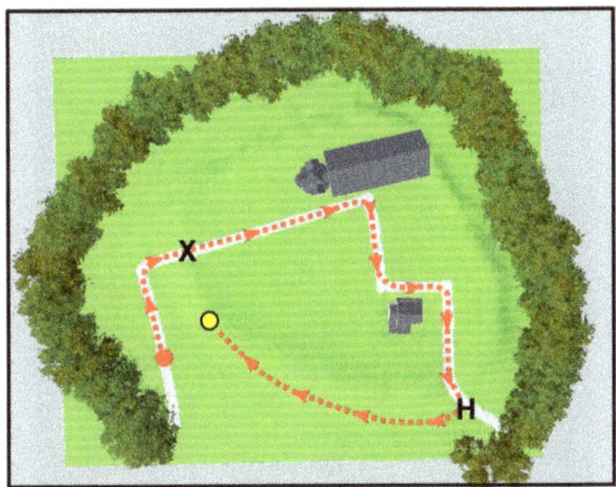

Figure 3. Layout of the virtual walk around the stronghold with the starting point (the red dot) and three viewing points (X, H, and the black dot).

4.3. Creating and Conducting a Questionnaire

According to the concept adopted, the survey was addressed to experts and gamers, representing different states of knowledge and skills, which was supposed to help obtain different opinions about the demonstrated way of becoming immersed in historio-geographical space including elements that were historically mutually exclusive.

The first group consisted of experts on the Middle Ages who derived their knowledge from the literature and their visits to Ostrów Lednicki. Nobody from that group was a gamer. The other group consisted of serious story game users via non-immersive screen-based presence, who had general knowledge of history from school and had never visited Ostrów Lednicki. Gamers declared to spend over 10 h per week, playing story games based on extended storylines. It was possible to invite 10 respondents (5 women and 5 men, aged roughly 35–60) to the first group; therefore, the second group also consisted of 10 members (men aged roughly 20–24).

We assumed that each participant would be constantly standing during the survey, basically right from the moment of putting on VR goggles and being handed the controller, then during "walking" down the trail and, finally, while completing the survey at the last viewing point. The difference in the course of the survey was mainly about the initial activity for the gamer, who received the printout of two photographs and was asked to take a look at them: the one with currently preserved remnants of the building's foundation and the area of the former stronghold in Ostrów Lednicki, and the one of a plaster model of the stronghold from the times of Mieszko I (Figure 1).

Before the participant put the goggles on, he/she was informed that, after putting the goggles on, he would be transferred to the virtual presentation of the stronghold in Ostrów Lednicki, in which the palatium with the chapel and the church from the times of Mieszko I had been reconstructed in the currently existing grass-covered area, along with the remnants of the rampart (Figure 4).

Having put the goggles on, the participant confirmed to see and hear everything correctly and to understand the functioning of the movement buttons placed on the controller that he was holding.

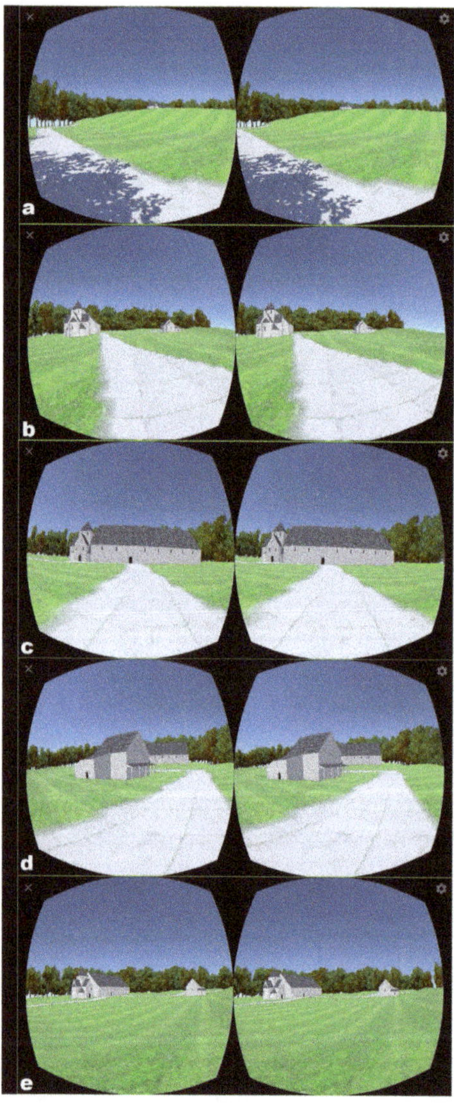

Figure 4. View in VR goggles during the virtual walk around the stronghold at the starting point (**a**), the first viewing point X (**b**), next to the palatium with the chapel (**c**), the second viewing point H (**d**), and from the last viewing point (**e**).

Then, the researcher conducting the survey informed the participant that he/she was located in Ostrów Lednicki, standing on the path before the grass-covered remnants of the rampart (Figures 3 and 4). The researcher asked the participant to stand by the black pale on the rampart and sharing his/her opinion on 5 main topics with 15 detailed issues concerning the entire visualization. With VR goggles on and without any time limits, the respondent answered each question separately and evaluated it. After reading out each question, the researcher would inform the respondent about the opportunity to share his/her own comments (Figure 5).

Figure 5. Researchers and an expert with VR goggles and with a controller in her right hand during the survey.

At the first main topic, the respondents would make the initial evaluation, and at the last, fifth main topic (fifteenth issue) they would provide their final evaluation. Providing evaluation concerning specific questions in three thematic groups was supposed to help obtain separate partial scores for the final evaluation.

Respondents were asked to share remarks (there were four possible evaluation types: very good, correctly, wrong, hard to define) on the following 5 main topics touching upon 15 issues:

1. The opportunity to see proportions of the palatium solid figures with the chapel and church in the state from the times of the first ruler of Poland compared to the current state,
2. Correctness of the reconstruction of the two pre-Romanesque buildings' elements: 2A: proportions of the building components, 2B: windows and portals, 2C: roofs and roofing tiles, 2D: plastered walls,
3. Functionality of this reconstruction method for two buildings in the current state of the stronghold in the following aspects: 3A: Obtaining knowledge of the features typical of buildings from the Mieszko I's stronghold by students, 3B: The method of studying historical features of buildings by experts, 3C: The understanding of links between the history of the beginning of the Polish state and the stronghold on the island, 3D: The opportunity to visualize two buildings on the remnants of foundations preserved to this day,
4. Parameters of immersion in VR used by researchers: 4A: movement exclusively along the trails and the rampart, 4B: Suggesting the viewing points for the observations of building features, 4C: Observation from the perspective of a pedestrian moving around the stronghold's area, 4D: Layout of trees with the animation of movement of the leaves, 4E: Hearing the sounds of wind and footsteps, and
5. General meaning of the demonstrated immersion method for virtual observation and interpretation of two reconstructed Romanesque buildings (derived from expertise) in the current state of the grass-covered rampart.

4.4. Graphically Juxtapose Opinions and Comments

According to our concept, we have presented the results of the survey, juxtaposed and with visual support. To facilitate the analysis and synthesis visually, three colors, associated with those of traffic lights, were employed for evaluation: green for "very good", yellow for "correct", red for "incorrect/wrong" and violet for "hard to define" (Figures 6 and 7). As empty boxes in tables were in colors symbolizing specific grades, symbols in black line were used to present the additional information in order not to conceal the colorful boxes. Simple emoticons for presenting three faces were used: "smiley face": positive feedback; "straight face": neutral feedback; and "frowning face": negative feedback. Additionally, the exclamation mark ("!") was placed in the table near the emoticon if, in his/her comment on the feature, the participant referred to knowledge from academic publications and/or his/her own experience from the visit to the stronghold. To facilitate the comparative analysis of the evaluation made by experts and gamers according to the group of issues and detailed topics, we prepared a comparative bar diagram in Figure 8, using analogical colors as in Figures 6 and 7.

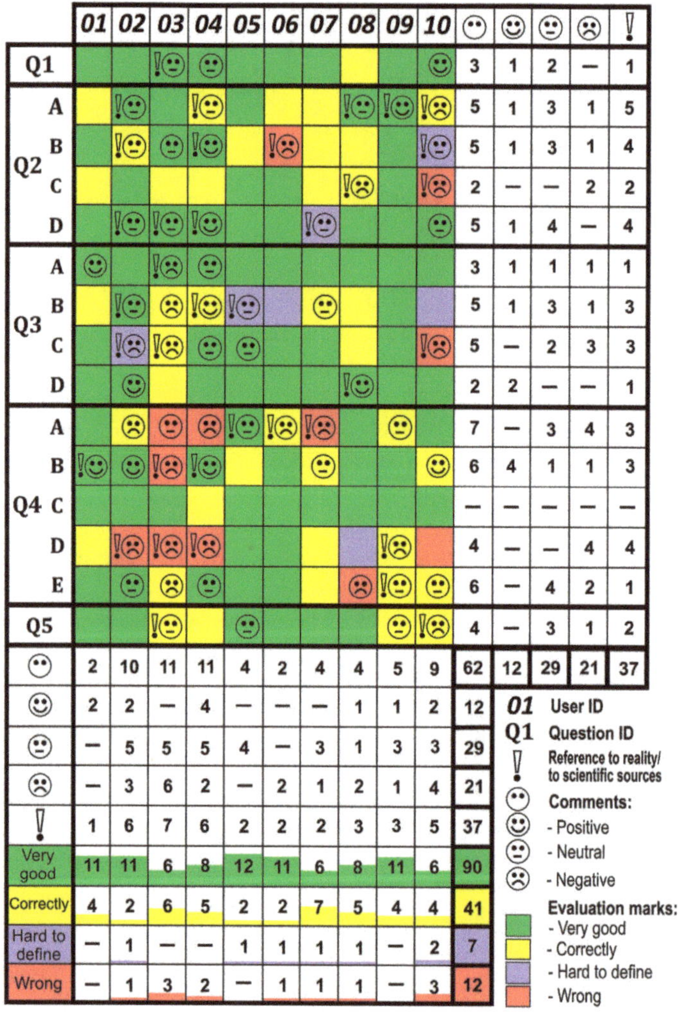

Figure 6. Evaluations and comments provided by ten experts.

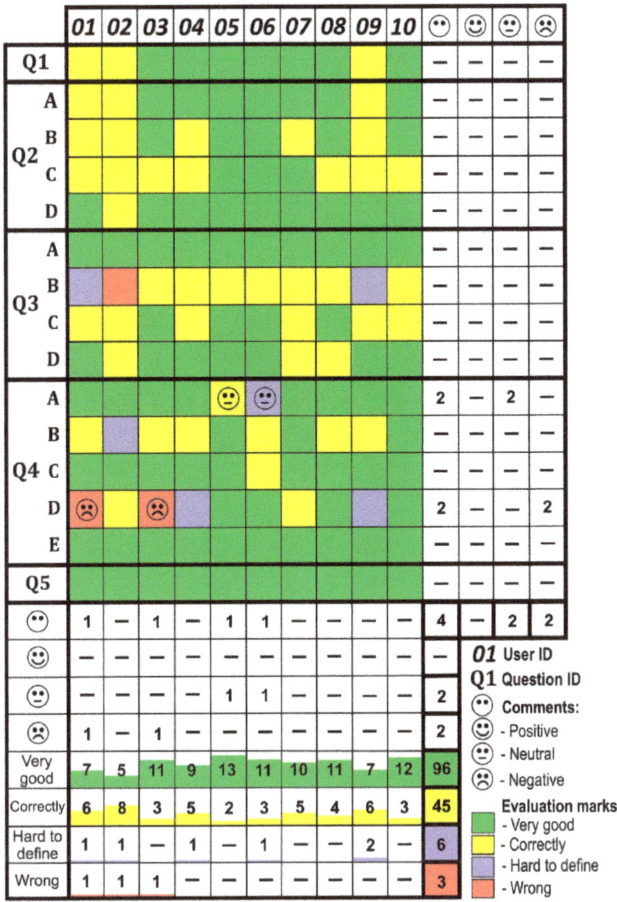

Figure 7. Evaluations and comments provided by ten gamers.

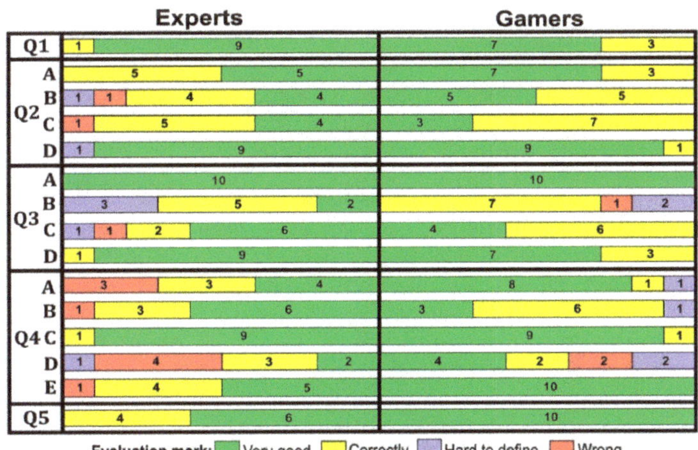

Figure 8. Comparative bar diagram: experts and gamers.

5. Results

After 20 questionnaires, it turned out that the time gamers needed to complete the experiment ranged between 12 and 19 min, with experts taking between 17 and 26 min, as they would frequently make comments. In the expert group, two women initially reported slight discomfort after putting their VR goggles on.

Figure 6 presents evaluations and comments provided by ten experts. Out of 150 evaluations, "very good" (60%) and "correctly" (27%) prevail. Each expert made a comment on selected answers to 15 questions at least twice. Two experts commented on 11 issues. The experts made 62 comments per 150 evaluated features. In 37 evaluations (60% comments), the experts referred to the knowledge derived from academic publications or their visits to the stronghold. Neutral comments definitely prevailed (47%), and the number of positive comments was the lowest (19%).

Only two experts answered to all the questions: "very good" and "correctly", and five of them used all four types of evaluation. Interestingly, only one negative answer was not supported with a comment, and, out of all 12 negative answers, 10 were supported with negative comments, and 1 comment was neutral. Surprisingly, in 7 cases, experts delivered negative comments on questions evaluated as "correctly" and, in one case, as "very good". Generally, experts would provide negative feedback along with negative evaluations, making references to expertise from academic publications or their knowledge of the stronghold from real life. The "hard to define" feedback was usually left without any comments. The majority of cases of positive evaluation were left without comments. Only 4C (Observation from the perspective of a pedestrian moving around the stronghold's area) was not commented upon by experts, receiving nine "very good" and one "correctly" answers. Other issues were commented on at least thrice, with 4A (Movement exclusively along the trails and the rampart) receiving the most comments. In general, each out of three topics received a similar number of comments.

Although topic 1 (The opportunity to see proportions of the reconstructed buildings) and topic 5, the last one (General meaning of the demonstrated immersion method), received mostly positive comments, the detailed questions from the topics: 2 (Correctness of the reconstruction), 3 (Functionality of this reconstruction method) and 4 (Parameters of the VR immersion) had diverse comments. 3D (The opportunity to visualize two buildings on the remnants of foundations preserved to this day: nine "very good" answers and one "correctly" answer) and 4C (Observation from the perspective of a pedestrian moving around the stronghold's area) were the exceptions. 4D (Artificial trees with the animation of movement of leaves) received the lowest evaluation. Topic 2 had the highest number of references to expertise and individual experience, which is understandable as it referred to correctness of reconstruction of some elements of two Romanesque buildings. Out of 62 comments made by experts, the following were made repeatedly:

- 1: in the literature, there are different hypotheses of reconstruction,
- 2A: consistency with the plaster model; correctness of reconstruction according to the academic versions remembered; the palatium was a bit taller; the roof cubature should be larger,
- 2B: portal should be rather in a different place; portals should be pre-Romanesque; coherent convention of the pre-Romanesque style should be maintained,
- 2C: shingle roofs rather than lead roofs,
- 2D: there are two parallel hypotheses that walls were either plastered or non-plastered,
- 3A: too simplified; very suggestive visualization to students,
- 3B: too few reconstructed elements for experts to use; no view of the lake; too few details,
- 3C: trees disrupted the view; trees are artificial, trees should be slender as there are tall alders growing there,
- 3D: consistency with other reconstructions,
- 4A: 5 comments expressing the urge to freely walk around the stronghold; 3 comments expressing the urge to walk round the palatium,

- 4B: 4 comments about accurate layout of the viewing points; 3 comments on the very well located last viewing point by the pale,
- 4D: better remove the trees as they were not there in the 10th century,
- 4E: the rustle of trees reflects the real sound; there is no twitter of birds, and
- 5: the structure of plaster on the walls should be changed; other elements of the stronghold can be added; time conflict resulting from the combination of the 10th and the 21st century; the rampart was taller in relation to buildings in the 10th century.

Figure 7 demonstrates the evaluation by gamers with the prevailing green (64%) and yellow (30%) boxes, comprising 94% of boxes in total. In general, all issues received high evaluation, 3B (The method of studying historical features of buildings by experts) and 4D (Artificial trees with the animation of movement of the leaves) being the exception. Three questions, 3A (Obtaining knowledge of the features typical of buildings from the Mieszko I's stronghold by students), 4E (Hearing the sounds of wind and footsteps) and 5 (General meaning of the demonstrated immersion method), received 10 "very good" evaluations. Four gamers made single comments on the fourth main topic (Parameters of immersion in VR used by researchers) in which they were highly experienced due to their "presence" in the world of computer games. Two negative comments concerned 4D (Layout of trees with the animation of movement of the leaves), as in the opinion of gamers, such artificial movement of leaves on trees disrupted the perception of the 3D view. The other two comments were made to express the willingness of gamers to freely move around the stronghold.

Comparing the way that the opinions were expressed by 10 experts and 10 gamers, one can draw a conclusion that the group of gamers is more homogenous as the same comments are made by most gamers on the same issues. On the other hand, a diverse combination of colors and faces in boxes denoting the same issues indicates individualized personalities of experts. All the respondents used mostly the "very good" or "correctly" evaluation, even though gamers used them slightly more often (94%) than experts (87%). Such high evaluation was generally connected with a positive attitude and interest in new technologies since no respondent had previously had the opportunity to become immersed in VR in goggles.

The "hard to define" answers constitute only 4% of all answers in both groups of respondents, which proves that all of them were highly committed and avoided ambiguous answers. We assumed that gamers would avoid unambiguous evaluation of issues addressed to historians. However, this assumption turned out to be wrong as gamers would willingly comment on issues connected with the knowledge of medieval history, which proves that one is willing to make judgements concerning the topics that one is no expert on, lacking expertise and experience in the field. On the other hand, experts on the history of the stronghold that were not familiar with the world of screen games would express their opinions on parameters of the application that rather required the experience of gamers. Skepticism, expressed more by gamers rather than experts, about the recognition of high functionality of the application in terms of studying historical features of buildings by experts, came as a surprise.

The marks to the third main topic referred to possibly open, informative and didactic functionality of the application to the wide audience. What is striking is that two issues (3A with 100% "very good" and 4C) received similarly positive feedback. The greatest discrepancy in marks occurred for 4A, as experts mostly reported the willingness to freely move around the palatium, whereas gamers approved of the designed way of moving around. It seems that the first-hand experience of experts resulted in their request to be able to move around the most significant building in the stronghold (Figure 8).

Experts were the most critical of the issues about VR parameters, referring to the real state of the area, e.g., criticizing the wrong shape of the artificial trees as they remembered that alders growing there were taller. A few gamers criticized the excessive movement of leaves on trees as they were used to more subtle movements of leaves in games. Interestingly, experts and gamers provided the lowest evaluation to the layout of trees with the animation of the leave movement.

Experts delivered more diverse evaluation with the higher number of comments, in most cases referring to their expertise and first-hand experience in terms of the second main topic. Gamers, in turn, evaluated these issues in a positive way and their lack of criticism, combined with lack of expertise, may be linked to their frequent "presence" in the world of games with buildings only imitating the original medieval architecture.

6. Discussion and Conclusions

Generally, one can conclude that the evaluation by experts in medieval strongholds and serious screen game users of the specific way of immersion in the VR of reconstructed buildings in the current area provided researchers with the extended view on its effectiveness and attractiveness, as well as with suggestions on further design process.

Analog cartographic and graphic materials constitute a solid foundation for creating specific immersive VR for cultural objects that have only remnants and a marked rampart in the relief left (Figure 1) [26]. Reliable reconstructions are saved as perspective drawings in academic publications or in the form of plaster models. Selecting the reconstruction version out of several clashing historical versions, as well as proper analysis and selection of the extensive source material, is what constitutes a problem in the concept of design. Analog and digital data management can be conducted in the orderly way thanks to parallel actions in several workspaces. Unfortunately, despite the assumption to use the lowest possible number of digital workspaces, four workspaces had to be used (Figure 2). A possible simple implementation has become crucial for creating a mobile VR application based on the spatial data game engine, geometry of solid figures and their graphic looks resulting from parallel modeling in three workspaces: 2D raster graphics, 3D buildings, and 3D relief [25].

Working out the same tasks and groups of questions for two different groups of users of the same virtual presentation of the historico-geographical space is essential to make a comparative analysis and draw coherent conclusions. Limiting the choice to only four answers, out of which the right one was supposed to be selected, allowed respondents to focus on delivering the answer and making the comments. It turned out to be relevant as each respondent asked for repeating the issues at least once, which probably resulted from the fact that it was the first time he/she was in the virtual stronghold with VR goggles on and listening to the questions of the researcher at the same time.

The fact that no gamer referred to photographs of the current state of the stronghold and the plastic model watched before putting their goggles on came as a surprise. On the other hand, experts would willingly refer to their expertise, the plastic model, and real-life experience in their comments. It may prove that only deeply engrained knowledge, supported by real-life experience, evokes appropriate reaction, whereas a short presentation of photographs without further explanation will not be remembered and will fail to provoke any comments by gamers, even though all gamers had general knowledge of the beginning of the Polish state and early styles in architecture.

A controversially designed "walking" down currently existing trails and grass-covered remnants of the rampart near the non-existing palatium and church that were 1000 years old was commented upon as a strange clash only by two experts. Thus, it can be thus adopted that users generally do not mind such a way of immersion.

On the basis of the evaluation obtained, one can draw a conclusion that the knowledge of the pre-Romanesque architecture and private first-hand experience had a positive impact on one designer, influencing the way the trail and viewing points were marked. However, experts, who had visited the stronghold more often, demanded that the twitter of birds should be added. It can be linked to synesthesia of the experience, just as it was the case for sounds in the process of cartographic communication [27–29]. A frequent use of games by another designer resulted in adding artificial footsteps that received a positive feedback from both user groups.

Graphic elements that enriched evaluation and comments allowed one to visually analyze and synthesize facts and capture links between them (Figures 6–8). The adaptation of the "Chernoff faces" method [12,21], known in traditional cartography, along with the employment of the exclamation mark, enabled one to link evaluation to comments, e.g., despite neutral and critical comments experts indicated positive features, and the majority of cases of negative evaluation were supported by negative comments and references to real-life experience. The issue of whether or not the employment of more types of faces would make the analysis more accurate remains open.

We are left with the issue of how to treat opinions from two different user groups. Both groups consisted of 10 members, and it was impossible to make them more numerous as it was necessary to formulate the criteria for experts with the real-life experience in a highly strict way. Moreover, we wanted to obtain opinions about the VR application that would confront the real world with the virtual one and juxtapose the comments provided by gamers who were transferred to the virtual reality. Hence, we incline toward making a complementary use of opinions provided by two user groups in the process of designing such applications. In our opinion, the choice of suggestions received from experts and gamers should be consistent with the concept of the multimedia presentation designers.

We are aware that there is no simple recipe for designing a good historical-geographic immersive application. On the one hand, applying the principles of designing cartographic signs facilitates the design process, but on the other hand, it limits the graphic creativity of the designer [30]. Therefore, the method of a special compilation of evaluation and comments in several design aspects proposed here is rather aimed at stimulating the design process in the directions preferred by designers of next applications.

Generally speaking, both groups demonstrated great openness to new opportunities to present the stronghold in VR, which proves high attractiveness of such medium. Informativeness in the historical and cultural context of the place, unique in terms of the birth of the Polish state, has also been evaluated very highly. However, defining the effectiveness of this medium in terms of visualizing historico-geographical space for public users requires further research that would also include other groups of respondents.

Author Contributions: Conceptualization, Beata Medyńska-Gulij; Methodology, Beata Medyńska-Gulij; Software, Beata Medyńska-Gulij and Krzysztof Zagata; Validation, Beata Medyńska-Gulij and Krzysztof Zagata; Formal Analysis, Beata Medyńska-Gulij and Krzysztof Zagata; Resources, Beata Medyńska-Gulij and Krzysztof Zagata; Data Curation, Beata Medyńska-Gulij and Krzysztof Zagata; Writing—Original Draft Preparation, Beata Medyńska-Gulij; All authors have read and agreed to the published version of the manuscript.

Funding: This paper is the result of research on visualization methods carried out within statutory research in the Department of Cartography and Geomatics, Faculty of Geographical and Geological Sciences, Adam Mickiewicz University in Poznań, in Poland.

Conflicts of Interest: The authors declare no conflict of interest.

References

1. Ramsey, E. Virtual Wolverhampton: Recreating the historic city in virtual reality. *Archnet-IJAR: Int. J. Arch. Res.* **2017**, *11*, 42–57. [CrossRef]
2. Edler, D. Where Spatial Visualization Meets Landscape Research and "Pinballology": Examples of Landscape Construction in Pinball Games. *KN-J. Cartogr. Geogr. Inf.* **2020**, *70*, 55–69. [CrossRef]
3. Walmsley, A.P.; Kersten, T.P. The Imperial Cathedral in Königslutter (Germany) as an Immersive Experience in Virtual Reality with Integrated 360° Panoramic Photography. *Appl. Sci.* **2020**, *10*, 1517. [CrossRef]
4. Kim, G.; Biocca, F. International Conference on Virtual, Augmented and Mixed Reality. In *Immersion Virtual Reality Can Increase Exercise Motivation and Physical Performance*; Springer: Cham, Germany, 2008; pp. 94–102.
5. Roettl, J.; Terlutter, R. The same video game in 2D, 3D or virtual reality—How does technology impact game evaluation and brand placements? *PLoS ONE* **2018**, *13*, e0200724. [CrossRef] [PubMed]
6. Slater, M. Immersion and the illusion of presence in virtual reality. *Br. J. Psychol.* **2018**, *109*, 431–433. [CrossRef] [PubMed]
7. Lorek, D. Multimedia integration of cartographic source materials for researching and presenting phenomena from economic history. *Geodesy Cartogr.* **2016**, *65*, 271–282. [CrossRef]

8. Medyńska-Gulij, B.; Lorek, D.; Hannemann, N.; Cybulski, P.; Wielebski, Ł.; Horbiński, T.; Dickmann, F. Die kartographische Rekonstruktion der Landschaftsentwicklung des Oberschlesischen Industriegebiets (Polen) und des Ruhrgebiets (Deutschland). *KN-J. Cartogr. Geogr. Inf.* **2019**, *69*, 131–142. [CrossRef]
9. Lerma, J.L.; Navarro, S.; Cabrelles, M.; Villaverde, V. Terrestrial laser scanning and close range photogrammetry for 3D archaeological documentation: The Upper Palaeolithic Cave of Parpalló as a case study. *J. Archaeol. Sci.* **2010**, *37*, 499–507. [CrossRef]
10. Van Elzakker, C.P.; Ooms, K. Understanding map uses and users. In *The Routledge Handbook of Mapping and Cartography*; Informa UK Limited: London, UK, 2017; pp. 55–67.
11. Hurni, L. *Multimedia Atlas Information Systems*; Springer Science and Business Media LLC: Berlin/Heidelberg, Germany, 2008; pp. 759–763.
12. Robinson, A.H.; Morrison, J.L.; Muehrcke, P.C.; Kimerling, A.J.; Guptill, S.C. *Elements of Cartography*; John Wiley & Sons: New York, NY, USA, 1995.
13. Kramers, E.R. Interaction with Maps on the Internet—A User Centered Design Approach for the Atlas of Canada. *Cartogr. J.* **2008**, *45*, 98–107. [CrossRef]
14. Cybulski, P.; Horbiński, T. User Experience in Using Graphical User Interfaces of Web Maps. *ISPRS Int. J. Geo-Inf.* **2020**, *9*, 412. [CrossRef]
15. Cybulski, P. Spatial distance and cartographic background complexity in graduated point symbol map-reading task. *Cartogr. Geogr. Inf. Sci.* **2020**, *47*, 244–260. [CrossRef]
16. Wielebski, Ł.; Medyńska-Gulij, B. Graphically supported evaluation of mapping techniques used in presenting spatial accessibility. *Cartogr. Geogr. Inf. Sci.* **2018**, *46*, 311–333. [CrossRef]
17. Halik, Ł.; Medyńska-Gulij, B. The Differentiation of Point Symbols using Selected Visual Variables in the Mobile Augmented Reality System. *Cartogr. J.* **2016**, *54*, 147–156. [CrossRef]
18. Horbiński, T.; Cybulski, P.; Medyńska-Gulij, B. Graphic Design and Button Placement for Mobile Map Applications. *Cartogr. J.* **2020**, 1–13. [CrossRef]
19. Medyńska-Gulij, B.; Wielebski, Ł.; Halik, Ł.; Smaczyński, M. Complexity Level of People Gathering Presentation on an Animated Map—Objective Effectiveness Versus Expert Opinion. *ISPRS Int. J. Geo-Inf.* **2020**, *9*, 117. [CrossRef]
20. Medyńska-Gulij, B.; Cybulski, P. Spatio-temporal dependencies between hospital beds, physicians and health expenditure using visual variables and data classification in statistical table. *Geodesy Cartogr.* **2016**, *65*, 67–80. [CrossRef]
21. Dent, B.D. *Cartography. Thematic Map Design*, 5th ed.; WCB/McGraw-Hill: Boston, MA, USA, 1996.
22. Tabaka, A. Historical Cultural Spaces—Adaptation and Functioning. The Case of the Museum of the First Piasts at Lednica. *Analecta Archaeologica Ressoviensia* **2018**, *13*, 287–308. [CrossRef]
23. Górecki, J. *Gród na Ostrowie Lednickim na tle Wybranych Ośrodków Grodowych Pierwszej Monarchii Piastowskiej*; Muzeum Pierwszych Piastów na Lednicy: Lednogóra, Poland, 2001; ISBN 83-903072-7-8.
24. Żurowska, K. *Ostrów Lednicki. U progu Chrześcijaństwa w Polsce*; Kraków, Instytut Historii Sztuki Uniwersytetu Jagiellońskiego: Kraków, Poland, 1993; ISBN 8386310014.
25. Banaszak, D.; Tabaka, A. *Ostrów Lednicki. Information Guide, Lednica*; Muzeum Pierwszych Piastów na Lednicy: Lednogóra, Poland, 2009; ISBN 978-83-61371-10-6.
26. Horbiński, T.; Medyńska-Gulij, B. Geovisualisation as a process of creating complementary visualisations: Static two-dimensional, surface three-dimensional, and interactive. *Geodesy Cartogr.* **2017**, *66*, 45–58. [CrossRef]
27. Imhof, E. *Thematische Kartographie*; De Gruyter: Berlin, Germany, 1972; ISBN 9783110021226.
28. Medyńska-Gulij, B. Topographische Manuskriptkarten der zweiten Hälfte des 18. Jahrhunderts als Aquarell- und Tuschezeichnung aus dem Haus Hannover. In *16. Kartographischehistorisches Colloquium, Marbach am Neckar 2012*; Heinz, M.A., Hüttermann, K.V.B., Eds.; Bonn: Kirschbaum Verlag, Germany, 2016; pp. 171–180. ISBN 978-3-7812-1928-1.

29. Edler, D.; Kühne, O.; Keil, J.; Dickmann, F. Audiovisual Cartography: Established and New Multimedia Approaches to Represent Soundscapes. *KN-J. Cartogr. Geogr. Inf.* **2019**, *69*, 5–17. [CrossRef]
30. Medyńska-Gulij, B. Point Symbols: Investigating Principles and Originality in Cartographic Design. *Cartogr. J.* **2008**, *45*, 62–67. [CrossRef]

International Journal of
Geo-Information

Article

Mini-Map for Gamers Who Walk and Teleport in a Virtual Stronghold

Krzysztof Zagata [1,*], Jacek Gulij [2], Łukasz Halik [1] and Beata Medyńska-Gulij [1]

[1] Department of Cartography and Geomatics, Faculty of Geographical and Geological Sciences, Adam Mickiewicz University, 61-712 Poznań, Poland; lhalik@amu.edu.pl (Ł.H.); bmg@amu.edu.pl (B.M.-G.)
[2] Faculty of Computing and Telecommunications, Poznan University of Technology, 60-965 Poznań, Poland; jacek.gulij@student.put.poznan.pl
* Correspondence: krzysztof.zagata@amu.edu.pl

Abstract: Studies of the effectiveness of multimedia cartography products may include mini-map design for navigation. In this study, we have touched upon designing gameplay to indicate the impact of the mini-map on the time effectiveness of a player that can walk or teleport himself/herself along marked out points in virtual topographic space. The eye-tracking examination of gamers' effectiveness in a non-complex game of collecting coins in a reconstructed stronghold on the holm provided us with a new perspective on the role of mini-maps. The more time gamers took to examine the mini-map, the more time they needed to finish the game, thus decreasing their effectiveness. The teleporting gamers had significantly higher time effectiveness than walking gamers, however, the data obtained showed only a minor difference between the proportions of the mini-map examination time to the total game time for walking and teleportation.

Keywords: mini-map; virtual stronghold; walking; teleportation; multimedia cartography; medium effectiveness; Unity; topographical space; gamer; gameplay; eye tracking

Citation: Zagata, K.; Gulij, J.; Halik, Ł.; Medyńska-Gulij, B. Mini-Map for Gamers Who Walk and Teleport in a Virtual Stronghold. *ISPRS Int. J. Geo-Inf.* **2021**, *10*, 96. https://doi.org/10.3390/ijgi10020096

Academic Editors: Wolfgang Kainz and Georg Gartner

Received: 29 December 2020
Accepted: 19 February 2021
Published: 22 February 2021

1. Introduction

To boost the usefulness of a map, one traditionally employs a location map, i.e., a map on a smaller scale that presents the geographical location of the area covered by the topographic map with reference to a larger administrative unit or physio-geographical unit [1,2]. The location map helps in geographical orientation at a more general level than the more detailed topographic map [3]. The concept of topographic orientation is related to the use of topographic maps directly in the field [4,5]. The content of topographic maps is compared to objects in the field to evaluate their actual location and spatial relationships (i.e., directions and distances) between them. In this study, the principles of cartographic design, especially layout construction, are of great importance. The layout includes the main, largest frame of cartographic content along with other frames of the map, with the legend being placed ideally on the right-hand side of the cartographic content or at the bottom of the map [6,7]. Global websites commonly use maps with mini-maps for car and pedestrian navigation. Their interface is evolving towards higher intuitiveness and quick navigation [8].

Navigation in a 3D computer game environment constitutes an essential element of the gameplay. Almost all the games use mini-maps to facilitate the character's movement in the game and the use of a virtual interface [9,10]. The mini-map has become standard not only in popular computer games with the open world, but also in visualizations and VR games (VR—Virtual Reality) [11–15]. The knowledge of the game space, one's location and individual stages of tasks may drastically change the chances of victory [16–19]. As suggested by game designers, mini-maps generally should not exceed 10% of the available display area [20]. Additionally, mini maps are sometimes referred to as a "corner map", although their display position is not standardized [21]. Walking is the most immersive locomotion technique in the movement around geographic space [22,23]. Multiple VR games that focus on the movement around virtual space employ walking at a natural pace [15,24,25]. Teleportation from one point to another by means of the hand-held controller, which significantly accelerates the movement in geographical space, is becoming increasingly popular in VR games [23,26–28].

The study of the effectiveness and attractiveness of multimedia cartography products may include subjective opinions of users, formulated in a descriptive way or in the form of specific marks [15,29,30]. Speed and/or correctness of task performance or interpreting the information about geographic phenomena constitute part of the objective characteristics of geomedia products that determine the effectiveness of use [31–37]. The speed of performing the spatial task with the use of the appropriate interactive tool is the most measurable factor of effectiveness. The effectiveness of multimedia cartography products is studied by means of: online questionnaires, directly supervised questionnaires, tasks performed in the field with the help of mobile devices, direct observations, observations with remote recording of the participant's movement and observations via eye tracking [36,38–42]. Homogenous groups of respondents with the proper number of representatives, e.g., 15–30 students or 5–15 experts per one respondent group, are usually invited to participate in research [15,36,43]. In multimedia cartography, statistical correlations are taken into consideration if homogeneous respondent groups participate in the research. For instance, the Spearman test is used to indicate the correlation between the respondents' answers and the times taken for tasks [36].

Designing mini-maps for navigation during walking and teleportation requires specific theoretical and technological procedures. The problem of the complementary employment of cartographic design rules with IT (information technology) notations (i.e., 3D topographic space and gameplay interaction with the possibility of recording the gamer's effectiveness by the method of eye tracking) has not been examined to date by any studies dealing with the effectiveness of multimedia cartography products.

In this analysis, we consider effectiveness as the total virtual game time of a single player, which means that the player that moves from the first to the last point faster is more effective. Thus, we touch upon designing gameplay to indicate the impact of the mini-map on the time effectiveness of a player that can walk or teleport himself/herself along marked out points in virtual topographic space.

The rest of the article is structured as follows. Section 2 presents the major aim and specific questions of the article analysis. In Section 3, we introduce our methodology for designing a VR application and experimental research process. Section 4 provides details of the experimental evaluation. Finally, in Section 5, we discuss the results and present our conclusions.

2. Aim and Questions

Considering complexity, different approaches to defining the concept, multiple possibilities of testing multimedia effectiveness and many other factors considered in various studies, we decided to focus on time effectiveness in virtual space as the most important factor. The major aim of the analysis is to examine the significance of mini-maps for the time effectiveness of a player walking around and teleporting in the virtual topographic space. Apart from this aim, we have also asked the following specific questions:

- What are the differences between the times of individual gameplay by walking and teleporting players, respectively?
- What impact does the mini-map examination time have on the total game time for walking and teleportation?
- What is the correlation between the mini-map examination time and the total game time while collecting the first coin and the last coin?
- Does complementary cartographic and IT game design in topographic space with mini-maps allow one to evaluate the gamer's effectiveness in the game?

3. Methodology

To meet the objective and answer the above questions, we have adopted four main research stages:

- To pinpoint the conceptual assumptions (Section 3.1);
- To create a game in a virtual stronghold following a scheme and a geographical layout of the gameplay elements (Section 3.2, Figures 1, 3 and 5);
- To prepare and carry out surveys among walking and teleporting players (Section 3.3, Figures 2 and 4).

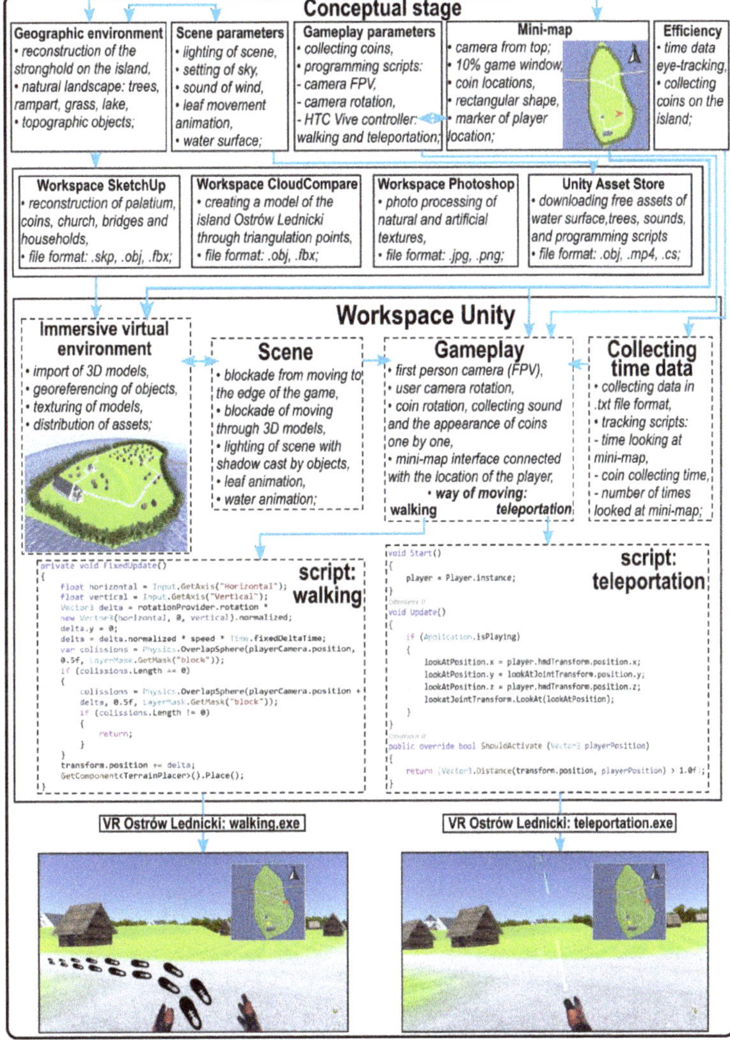

Figure 1. Scheme of working out the VR application for two game versions: with walking and teleportation.

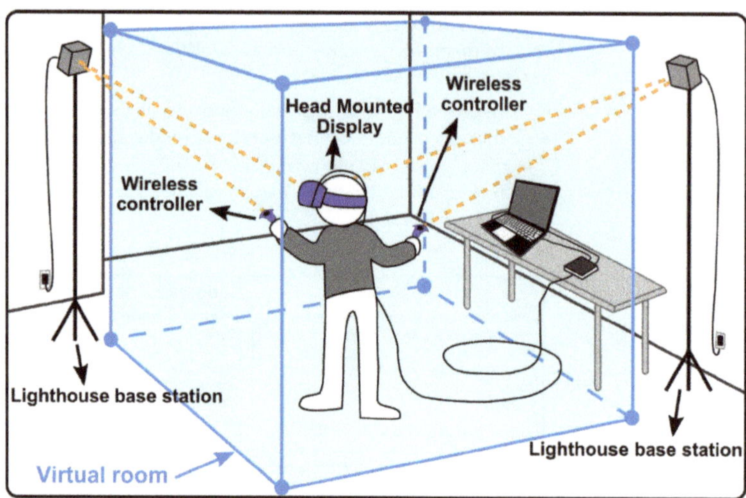

Figure 2. Equipment and room installation progress with HTC goggles (based on https:www.vive.com, accessed on 20 February 2021).

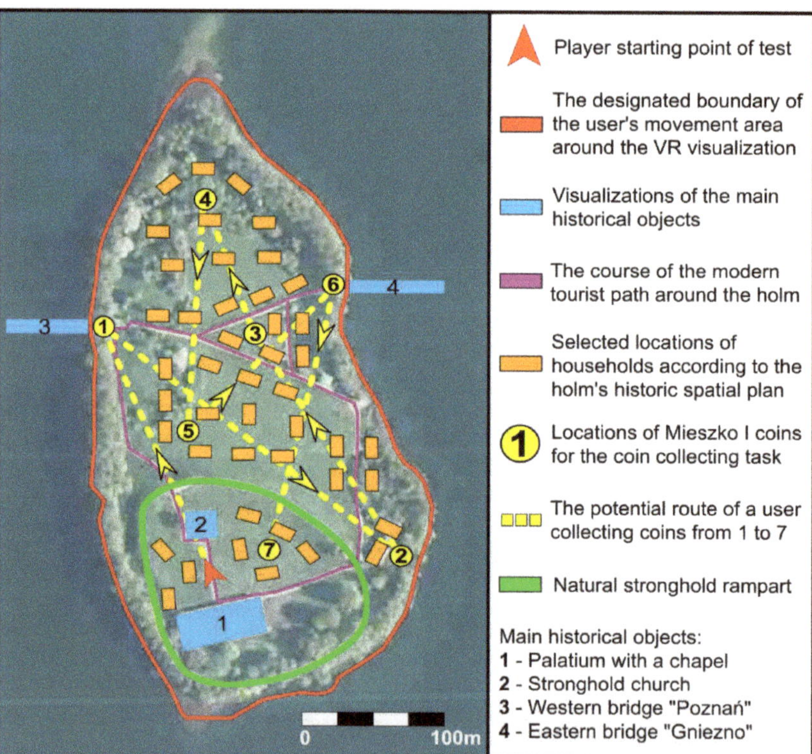

Figure 3. Layout of the application with the placement of individual game elements on the holm.

START
COIN 7
COIN 1
COIN 6
COIN 2
COIN 5
COIN 3
COIN 4

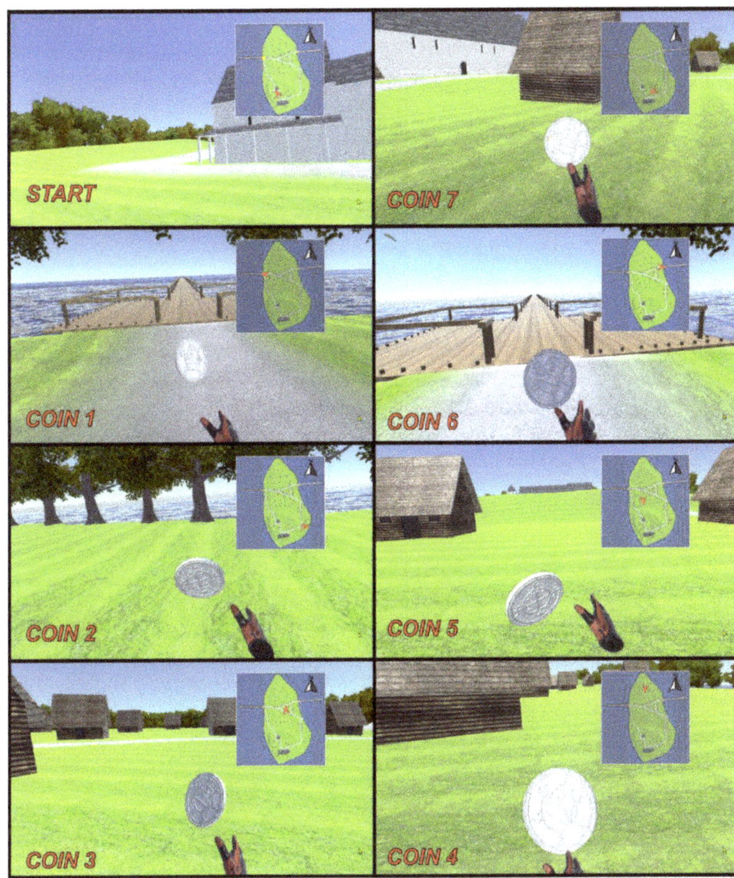

Figure 4. View in HTC Vive Pro Eye goggles during the collection of coins at the starting point and at each coin.

```
public float speed = 3.5f;
public Transform playerCamera;
private void Update()
{
    var trackpad = TrackPad.main.Trackpad;
    var magnitude = trackpad.magnitude;
    var delta = playerCamera.rotation * new Vector3(trackpad.x, 0, trackpad.y);
    delta.y = 0;
    delta = delta.normalized * magnitude * speed * Time.deltaTime;

    var colissions = Physics.OverlapSphere(playerCamera.position, 0.5f,
      LayerMask.GetMask("block","terrain"));
    if(colissions.Length == 0)
    {
        colissions = Physics.OverlapSphere(playerCamera.position + delta, 0.5f,
           LayerMask.GetMask("block", "terrain"));
        if(colissions.Length != 0)
        {
            return;
        }
    }
    transform.position += delta;
    GetComponent<TerrainPlacer>().Place();
}
```

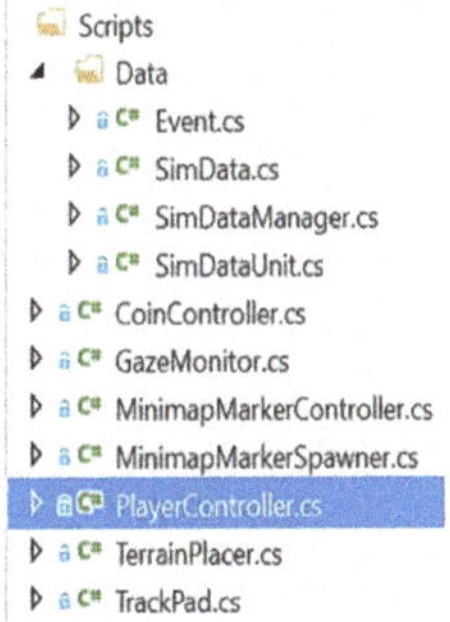

Scripts
Data
Event.cs
SimData.cs
SimDataManager.cs
SimDataUnit.cs
CoinController.cs
GazeMonitor.cs
MinimapMarkerController.cs
MinimapMarkerSpawner.cs
PlayerController.cs
TerrainPlacer.cs
TrackPad.cs

Figure 5. Script PlayerController.cs and list of written programming scripts for running the game.

3.1. Concept

Considering the cartographico-geographic attitude (cartographic design of topographic space and mini-maps), gamers' habits (typical game actions) and IT aspects (software in game engines), we adopted a concept embracing the following assumptions:

- Type of multimedia application: VR coin collecting game in specified locations in a small, limited, topographic space, which is possible to present in a single mini-map view;
- Multimedia design: the geographic part (creating the island's topography), the cartographic part (designing the trail and the mini-map) and the IT part (designing the game and eye tracking);
- Medium: full-immersive VR—presence and movement, supported with sound and animated effects;
- Software and equipment: graphic software (Photoshop), architectural software (SketchUp), geographical modeling software (CloudCompare), game engine (Unity), survey (HTC Vive Pro Eye goggles and HTC controllers);
- Technological process: managing transformation and data integration in a geomatic process in several workspaces;
- Geographic space: stronghold on the island, elements of natural landscape (trees, rampart, grass, lake) and historical buildings (church, palace, huts, bridges);
- Parameters of the scene: natural lighting, cloudless sky, sounds of wind, animation of leaves and water surface;
- Parameters of the gameplay: collecting coins, programming scripts, programming HTC Vive controllers for two types of movement: walking and teleportation;
- Mini-map: the view of the entire island in the stronghold's graphics; 10% of the game window view; in the right top corner; location of coins marked with yellow dots; northern orientation, rectangular shape; gamer's location indicated by the red arrow pointing to the direction of looking;
- Parameters of effectiveness: collecting time data scripts; individual and synthetic analysis; analysis of gamers' effectiveness;
- Respondents: screen-based video game users; lacking experience in immersive VR environments; playing a minimum of 10 h per week;
- The way of conducting the research: each gamer stays in the virtual room; eye calibration of the position of goggles and controllers; eye-tracking study with HTC goggles; the same task for each gamer: to collect 7 coins, time for the task: approximately 15 min; obtaining data on the total game time and the mini-map examination time;
- Expected research results: statistical and graphic specification of time effectiveness of a gamer that walks and teleports by means of the mini-map.

3.2. Creating a VR Application

The planned VR application was created as a part of four workspaces according to the following order situated in the scheme in Figure 1. According to Medyńska-Gulij [6], a workspace is an area in the application (i.e., digital work environment dedicated to the specific application) that allows one to perform digital operations (framing, cutting, rotation, adding points, georeferencing) for various data types (raster and vector) and formats (.png, .svg, .obj, .shp), operating on both desktop and mobile systems according to the interface and programming scripts.

It is worth mentioning that in the research, we used a virtual presentation of the stronghold on the holm of Ostrów Lednicki (Poland) worked out in four workspaces: SketchUp, CloudCompare, Photoshop and Unity. Hence, we would like to focus specifically on cartographic design (mini-maps), GIS design (scene and gameplay for walking and teleportation) and IT design (scripts allowing one to obtain time data during the eye-tracking study) [15].

Cartographic attitude included the design and creation of the geographic space of the stronghold, with preservation of its natural topography, and designing a mini-map. The

IT attitude encompassed the implementation of programming scripts for the game and programming scripts that allowed one to collect time data for the eye-tracking examination.

The first stage of design in Unity was to import a 3D model of natural relief and a georeference of modeled historical objects and to add landscape assets and task coins (Figure 3). Then, individual scene parameters were worked out, following the initially assumed concept. Box colliders were implemented at both ends of the holm, and several of them were also used to prevent the gamers from walking through the building walls, to create the realistic space of the holm. Scene light was directed toward the north to make shadows of objects and create a natural 3D impression of the holm landscape. Leaves and the surface of the water were animated, and wind sounds, as well as footstep sounds, were added to boost the gamers' level of immersion.

To make the design of a rectangular mini-map reflect the space of the holm as seen from a pedestrian's point of view, we used an extra camera for the bird's eye view (Figure 3). Following the cartographic rules, the mini-map was north-oriented and received a north arrow. Using the "Canvas" object (the area inside of which all user interface elements should be), the mini-map was set in the top right corner of the gamers' view. The position of the gamer is represented by the red arrow, and the location of the next coin to be collected is symbolized by the yellow ring. In this study, the character is not bound to the center of the map field. The animated symbol representing a player constantly changes its position on the mini-map, and the map always includes the same reference objects (world-oriented mini-map) [20].

SteamVR, a Unity plugin, was responsible for the main interface between the gamer and the virtual stronghold. The plugin solves the problem of configuring the first-person camera in such a way that it displays a reliable view in goggles adapted to different perspectives from both eyes. Additionally, it includes ready implementation of the reflection of the gamer's head and hands in the world and facilitates programming behaviors related to the use of the touch panel and controller buttons.

The next step was to write programming scripts needed for the correct operation of the game and the development of conceptual assumptions (Figure 5). "CoinController", responsible for presenting the coin by rotating it and creating audio-visual effects on collection, was the first script to be written. It was also responsible for managing the order in which the coins appeared, and it also logged the time of coin collection for further processing. The next script was "GazeMonitor", which checked whether the object was in the center of the player's vision. Its functionality was supplied by eye tracking and the Tobii plugin. Tobii XR SDK for Unity offers simple add-on-independent HTC Vive Pro Eye methods of accessing eye-tracking data and tools for scene development. The following scripts "MinimapMarkerController" and "MinimapMarkerSpawner" were responsible for creating and controlling mini-map markers for the position of the gamer and coins. While the script responsible for navigating through the teleportation was included in the SteamVR plugin, the script responsible for walking in geographic space had to be rewritten from scratch. The script "PlayerController", depicted in Figure 5, was responsible for the user's walking and collisions with buildings. The "TerrainPlacer" script, responsible for aligning objects to relief elements, was used to glue not only buildings but also coins and the player to the ground. The next step of the correct operation of the "PlayerController" script was to write the "TrackPad" script, which was responsible for collecting and storing HTC controller trackpad touch data. The walking speed option with the controller was set to 4.8 meters per second to represent the natural movement in virtual terrain and to match the speed of movement through teleportation. HTC documentation shows that the distance of teleporting with a controller is a maximum of 45 degrees of a parabolic arc from the player's elevation of the controller.

The last step in the IT approach was to create a package of scripts that would be responsible for collecting time and eye-tracking data (Figure 5). The first script was "SimData Manager", responsible for enabling other scripts to log events, collect data and process collected data into a readable TXT log file. This manager was accompanied by models, i.e., scripts that contain only data structures and have no behavior. "SimData" was such a model and it contained a collection of "Event" entries and a list of "SimDataUnit" records. The "Event" model was supposed to contain the information about the user's action, or a specific system-generated event at a specific time, such as coin collection. Another model, "SimDataUnit", was responsible for storing a temporary simulation status contained within one frame at a specific time.

Eventually, after all operations, the created applications were exported. The walking.exe application, including the way of moving by holding the touchpad, and teleportation.exe, including teleportation by clicking on the touchpad, were exported.

3.3. *Participants and Experimental Process*

We invited 40 game users, selected at random among students, who declared to spend over 10 h per week on screen games but lacking any experience with VR games, to participate in the research. Twenty game users collected coins by means of teleportation, and the other half by walking around the holm. Game users, aged 17–26, participated in the game voluntarily, without any financial gratification, and could resign from further gameplay at any moment.

The research was conducted on a laptop with Windows 10 and two applications created in Unity: walking.exe and teleportation.exe (Figure 1). To carry out the research, we used HTC Vive Pro Eye goggles with a definition of 2880 × 1600 px. In Steam VR, we prepared a virtual room for the gamer, configured the laser tracking of base stations for establishing the position of goggles and controllers and calibrated the floor's position (Figure 2). To capture the gamer's gaze, we used the in-built device for eye tracking in the HTC goggles.

Before each gamer put on VR goggles, he was informed that after putting them on, he would find himself in the virtual medieval stronghold on the holm (Figure 4). The gamer was informed how to set the focus in the VR goggles and how to use the wireless controllers. Then, each gamer was introduced to the main goal of the game, which was to collect seven coins. Each of them appeared on the mini-map in the form of a yellow dot, whereas their location was marked with a red arrow pointing in the direction the gamer was looking.

After putting on the goggles, the gamer confirmed that he had moved to the virtual stronghold, could hear sounds and understood the functioning of the controllers, as well as could see his own location on the mini-map. The confirmation by the gamer that he had started to move towards the first coin noticed on the mini-map initiated measuring the total game time (START in Figure 4), and collecting the 7th coin ended measuring the game time (Figure 4).

4. Results

Obtained time data were placed in two tables: walking gamers (Table 1) and teleporting gamers (Table 2). Data in tables were divided into three major data categories: total time, individual time—from the start to the collection of the first coin—and individual time—from the collection of the 6th coin to the collection of the 7th coin. For each category, we distinguished the three most significant data subcategories for each type of movement: time-Walking-Space (tWS)—the total game time for walking gamers, time-Teleportation-Space (tTS)—the total game time for teleporting gamers, time-Walking-Mini-map (tWM)—the mini-map examination time for walking gamers, time-Teleportation-Mini-map (tTM)—the mini-map examination time for teleporting gamers and tWM/tWS and tTM/tTS, i.e., the ratio in percent. To present the data from Tables 1 and 2, we used column charts and a line chart (Figures 6 and 7).

Table 1. Time data for walking (20 gamers). The total game time for walking gamers—tWS (time-Walking-Space), the mini-map examination time for walking gamers—tWM (time-Walking-Mini-map).

	A. Total Time—Walking			B. Individual Time (Start–1 Coin)—Walking			C. Individual Time (6–7 Coins)—Walking		
Id	tWS:	tWM:	tWM/tWS:	t_1WS:	t_1WM:	t_1WM/t_1WS:	t_2WS:	t_2WM:	t_2WM/t_2WS:
1	353.2	177.9	50%	56.8	16.3	29%	56.0	27.6	49%
2	419.0	208.6	50%	59.6	29.0	49%	80.1	47.0	59%
3	386.5	181.0	47%	87.5	41.8	48%	54.5	19.9	36%
4	453.4	178.7	39%	79.7	20.0	25%	68.4	25.2	37%
5	544.8	208.5	38%	64.1	6.7	10%	67.6	38.1	56%
6	530.4	200.3	38%	71.5	17.7	25%	67.4	32.8	49%
7	377.0	130.7	35%	55.7	8.3	15%	55.6	29.7	53%
8	392.7	129.4	33%	65.6	19.2	29%	59.0	18.8	32%
9	380.1	119.5	31%	59.7	13.4	22%	54.5	16.5	30%
10	371.0	113.8	31%	59.2	13.2	22%	60.2	21.0	35%
11	393.2	108.7	28%	49.8	13.2	27%	80.5	20.9	26%
12	375.3	100.3	27%	69.4	13.8	20%	54.5	18.1	33%
13	412.4	110.2	27%	56.2	8.1	14%	81.9	25.9	32%
14	366.6	94.3	26%	62.8	10.6	17%	59.6	17.8	30%
15	378.8	91.4	24%	66.8	12.8	19%	60.1	15.6	26%
16	499.2	114.3	23%	77.8	8.0	10%	67.3	24.2	36%
17	349.2	59.6	17%	61.6	8.8	14%	56.1	7.9	14%
18	302.3	49.7	16%	49.4	5.9	12%	49.2	8.7	18%
19	389.2	60.0	15%	66.6	6.2	9%	66.0	11.4	17%
20	379.0	56.2	15%	55.0	5.4	10%	88.9	11.2	13%
Median	383.3	114.0	29%	62.2	13.0	19%	60.2	20.4	33%

Table 2. Time data for teleportation (20 gamers). The total game time for teleporting gamers—tTS (time-Teleportation-Space), the mini-map examination time for teleporting gamers—tTM (time-Teleportation-Mini-map).

	A. Total Time—Teleportation			B. Individual Time (Start–1 Coin)—Teleportation			C. Individual Time (6–7 Coins)—Teleportation		
Id	tTS:	tTM:	tTM/tTS:	t_1TS:	t_1TM:	t_1TM/t_1TS:	t_2TS:	t_2TM:	t_2TM/t_2TS:
1	201.5	79.2	39%	32.6	5.4	17%	19.7	12.4	63%
2	218.4	77.9	36%	45.5	8.1	18%	30.5	14.2	47%
3	273.3	96.5	35%	42.0	16.4	39%	44.2	11.8	27%
4	148.5	52.0	35%	27.1	10.9	40%	24.5	6.3	26%
5	365.1	123.0	34%	74.6	19.7	26%	55.3	27.6	50%
6	164.7	55.3	34%	34.9	7.9	23%	15.5	5.8	37%
7	166.3	55.6	33%	26.3	12.8	49%	24.9	6.5	26%
8	323.9	106.8	33%	44.5	10.9	24%	44.4	21.5	48%
9	222.2	73.0	33%	39.3	18.7	48%	25.6	10.0	39%
10	196.6	60.7	31%	32.9	4.5	14%	33.6	13.6	41%
11	302.2	93.0	31%	31.0	5.2	17%	64.0	23.8	37%
12	418.6	123.6	30%	50.0	19.7	39%	38.5	24.4	63%
13	213.8	63.1	30%	37.7	11.6	31%	27.3	9.8	36%
14	268.3	77.4	29%	33.0	2.9	9%	60.7	25.8	43%
15	315.8	87.2	28%	48.9	14.0	29%	33.8	5.6	17%
16	203.9	56.3	28%	30.2	9.7	32%	25.3	4.7	18%
17	331.7	81.1	24%	49.9	8.9	18%	47.3	9.0	19%
18	237.1	38.8	16%	33.9	4.7	14%	27.6	6.0	22%
19	255.2	41.6	16%	47.5	5.6	12%	28.4	6.2	22%
20	331.5	48.6	15%	31.2	5.5	18%	62.8	5.6	9%
Median	246.2	75.2	31%	36.3	9.3	24%	32.0	9.9	37%

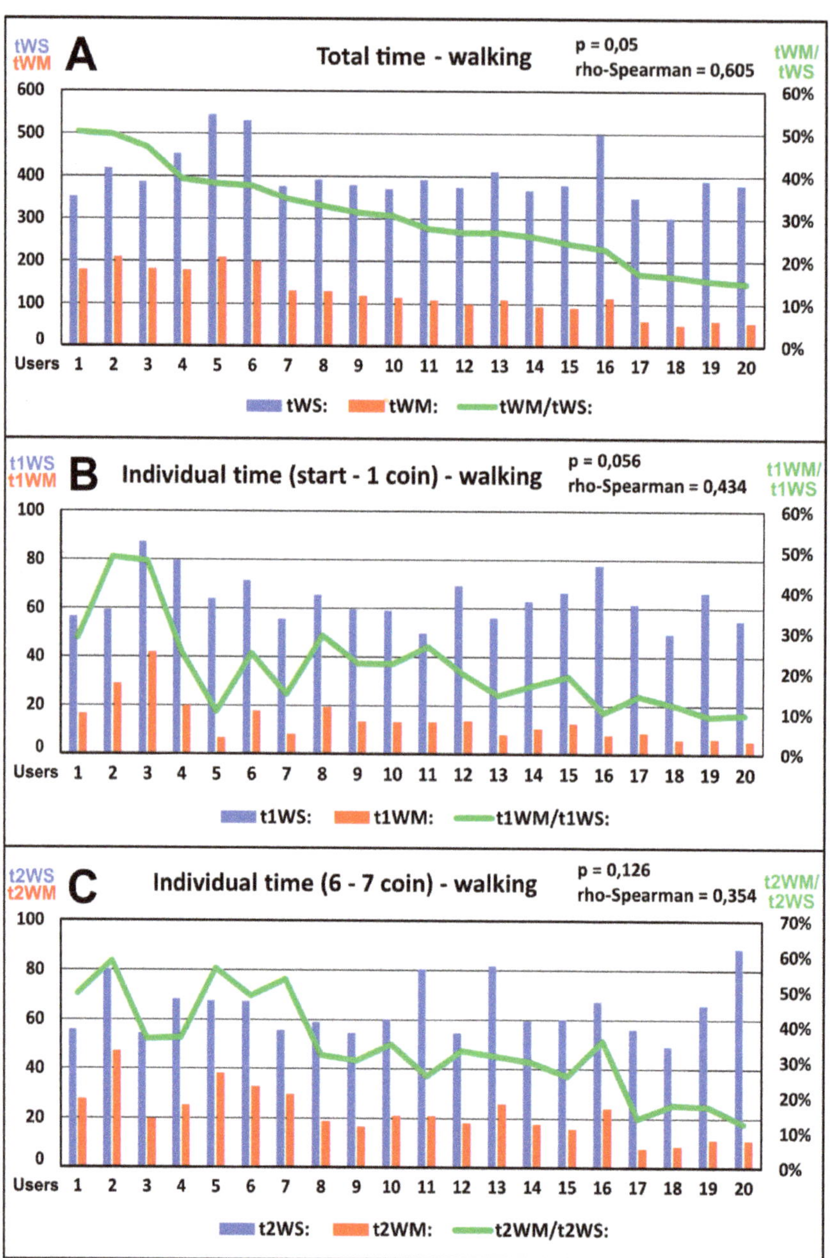

Figure 6. Visualization of time data for walking.

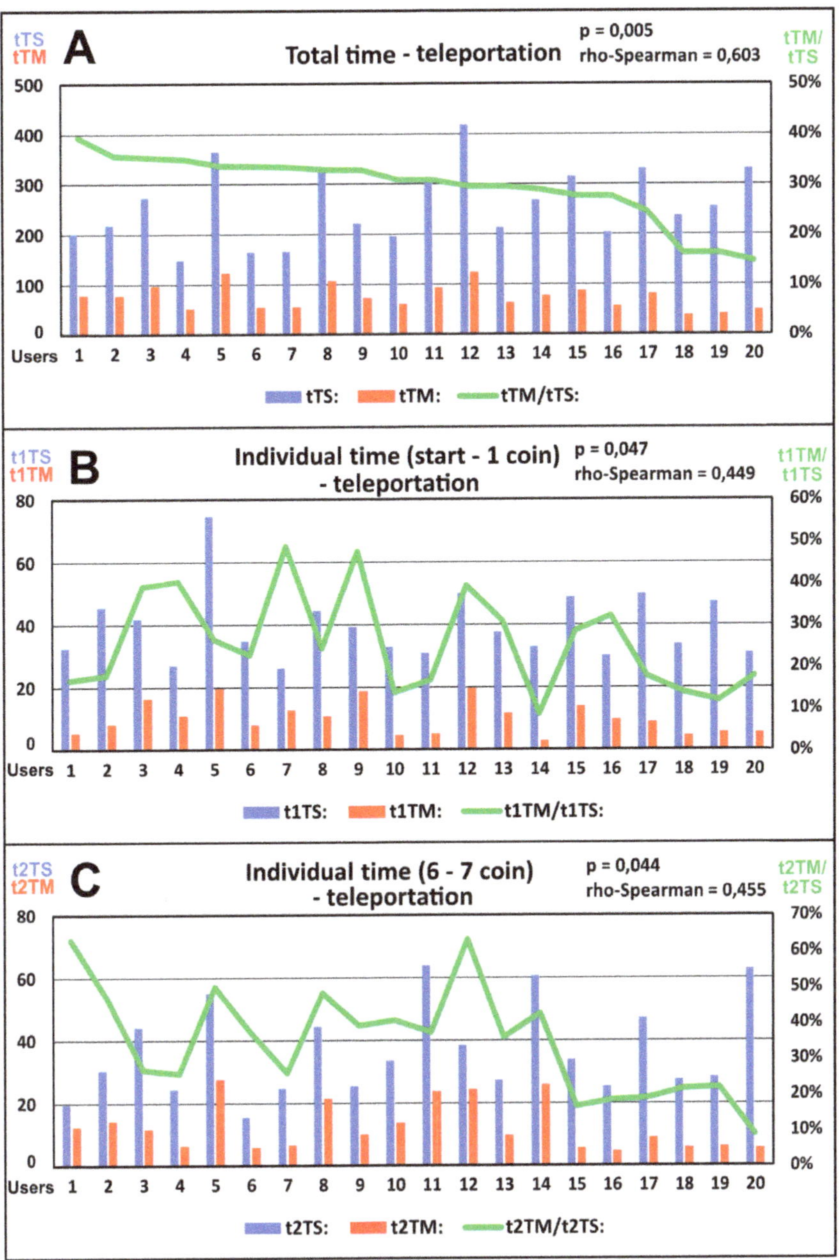

Figure 7. Visualization of time data for teleportation.

In analytical comparisons, we wanted to demonstrate the level of time effectiveness in individual games. Effectiveness is understood as the total game time achieved by a single gamer. In this research, a gamer that needs less time to collect all the coins has greater effectiveness.

In the analysis of numerical data, we assumed that the measurements of the mini-map examination time that are below 0.3 s and appeared after collecting any coins until the next correct mini-map examination time measurement that exceeded 0.3 s would be deleted. For a middle-aged person, the reaction time is around 0.2–0.4 s, hence, we rounded the number down, as multiple young people participated in the research [44]. Errors may have resulted from the incorrect refreshing of display frames during teleportation or too rapid head movements. We recorded a dozen or so such erroneous time measurements.

We sorted out the results for walking gamers, with the greatest ratio of mini-map examination time to the total game time (Table 1) as our point of reference. Gamer 5 had the longest total game time (544.8 s), and gamer 18 had the shortest (302.3 s) (Table 1A). The time difference was 242.5 s, which equals an effectiveness discrepancy of 55% between the two gamers. The average total game time (the median) for the entire group was 383.3 s, which marks the average effectiveness level for walking gamers. Comparing the shortest and the longest time to the median, gamer 18 showed an effectiveness increase of 21% (81.0 s), whereas gamer 5 showed an effectiveness decrease of 42% (161.5 s).

The longest mini-map examination time was recorded for gamer 2 (208.6 s), and the shortest for gamer 18 (49.7 s) (Table 1A). For most of the walking users, the ratio of the total mini-map examination time to the total game time exceeded 25%, which means that over $\frac{1}{4}$ of their total walking time was spent on looking at the navigation tool. Collecting the first coin was easier for most gamers without the excessive use of the mini-map (15 gamers below 25%), whereas collecting the last coin required longer examination of the mini-map (16 gamers above 25%).

When coming up to the first coin, gamer 3 was the one that examined the mini-map the longest (41.8 s), and gamer 20 the shortest (5.4 s) (Table 1B). At the end of the task, right before collecting the last coin, gamer 2 took the most time to examine the mini-map (47.0 s), and gamer 17 the least (7.9 s) (Table 1C). A significant majority of walking users examined the mini-map longer at the last coin than at the first one, even though the average time of completing the task was comparable. Only gamers 3, 8 and 17 took more time prior to the first coin and less time prior to the last coin, which means that only these three gamers, despite similar total game time, similar route and obstacles in the form of the rampart and buildings, had higher effectiveness of using the mini-map during the game.

To examine the correlation between the total game time and the mini-map examination time, we used the Spearman correlation test. There is a correlation between the use of the mini-map and the total game time ($r = 0.605$) (Figure 6A), however, only for the first diagram. This means that the more walking users used the mini-map during the game, the more time they needed for collecting all coins.

The results of teleporting gamers were sorted out, using the largest ratio of the mini-map examination to the total game time as a point of reference (Table 2). Gamer 12 was the one with the longest total game time (418.6 s), and gamer 4 had the shortest (148.5 s) (Table 2A). The time difference is 270.1 s, which gives a disparity of 35% between game users. The average total game time for the entire group was 246.2 s, which constitutes the average effectiveness level for teleporting gamers. Comparing the shortest and the longest time to the median, gamer 4 showed an effectiveness increase of 40% (97.6 s), and gamer 12 showed an effectiveness decrease of 70% (172.5 s).

The longest mini-map examination time was registered for gamer 12 (123.6 s), and the shortest for gamer 18 (38.8 s) (Table 2 A). For a significant majority of teleporting gamers, the ratio of the total mini-map looking time to the game finishing time was higher than 25% but did not exceed 40%, which means that they spent over $\frac{1}{4}$ of their time looking at their location and the location of coins. Collecting the first coin for almost half of the game users was quicker when they did not use the mini-map for too long (11 gamers below 25%), whereas collecting the last coin seemed more difficult and required longer examination of the mini-map (14 people above 25%).

Gamers 5 and 12 needed the most time to examine the mini-map (19.7s), and gamer 14 the least time (2.9 s) (Table 2B). During the last walk, gamer 5 needed the most time to look at the mini-map (27.6 s), and gamers 15 and 20 the least time (5.6 s) (Table 2C). During teleportation, most gamers took more time to look at the mini-map when collecting the last coin, compared to the time they took when collecting the first coin.

According to the Spearman correlation test for the first diagram, there is a significant correlation between the use of the mini-map and the total game time (r = 0.603) (Figure 7A). This shows that gamers using the mini-map for longer during the game needed more time to collect all coins. The second and third diagram depict similar correlations, however, they have lower statistical power (Figure 7B,C), which means that for collecting the first and the seventh coin, the more time the gamer took to examine the mini-map, the longer it took for him to finish the game.

To assess the effectiveness of walking and teleportation users, we juxtaposed the medians for individual categories in Table 3, and in diagrams in Figure 8. Walking gamers needed significantly more time to finish the game (383.3 s) than teleporting gamers (246.2 s) (Table 3 A). When the game concept was created, it was assumed that the walking speed was 4.8 m/s, in accordance with the natural feeling of the gamer's movement in topographic space. Reading the default teleportation settings by HTC, one can observe that the maximum teleportation distance depends on the movement of the controller by the gamer (with a maximum of 45 degrees of a parabolic arc).

Table 3. Comparison of the median time for walking and teleportation.

	A. Median for Total Time			B. Median for Individual Time (Start–1 Coin)			C. Median for Individual Time (6–7 Coins)		
Id	tS:	tM:	tM/tS:	t_1S:	t_1M:	t_1M/t_1S:	t_2S:	t_2M:	t_2M/t_2S:
Walking	383.3	114.0	29%	62.2	13.0	19%	60.2	20.4	33%
Teleportation	246.2	75.2	31%	36.3	9.3	24%	32.0	9.9	37%

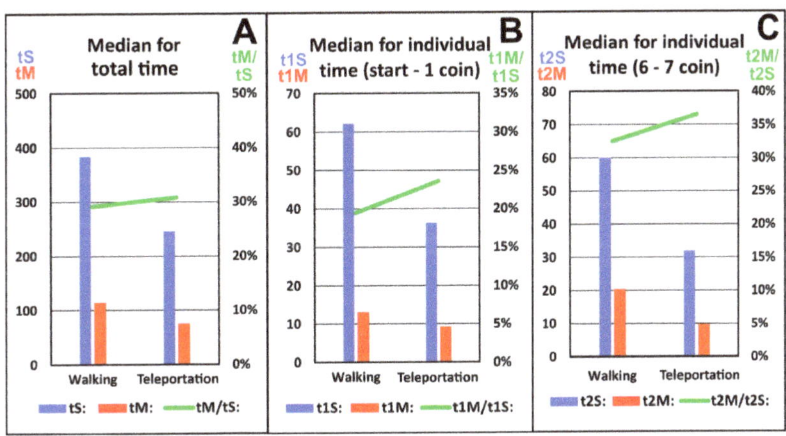

Figure 8. Comparative visualization of time for walking and teleportation.

The comparison of two medians of the total game time reveals that the difference is 137.1 s, which means that the teleporting gamers were 36% more effective than the walking gamers. In proportion to the total game time, the mini-map examination time was also longer for walking (114.0 s) than for teleportation (75.2 s) (Table 3A). Juxtaposing the proportion of the mini-map examination time to the total game time, one can conclude that these proportions are highly similar, i.e., 31% for teleporting gamers and 29% for walking gamers. The difference of 2% constitutes around 3 s, which demonstrates just a minor difference between gamers. Both walking and teleporting gamers took significantly more time to use the mini-map while collecting the last coin (33% and 37%) than while collecting the first one (19% and 24%) (Table 3B,C).

5. Discussion and Conclusions

Generally speaking, it can be said that the analysis of gamers' time effectiveness in a non-complex game of collecting coins in the reconstructed stronghold on the holm provided us with a new perspective on the role of mini-maps in movement in virtual topographic space.

Complementary cartographic and IT design employed in the research is, in the case of VR with mini-maps, consistent with the rules of map design that have always been determined by the technology of publishing maps and other products of multimedia cartography. We designed a game in the closed space of the holm with a relatively small number of topographic objects in such a way that the gamer walks the previously used paths on his way to the next coin. Such a strategy was supposed to accustom the gamer to the virtual topographic space. Each coin was situated in such a place that the gamer could not see it immediately after obtaining the previous one, which resulting in him having to use the mini-map.

The appearance, location and size of the mini-map became a problem in the concept of cartographic design. A traditional location map has generalized content, compared to the highly detailed main cartographic content. Our mini-map was designed as a photographically decreased view of the entire holm with all topographic elements. It was supposed to combine the perception of objects seen horizontally on the mini-map with their VR equivalents. Moreover, we employed graphic enhancement with intense coloring of the coin locations and of the gamer, so that they became the gamer's main focus [1,45,46]. It was debatable to place the north arrow traditionally on the map, as it could cause visual chaos. Even experts consider the north arrow an unnecessary element of multimedia cartography [43]. The size of our mini-map was adapted, according to the suggestions of designers and computer game developers, to occupy approximately 10% of the display screen available [9].

It became necessary to prepare the same task for two gamer groups in a single game to conduct a comparative analysis and draw synthetic conclusions on walking and teleportation. The total game time and the mini-map examination time were the two most relevant datasets of the research (Tables 1–3; Figures 6–8). The total game time was adopted as a factor determining the gamer's effectiveness. We assumed that an effectiveness increase occurs when the gamer finishes the game faster than the median of 20 gamers' times, and an effectiveness decrease occurs when it takes more time for the gamer to collect coins. In this analysis, our groups consisted of 20 people and both were homogeneous, in keeping with the assumptions of other multimedia cartography studies, but we recognize that more people would have to be considered for more detailed statistical research.

Optimally effective gamer's behavior would occur if the gamer examined the mini-map once to identify his location and the coin, and then continued the walking or teleportation without using the mini-map anymore. This assumption turned out to be correct, as the research revealed that the more time gamers took to examine the mini-map, the more time they needed to finish the game, thus decreasing their effectiveness. On the other hand, without the mini-map, the effectiveness level would drop significantly, as gamers would have to rely solely on their intuition to find coins.

The research proved that teleporting gamers had significantly higher time effectiveness than walking gamers, however, the data obtained showed only a minor difference between the proportions of the mini-map examination time to the total game time for walking and teleportation. A higher percentage obtained for teleportation rather than walking means that a minimally longer time is needed to check one's own location and the location of coins when one moves faster. Comparing the ratio of the mini-map examination time to the total game time prior to collecting the first coin and prior to collecting the last coin came as the biggest surprise. We assumed that prior to the last coin, gamers would take less time to examine the mini-map since they already had had time to accustom themselves to the topographic space and the mini-map function when they had been collecting previous coins. It turned out gamers took much more time to examine the mini-map prior to collecting the last coin rather than the first one, even though the routes to the first and the last coin were highly similar. That means that the learning effect failed to appear for both gamer groups and the effectiveness level in the use of the mini-map at the beginning and at the end of the game was not related to learning the game's topography by gamers. Interestingly, the proportions are larger for teleportation, i.e., gamers would take significantly more time to examine the mini-map prior to collecting the last coin than prior to collecting the first one.

The data obtained according to the research concept adopted became the foundation for drawing the above conclusions, however, the authors of the research realize that their suggestion to use eye tracking to determine the gamer's effectiveness in moving in virtual topographic space should continue to be followed to search for guidelines on how to design parameters of the mini-map.

In our future studies, we are planning to compare different types of mini-maps, dynamic and static ones, to extend this analysis. We are also planning to compute a model that would predict user cognitive performance in virtual reality. Thus, our future research should focus on statistical analysis of different cognitive strategies of processing spatial information.

Author Contributions: Conceptualization, Krzysztof Zagata and Beata Medyńska-Gulij; Data curation, Krzysztof Zagata; Formal analysis, Krzysztof Zagata and Beata Medyńska-Gulij; Investigation, Krzysztof Zagata, Jacek Gulij and Beata Medyńska-Gulij; Methodology, Krzysztof Zagata, Jacek Gulij and Beata Medyńska-Gulij; Resources, Krzysztof Zagata, Jacek Gulij, Beata Medyńska-Gulij and Łukasz Halik; Software, Krzysztof Zagata and Jacek Gulij; Supervision, Beata Medyńska-Gulij; Validation, Krzysztof Zagata and Beata Medyńska-Gulij; Visualization, Krzysztof Zagata; Writing—original draft, Krzysztof Zagata, Jacek Gulij and Beata Medyńska-Gulij; Writing—review and editing, Krzysztof Zagata and Beata Medyńska-Gulij. All authors have read and agreed to the published version of the manuscript.

Funding: This paper is the result of research on visualization methods carried out within statutory research in the Department of Cartography and Geomatics, Faculty of Geographical and Geological Sciences, Adam Mickiewicz University in Poznań, Poland.

Conflicts of Interest: The authors declare no conflict of interest.

References

1. Dent, B.D. *Cartography: Thematic Map Design*, 5th ed.; WCB/McGraw-Hill: Boston, MA, USA, 1999.
2. Kraak, M.-J.; Ormeling, F. *Cartography: Visualization of Geospatial Data*, 4th ed.; CRC Press: Boca Raton, FL, USA, 2020.
3. Robinson, A.H.; Morrison, J.L.; Muehrcke, P.C.; Kimerling, A.J.; Guptill, S.C. *Elements of Cartography*, 6th ed.; Wiley: New York, NY, USA, 1995.
4. Forrest, D.; Pearson, A.; Collier, P. The Representation of Topographic Information on Maps—The Coastal Environment. *Carto. J.* **1997**, *34*, 77–85. [CrossRef]
5. Wielebski, Ł.; Medyńska-Gulij, B.; Halik, Ł.; Dickmann, F. Time, Spatial, and Descriptive Features of Pedestrian Tracks on Set of Visualizations. *ISPRS Int. J. Geo-Inf.* **2020**, *9*, 348. [CrossRef]
6. Medyńska-Gulij, B. *Kartografia i Geomedia*; Wydawnictwo Naukowe PWN: Warsaw, Poland, 2021.
7. Edler, D.; Keil, J.; Tuller, M.-C.; Bestgen, A.-K.; Dickmann, F. Searching for the 'Right' Legend: The Impact of Legend Position on Legend Decoding in a Cartographic Memory Task. *Cartogr. J.* **2018**, *57*, 6–17. [CrossRef]
8. Horbiński, T.; Cybulski, P.; Medyńska-Gulij, B. Graphic Design and Button Placement for Mobile Map Applications. *Cartogr. J.* **2020**, *57*, 196–208. [CrossRef]
9. Edler, D.; Keil, J.; Wiedenlübbert, T.; Sossna, M.; Kühne, O.; Dickmann, F. Immersive VR Experience of Redeveloped Post-industrial Sites: The Example of "Zeche Holland" in Bochum-Wattenscheid. *KN J. Cartogr. Geogr. Inf.* **2019**, *69*, 267–284. [CrossRef]
10. Hruby, F.; Castellanos, I.; Ressl, R. Cartographic Scale in Immersive Virtual Environments. *KN J. Cartogr. Geogr. Inf.* **2020**, *21*, 1–7. [CrossRef]
11. Wessels, S.; Ruther, H.; Bhurtha, R.; Schroeder, R. Design and creation of a 3D virtual tour of the world heritage site of Petra, Jordan. In Proceedings of the AfricaGEO, Cape Town, South Africa, 1–3 July 2014.
12. Callieri, M.; Dellepiane, M.; Scopigno, R. Remote visualization and navigation of 3d models of archeological sites. *ISPRS Int. Arch. Photogramm. Remote Sens. Spat. Inf. Sci.* **2015**, *5*, 147–154. [CrossRef]
13. Yuqiang, B.; Niblock, C.; Bonenberg, L. Lincoln Cathedral Interactive Virtual Reality Exhibition. In Proceedings of the 17th International Conference on Computer Aided Architectural Design Futures, Istanbul, Turkey, 12–14 July 2017.

14. Khan, N.; Rahman, A.U. Rethinking the Mini-Map: A Navigational Aid to Support Spatial Learning in Urban Game Environments. *Int. J. Hum.-Comput. Interact.* **2018**, *34*, 1135–1147. [CrossRef]
15. Medyńska-Gulij, B.; Zagata, K. Experts and Gamers on Immersion into Reconstructed Strongholds. *ISPRS Int. J. Geo-Inf.* **2020**, *9*, 655. [CrossRef]
16. Si, C.; Pisan, Y.; Tan, C.T.; Shen, S. An initial understanding of how game users explore virtual environments. *Entertain. Comput.* **2017**, *19*, 13–27. [CrossRef]
17. Johanson, C.; Gutwin, C.; Mandryk, R.L. The Effects of Navigation Assistance on Spatial Learning and Performance in a 3D Game. In Proceedings of the Annual Symposium on Computer-Human Interaction in Play, Amsterdam, The Netherlands, 15–18 October 2017; Association for Computing Machinery (ACM): New York, NY, USA, 2017; pp. 341–353.
18. Kłosiński, M. *Hermeneutyka Gier Wideo. Interpretacja, Immersja, Utopia*; Wydawnictwo IBL PAN: Warsaw, Poland, 2018.
19. Dominic, J.; Robb, A. Exploring Effects of Screen-Fixed and World-Fixed Annotation on Navigation in Virtual Reality. In Proceedings of the 2020 IEEE Conference on Virtual Reality and 3D User Interfaces (VR), Atlanta, GA, USA, 22–26 March 2020; pp. 607–615.
20. Adams, E. *Fundamentals of Game Design*, 3rd ed.; New Riders: Thousand Oaks, CA, USA, 2014.
21. Edler, D.; Keil, J.; Dickmann, F. Varianten interaktiver Karten in Video-und Computerspielen—Eine Übersicht. *KN J. Cartogr. Geogr. Inf.* **2018**, *68*, 57–65. [CrossRef]
22. Slater, M.; Usoh, M.; Steed, A. Taking steps. *ACM Trans. Comput. Interact.* **1995**, *2*, 201–219. [CrossRef]
23. Funk, M.; Müller, F.; Fendrich, M.; Shene, M.; Kolvenbach, M.; Dobbertin, N.; Günther, S.; Mühlhäuser, M. Assessing the Accuracy of Point & Teleport Locomotion with Orientation Indication for Virtual Reality using Curved Trajectories. In Proceedings of the 2019 CHI Conference on Human Factors in Computing Systems, Glasgow, Scotland, UK, 4–9 May 2019; Association for Computing Machinery (ACM): New York, NY, USA, 2019; p. 147.
24. Llorach, G.; Evans, A.; Blat, J. Simulator sickness and presence using HMDs. In *Proceedings of the 20th ACM Symposium on Access Control Models and Technologies, Edinburgh, Scotland, UK, November 2014*; Association for Computing Machinery (ACM): New York, NY, USA, 2014; pp. 137–140.
25. Langbehn, E.; Lubos, P.; Steinicke, F. Evaluation of Locomotion Techniques for Room-Scale VR. In Proceedings of the Virtual Reality International Conference—Laval Virtual, Laval, France, 4–6 April 2018; Association for Computing Machinery (ACM): New York, NY, USA, 2018; p. 4.
26. Langbehn, E.; Steinicke, F. Redirected Walking in Virtual Reality. In *Encyclopedia of Computer Graphics and Games*; Springer International Publishing: Berlin, Germany, 2018; pp. 1–11.
27. Lütjens, M.; Kersten, T.P.; Dorschel, B.; Tschirschwitz, F. Virtual Reality in Cartography: Immersive 3D Visualization of the Arctic Clyde Inlet (Canada) Using Digital Elevation Models and Bathymetric Data. *Multimodal Technol. Interact.* **2019**, *3*, 9. [CrossRef]
28. Walmsley, A.P.; Kersten, T.P. The Imperial Cathedral in Königslutter (Germany) as an Immersive Experience in Virtual Reality with Integrated 360° Panoramic Photography. *Appl. Sci.* **2020**, *10*, 1517. [CrossRef]
29. Medynska-Gulij, B. The Effect of Cartographic Content on Tourist Map Users. *Cartography* **2003**, *32*, 49–54. [CrossRef]
30. Wielebski, Ł.; Medyńska-Gulij, B. Graphically supported evaluation of mapping techniques used in presenting spatial accessibility. *Cartogr. Geogr. Inf. Sci.* **2019**, *46*, 311–333. [CrossRef]
31. Çöltekin, A.; Heil, B.; Garlandini, S.; Fabrikant, S.I. Evaluating the Effectiveness of Interactive Map Interface Designs: A Case Study Integrating Usability Metrics with Eye-Movement Analysis. *Cartogr. Geogr. Inf. Sci.* **2009**, *36*, 5–17. [CrossRef]
32. Smith, S.P.; Du'Mont, S. Measuring the effect of gaming experience on virtual environment navigation tasks. In Proceedings of the 2009 IEEE Symposium on 3D User Interfaces, Lafayette, LA, USA, 14–15 March 2009; pp. 3–10.
33. Cybulski, P.; Wielebski, Ł. Effectiveness of Dynamic Point Symbols in Quantitative Mapping. *Cartogr. J.* **2018**, *56*, 146–160. [CrossRef]
34. Gkonos, C.; Enescu, I.I.; Hurni, L. Spinning the wheel of design: Evaluating geoportal Graphical User Interface adaptations in terms of human-centred design. *Int. J. Cartogr.* **2018**, *5*, 23–43. [CrossRef]
35. Cybulski, P. Spatial distance and cartographic background complexity in graduated point symbol map-reading task. *Cartogr. Geogr. Inf. Sci.* **2020**, *47*, 244–260. [CrossRef]
36. Cybulski, P.; Horbiński, T. User Experience in Using Graphical User Interfaces of Web Maps. *ISPRS Int. J. Geo-Inf.* **2020**, *9*, 412. [CrossRef]
37. Mccoleman, C.; Thompson, J.; Anvari, N.; Azmand, S.J.; Barnes, J.; Barrett, R.C.A.; Byliris, R.; Chen, Y.; Dolguikh, K.; Fischler, K.; et al. Digit eyes: Learning-related changes in information access in a computer game parallel those of oculomotor attention in laboratory studies. *Atten. Percept. Psychophys.* **2020**, *82*, 2434–2447. [CrossRef]
38. Medynska-Gulij, B.; Halik, Ł.; Wielebski, Ł.; Dickmann, F. Mehrperspektivische Visualisierung von Informationen zum räumlichen Freizeitverhalten—Ein Smartphone-gestützter Ansatz zur Kartographie von Tourismusrouten. *KN J. Cartogr. Geogr. Inf.* **2015**, *65*, 323–329. [CrossRef]
39. Halik, Ł.; Medyńska-Gulij, B. The Differentiation of Point Symbols using Selected Visual Variables in the Mobile Augmented Reality System. *Cartogr. J.* **2017**, *54*, 147–156. [CrossRef]
40. Van Elzakker, C.P.; Ooms, K. Understanding map uses and users. In *The Routledge Handbook of Mapping and Cartography*; Informa UK Limited: London, UK, 2018; pp. 55–67.

41. Alghofaili, R.; Sawahata, Y.; Huang, H.; Wang, H.-C.; Shiratori, T.; Yu, L.-F. Lost in Style. In Proceedings of the 2019 CHI Conference on Human Factors in Computing Systems, Glasgow, Scotland, UK, 4–9 May 2019; Association for Computing Machinery (ACM): New York, NY, USA, 2019; p. 348.
42. Burch, M. Teaching Eye Tracking Visual Analytics in Computer and Data Science Bachelor Courses. In Proceedings of the Symposium on Eye Tracking Research and Applications, Stuttgart, Germany, 2–5 June 2020; Association for Computing Machinery (ACM): New York, NY, USA, 2020.
43. Medyńska-Gulij, B.; Wielebski, Ł.; Halik, Ł.; Smaczyński, M. Complexity Level of People Gathering Presentation on an Animated Map—Objective Effectiveness Versus Expert Opinion. *ISPRS Int. J. Geo-Inf.* **2020**, *9*, 117. [CrossRef]
44. Kay, M.; Rector, K.; Consolvo, S.; Greenstein, B.; Wobbrock, J.O.; Watson, N.F.; Kientz, J.A. PVT-Touch: Adapting a Reaction Time Test for Touchscreen Devices. In Proceedings of the 2013 7th International Conference on Pervasive Computing Technologies for Healthcare and Workshops, Venice, Italy, 5–8 May 2013.
45. Cybulski, P.; Wielebski, Ł.; Medyńska-Gulij, B.; Lorek, D.; Horbiński, T. Spatial visualization of quantitative landscape changes in an industrial region between 1827 and 1883. Case study Katowice, southern Poland. *J. Maps* **2020**, *16*, 77–85. [CrossRef]
46. Smaczyński, M.; Medyńska-Gulij, B.; Halik, Ł. The Land Use Mapping Techniques (Including the Areas Used by Pedestrians) Based on Low-Level Aerial Imagery. *ISPRS Int. J. Geo-Inf.* **2020**, *9*, 754. [CrossRef]

International Journal of
Geo-Information

Article

The Land Use Mapping Techniques (Including the Areas Used by Pedestrians) Based on Low-Level Aerial Imagery

Maciej Smaczyński *, Beata Medyńska-Gulij and Łukasz Halik

Department of Cartography and Geomatics, Faculty of Geographical and Geological Sciences, Adam Mickiewicz University in Poznań, 61-712 Poznań, Poland; bmg@amu.edu.pl (B.M.-G.); lhalik@amu.edu.pl (Ł.H.)

* Correspondence: maciej.smaczynski@amu.edu.pl

Received: 9 November 2020; Accepted: 12 December 2020; Published: 16 December 2020

Abstract: Traditionally, chorochromatic maps with a qualitative measurement level are used for land use presentations. Along with the use of UAV (Unmanned Aerial Vehicles), it became possible to register dynamic phenomena in a small space. We analyze the application of qualitative and quantitative mapping methods to visualize land use in a dynamic context thanks to cyclically obtained UAV imaging. The aim of the research is to produce thematic maps showing the actual land use of the small area urbanized by pedestrians. The research was based on low-level aerial imagery that recorded the movement of pedestrians in the research area. Additionally, based on the observation of pedestrian movement, researchers pointed out the areas of land that pedestrians used incorrectly. For this purpose, the author will present his own concept of the point-to-polygon transformation of pedestrians' representation. The research was an opportunity to demonstrate suitable mapping techniques to effectively convey the information on land use by pedestrians. The results allowed the authors of this article to draw conclusions on the choice of suitable mapping techniques during the process of thematic land use map design and to specify further areas for research.

Keywords: thematic map; mapping techniques transformation; land use; UAV; geometric representation; pedestrian; chorochromatic maps; dot map; heat map

---

1. Introduction

The employment of Unmanned Aerial Vehicles (UAV) for obtaining data about changes in land use in the form of images occurs in small areas in which various changes can be observed. Thanks to their capability to position themselves at a constant height and remain stationary for a specific period, multirotor UAVs can provide opportunities to observe the phenomenon that occurs in each area. The ability to start air raids every couple of minutes over the area of several hectares allows one to record people moving around. In our research, we have touched upon the problem of producing thematic maps of the land cover, illustrating official land use, including places whereby the use of which has been changed by pedestrians.

Methods of obtaining UAV data require creating a digital elevation model (DEM) and a point cloud based on the images obtained [1–3]. The actual reflection of geometry of the recorded area or object, including defining its actual location, is particularly interesting. To do so, researchers focus on GCPs that serve as reference points used in the aerial triangulation of imagery [4–6]. Using geodetic techniques of measurement allows one to have accurate coordinate values and accurate heights of the photogrammetric models [7–11].

It becomes crucial to properly adjust mapping methods of presenting raster data transformations from images into qualitative and quantitative vector data. We can see the necessity to develop well-known methods of cartographic presentation for visualizing data obtained from UAVs. Obtaining static data based on images is relatively simple, but the observation of and recording changes in dynamic phenomena in the form of images requires a particular approach [12]. Recording space used by pedestrians outside designated sidewalks is one of such phenomena. We assume that actual land cover also includes also the areas occupied by pedestrians outside designated sidewalks, e.g. worn shortcut paths or areas along sidewalks used when the sidewalk is too narrow for all pedestrians.

The aim of the research on cartographic visualization is to obtain highly effective spatial information. The effectiveness of cartographic visualisations connect to the creation of maps that enable the simple and unequivocal reading of features of geographical phenomena [13]. The effectiveness of visualisation is defined as both the efficiency and effectiveness of the communication of information [14]. It is directly proportional to the quantity of information correctly received and is inversely proportional to the time used to obtain it [15].

The increase in cartographic effectiveness significantly improves the usefulness of maps [16] by suitable mapping methods and graphic variables, with which we can capture and demonstrate changes occurring in the area [17–19]. The choice of the mapping technique is key to thematic map design. Mapping techniques are specific methods applied to the cartographic presentation of spatial phenomena and the relations existing there between. The following factors are decisive for the effectiveness of the obtained map: measurement level, geometric form of objects, and graphic variables. The classification of attributes is effected on specific measurement levels by the assignment of numbers or categories to objects, which is tantamount to the presentation of the relations existing between these objects by suitably selected measurement scales. Each map comprises a set of graphical elements that present spatial data [20]. First and foremost, general geographical and topographical maps provide nominal information, and to a lesser extent, quantitative data. Several mapping methods may be used in any one map. Quantitative data are presented applying a statistical approach, and the end-product of the utilization of this specific mapping method are statistical maps. The geometrical form of objects requires the adoption of an appropriate mapping method. Point phenomena may be represented to emphasise qualitative or quantitative attributes.

For the representation of areal phenomena, use is made of the following: chorochromatic maps, choropleth map, dasymetric maps, areal cartodiagrams, scope maps, isoline maps and statistical surfaces. Apart from the division into qualitative and quantitative data, it is worth mentioning the regular or irregular nature of the distribution of areal phenomena, which frequently entails the selection of basic fields for data [21]. Satellite or UAV images, which show types of land cover for individual grids, are called chorochromatic graticule maps, where each grid cell has a different colour.

The dot method is considered as a variant of the point symbol method, but on the dot map, a small circle becomes the sign, the size of which enables the more precise location of the phenomenon. The most logical method of location entails introducing the assumption that one dot presents one object, but in practice a specific value is given to a single dot; this is known as the dot ratio [22].

Heat maps (specific variant of the choropleth map), with various color scales used for areas with different intensity of points that represent magnitude of a given phenomenon, are particularly beneficial for visualizations and the exploration of large quantitative data sets [23–26].

The combination and transformations of mapping methods of point and areal phenomena require appropriate graphical procedures that make it possible to maintain the legibility of the cartographic visualisation effect, i.e., enabling the unequivocal interpretation of spatial relations.

The main objective of our work was to work out thematic maps based on UAV imagery that would present the actual use of the small, urbanized area, including places for which pedestrians changed the form of use. In this research, we pinpointed a few intermediary goals:

- To obtain data on land use and the location of pedestrians in the research area based on low-level aerial imagery.
- To adjust mapping methods to the point-to-polygon transformation of pedestrians' representation.

- To use the dot map, buffer map, chorochromatic map and the so-called *heat map* to present quantitative data by means of the area method, and for cartographic presentation of the actual land use by pedestrians.
- To demonstrate the effectiveness of the thematic maps created, presenting the actual land use by pedestrians.

2. Study Area

The inclusion of pedestrian movement in the identification of the actual land use has become the main determinant in the process of pinpointing conditions the research area should be characterized by. We have assumed that the use of the research area by pedestrians should be as intense as possible. For the purposes of the research, we concluded that urbanized areas would be most intensely used by pedestrians. A relatively dense and diverse form of urbanized areas, as well as the nature of the phenomenon analyzed, and technology employed for observation, should help one reduce the size of the research area. In this study, we assumed that the maximum research area, depending on the specificity of construction of selected areas, would be no larger than 10 ha. The construction of the research area should be balanced and include both buildings and structures, such as corridors (roads, sidewalks, parking lots), objects of the so-called small architecture and green spaces. The study was conducted in the western part of Poland, the Greater Poland province, in the city of Poznań. In accordance with the standards adopted in terms of the research area characteristics, we selected the area that was a part of the campus of Adam Mickiewicz University in Poznań. The area is 9 ha and met all the selected criteria (Figure 1). A chosen fragment is one of the parts most intensely used by pedestrians, as it is located between the main part of the campus and the railway station. The area selected for the research is used daily by thousands of students and university employees, as well as the residents of the nearby housing estates.

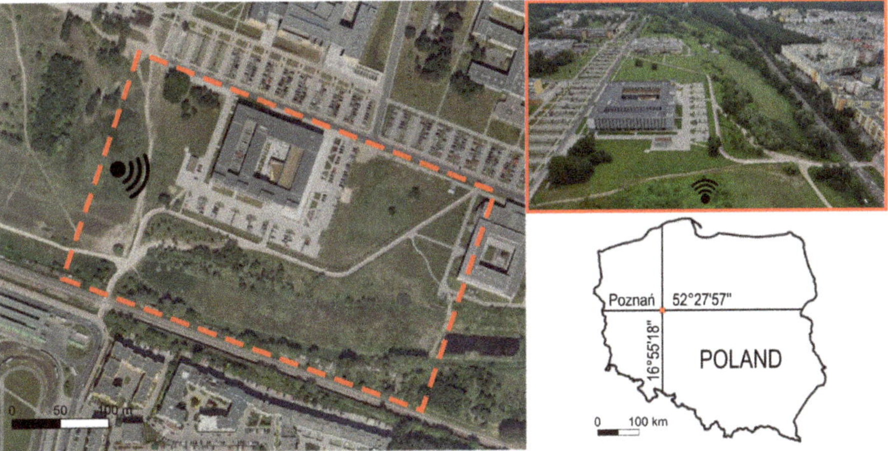

Figure 1. Location of the studied area (background layer orthophotomap from national geoportal www.geoportal.gov.pl).

Then, we analyzed the research area in terms of transport infrastructure (Figure 2). Our priority was to mark out areas for pedestrians, i.e., sidewalks, routes for pedestrians and cyclists, parking lots and access roads. These elements were included in the research as areas in which pedestrian movement was allowed (Figure 2 Transport infrastructure). Additionally, we presented built-up areas (Figure 2 Buildings). The rest of the research area consisted of green areas, classified as nonpedestrian zones. The areas marked out constituted qualitative data. A map presenting the infrastructure of the research area (Figure 2) was worked out with the application of the chorochromatic mapping method, traditionally used for presenting land use.

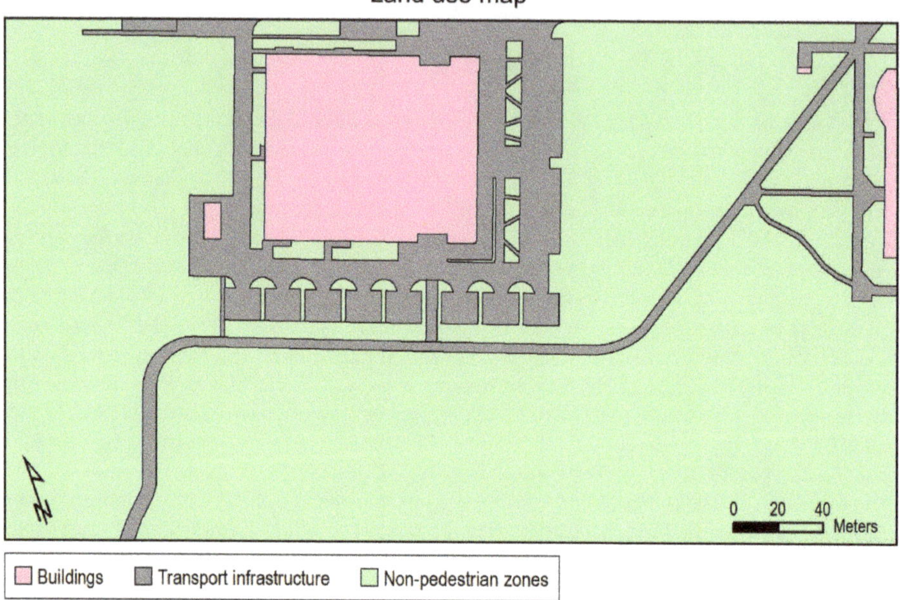

Figure 2. The map of the research area in terms of transport infrastructure: the chorochromatic method (qualitative data).

3. Methodology

To achieve our objectives, we adopted four main stages of research:

- Concept of the point-to-polygon transformation of pedestrian representation (Section 3.1; Figure 3),
- Obtaining low-level imagery (Section 3.2),
- Identification and vectorization of pedestrians (Section 3.3; Figure 4),
- Cartographic visualization of pedestrians' location (Section 3.4; Figures 5–7),
- Results: working out final maps (Section 3.5; Figures 8 and 9).

3.1. Concept of the Point-to-Polygon Transformation of Pedestrian Representation

The employment of UAVs for recording the same area at different times makes it possible to scrutinize changes between individual images. Pedestrian movement is a highly dynamic phenomenon; hence, we have decided to use short time intervals between individual images of the research area. We adopted the interval of 10 seconds that would help us record pedestrian movement. The process of aerial triangulation of the images obtained, and their subsequent vectorization, allowed us to specify the location of individual pedestrians, which, in turn, made it possible to represent pedestrians by means of point objects and to identify each of them in the coordinate system (X; Y).

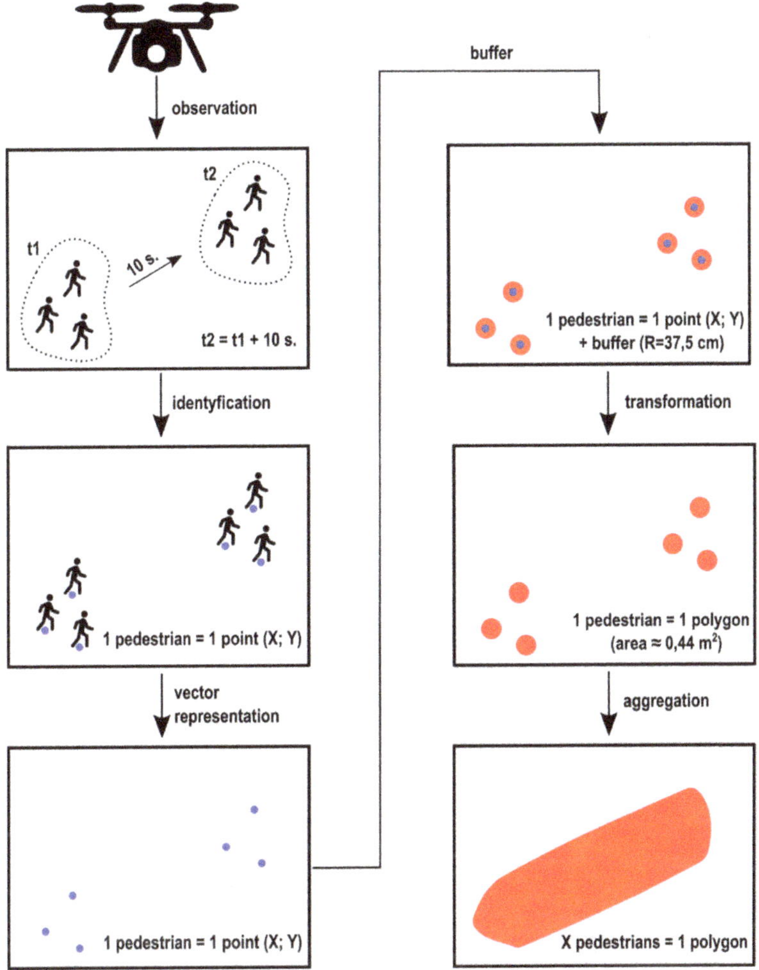

Figure 3. The concept of point-to-polygon transformation of pedestrians' representation.

Figure 4. Vectorization of pedestrians from the research area.

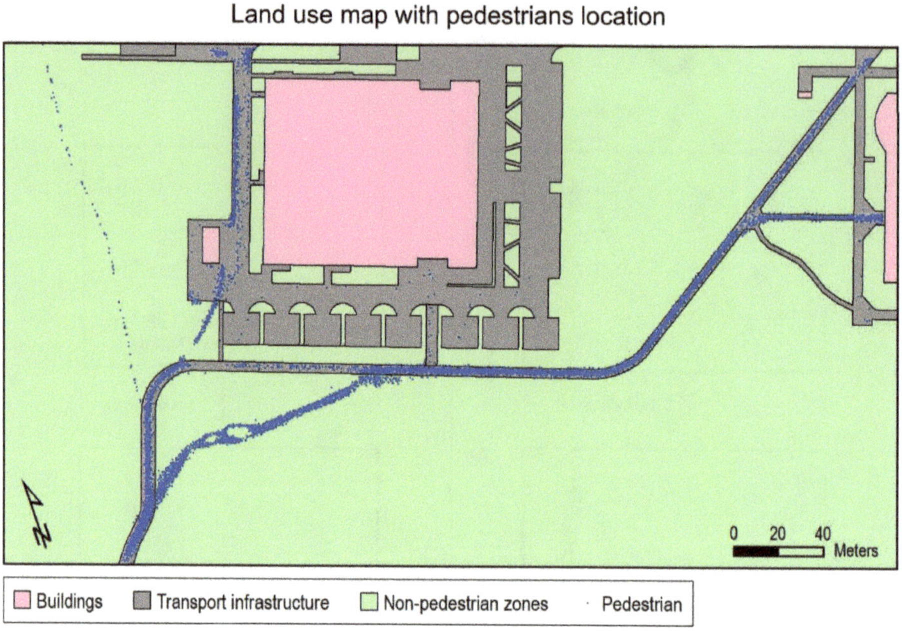

Land use map with pedestrians location

0 20 40 Meters

Buildings Transport infrastructure Non-pedestrian zones · Pedestrian

Figure 5. The map of land use with the location of pedestrians in the research area: the chorochromatic method (qualitative data) and the dot method (quantitative data).

Land use map with pedestrians location outside of the transport infrastructure

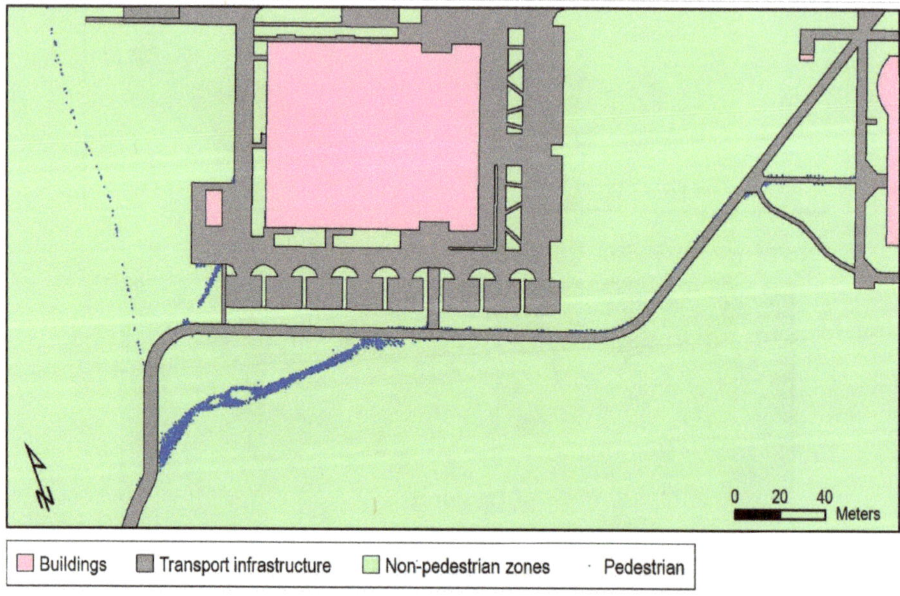

Figure 6. The map of land use with the location of pedestrians outside the transport infrastructure: the chorochromatic method (qualitative data) and the dot method (quantitative data).

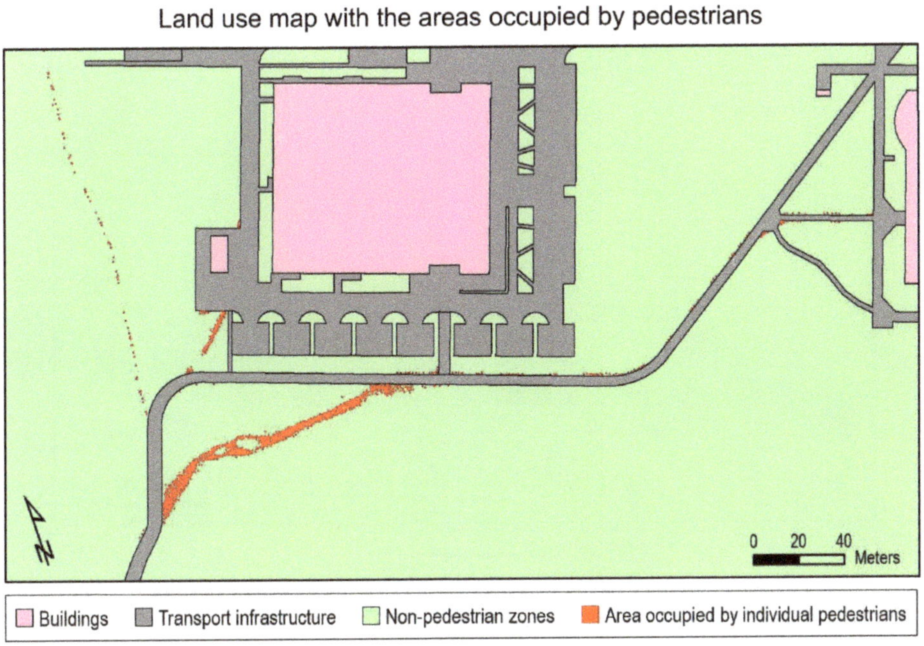

Land use map with the areas occupied by pedestrians

0 20 40 Meters

Buildings Transport infrastructure Non-pedestrian zones Area occupied by individual pedestrians

Figure 7. The map of land use with the areas occupied by pedestrians: the chorochromatic method (qualitative data).

Land use map with informal areas used by pedestrians

0 20 40 Meters

Buildings Transport infrastructure Non-pedestrian zones Informal areas of land used by pedestrians

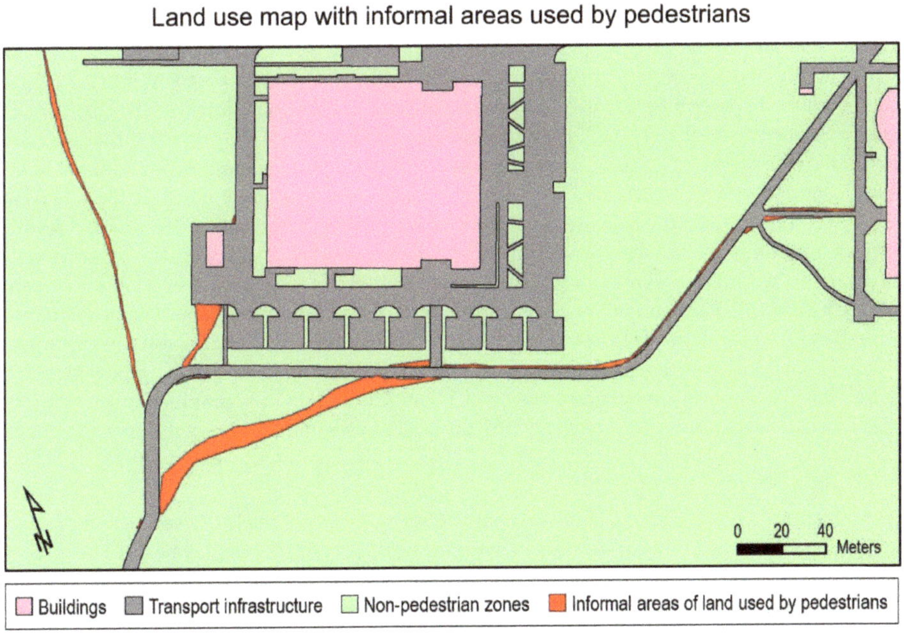

Figure 8. The map of informal areas of land used by pedestrians: the chorochromatic method (qualitative data).

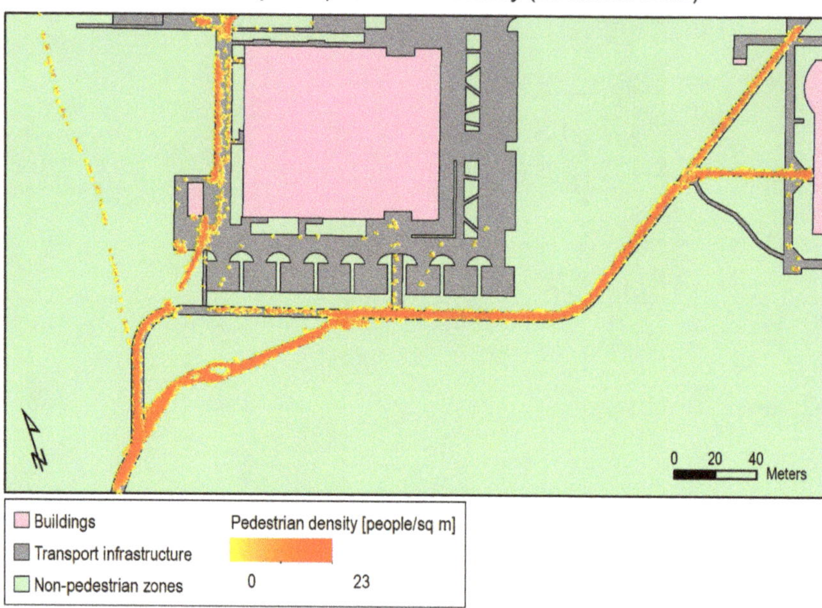

Figure 9. The map of pedestrian density in the research area (continuous scale): the chorochromatic method (qualitative data).

The ability to specify the accurate location of pedestrians in the images obtained and a large scale of the map resulted in the conclusion that the surface object would make the more accurate representation, allowing us to present the area occupied by an individual person in the way that resembles the actual state. To do that, it was necessary to transform the point representation of pedestrians into the polygon representation. To determine the area occupied by an individual pedestrian, we have specified the value of a buffer by means of which transformation would take place. We assumed that the buffer value of a single point would be 37.5 cm (diameter of 75 cm), which corresponded with the largest length of the pedestrian's footstep (62.5–75 cm) [27]. Then, we marked out the area of the actual land use based on the obtained polygon objects representing pedestrians. The aim of this was to determine the areas informally used by pedestrians. With recording pedestrians at intervals, it is not possible to determine the location of individual pedestrians between the images obtained. In the research, we have assumed that obtaining images at short intervals and the average speed of pedestrians of 5 km/h [28–30] would allow us to determine the aggregate distance between pedestrians captured in individual images. Figure 3 depicts a concept of point-to-polygon transformation of the pedestrian's representation. It shows the different stages of identification and cartographic representation of pedestrians in the research area, as well as the final result presenting the area of the actual land use.

3.2. Obtaining Low-Level Imagery

In this research, we decided to obtain spatial data by means of the low-level imagery method. The time of recording pedestrians' movement in the research area with the employment of UAVs strictly correlated with the maximum pedestrian movement intensity. A field survey allowed us to determine that pedestrian movement in the selected research area was the most intense between 7:30 and 8:15 on weekdays during the academic year. Thus, we decided to conduct the research with the employment of UAVs specifically during this time. Having considered the size of the research area and the necessity to observe pedestrian movement, we concluded that the multirotor UAV would be the best choice in terms of obtaining visual data. We used the Tarot X6 platform, equipped with a camera with 16.1 Mpx matrix. Although the platform was equipped with the Global Navigation Satellite Systems (GNSS), we decided that the aerial triangulation process of the imagery obtained would be carried out with the employment of GCPs located in the research area, which allowed us to make the process as accurate as possible [31,32]. The GNSS RTK technique is one of the methods of measuring GCPs [33]. For the measurement, we used the GNSS Trimble R4 Receiver, model 3, with differential corrections provided by the permanent reference station network. To create as accurate photogrammetric analysis of the research area as possible, it was necessary to record it from all sides and to work out the 3D model, using the Stucture-from-Motion (SfM) algorithm [34,35].

It was important to provide security during UAV raids. The observation of pedestrian movement was conducted in accordance with the existing provisions of law. In the study, the priority was to ensure safety for the operator, the UAV platform, and most of all for people in the vicinity during the flight mission using UAV technology. Bearing in mind safety and legal regulations, it became impossible to fly directly over pedestrians. This was mainly due to the inability to obtain images using the classic photogrammetric flight path based on photogrammetric series. In order to achieve the aim of the study, we have determined that it is necessary to register all pedestrians moving in the study area at the same time. We also assumed that the registration of pedestrian traffic would be carried out from one observation post. Such an approach in the conducted study forced the necessity to obtain oblique photos. We used an unmanned aerial vehicle (UAV)—a multirotor platform—which ensured stable registration of the research area from one observation position. The photogrammetric images obtained were processed in the Agisoft Metashape Professional software. The GCP-based aerial triangulation process resulted in obtaining the average RMSE value for all GCPs at the level of 0.27 m. Orthophotomaps were exported from Agisoft Metashape Professional for all the images obtained in the GeoTIFF format to identify pedestrians recorded in the research area. During the recording of pedestrian movement in the research area, we obtained 275 images in total.

3.3. Identification and Vectorization of Pedestrians

The orthophotomaps, created for all images, were then imported to the GIS software. In the study, we will use the ESRI ArcMap software to process spatial data and develop thematic maps. In this software, we vectorized pedestrians recorded in individual images to determine their exact situational coordinates (X; Y). It was difficult to determine the accurate vectorization place of each pedestrian, as pedestrians were constantly moving. We specified that the place of projecting the center of gravity of each pedestrian in the research area plane would constitute the vectorization point (Figure 4); this was conducted for the 275 images obtained at 10-s intervals.

3.4. Cartographic Visualization of Pedestrians' Location

In our attempt to meet research objectives on designing maps that use adequate mapping techniques for presenting the real land use by pedestrians, we adopted the following order of creating intermediary maps:

- The map depicting land use with the location of pedestrians in the research area: the area method (qualitative data) and the dot method (quantitative data) (Figure 5).
- The map depicting land use with the location of pedestrians outside the transport infrastructure: the area method (qualitative data) and the dot method (quantitative data) (Figure 6).
- The map depicting land use with the areas occupied by pedestrians: the area method (qualitative data) (Figure 7).

Mapping techniques were listed in terms of data type (geometry of objects).

Vectorization conducted in the research allowed one to obtain quantitative data presenting the location (X; Y) of pedestrians in the research area. The most natural way to present such data is to use dots [36]. For presenting quantitative data we used the dot method. Points as 0-dimensional objects are represented with a pair of coordinates [24], as presenting a point object on the map through direct physical representation of point geometry would make such object invisible. To read its location, it was necessary to use the additional attribute, the dot ratio. When designing our dot map, we assumed the dot width (diameter) of 0.5 mm, because it is the smallest symbol size recognizable by the map user [37]. The dot created is a cartographic representation of a point feature, and the geometric center of the dot corresponds to the coordinates of the point it represents. On the previously prepared map (Figure 5) pedestrians were represented as objects in the shape of a small circle, so that it is visible to map users. This is only the cartographic representation of a point object on the map. Quantitative data were presented in the background of the map showing the infrastructure of the research area (Figure 2). Such combination of methods of cartographic presentation made it possible to present both qualitative and quantitative data. As a result, researchers obtained a map for the creation of which two methods, the dot method and the chorochromatic map were used. In Figure 5 we demonstrated the spatial layout of all the recorded pedestrians (quantitative data) in the background of the infrastructure of the research area (qualitative data).

In the research. we decided to mark out the areas incorrectly used by pedestrians, i.e., informal land use. Pedestrians, who were moving outside the specified transport infrastructure, were considered informal land users. To single such people out of all recorded pedestrians, we carried out subtraction on the quantitative data set. Such activity was possible thanks to the use of the *Erase* tool, which allowed to localize only those pedestrians outside the designated transport infrastructure. In Map 6 we marked pedestrians using land in the informal way. As in Figure 5, for presenting quantitative (pedestrians) and qualitative (infrastructure) data, we used two mapping methods, i.e., the dot method and chorochromatic mapping.

A small research area and the UAV technology employed allowed us to obtain accurate location of individual pedestrians. We concluded that the representation of pedestrians by means of the dot mapping method was not adequate. It was related to the opportunity to work out large-scale cartographic visualizations that can faithfully reflect pedestrians in the research area. The intermediary objective of the research was to determine the area of the land that pedestrians used incorrectly, i.e., informal areas. We specified that it was necessary to carry out the point-to-polygon transformation of pedestrians' representation.

Transformation was understood as a change of the mapping technique, i.e., the transformation of the geometry of the element used in cartography, which would allow us to determine the area that pedestrians occupy outside the transport infrastructure specified. The average length of a human footstep is between 62.5 to 75 cm, depending on sex and type of walk [27]. We assumed that the area occupied by an individual moving pedestrian would be equal to a circle that is 75 cm (the maximum length of a human footstep) in diameter. That meant that each point object representing an individual pedestrian was been surrounded by a buffer of 37.5 cm, counting from the values of coordinates describing the location of the point. After transformation, one person was represented by means of a polygon object with the geometry of a circle occupying the area of 0.44 m^2 (Figure 7). The ArcMap *Buffer* tool was used to transform point features into area features. Transformation of the point representation of a pedestrian into the polygon representation resulted in the change in data type (from quantitative to qualitative data). The map presenting the area occupied by pedestrians outside the transport infrastructure (Figure 7) was worked out with the use of the chorochromatic mapping method. The buffer value adopted in the research constitutes just an example and can be modified, depending on needs and the assumptions of the research.

3.5. Results: Working out Final Maps

Working out intermediary maps (Figures 5–7) made it possible to create additional cartographic analyses that effectively demonstrate the actual land use by pedestrians. Since we distinguished two mapping techniques, the following map types were worked out:

- The map presenting informal areas of land used by pedestrians: the area method (qualitative data) (Figure 8),
- The map presenting pedestrian density in the research area: the area method (qualitative data) (Figure 9).

The transformation of point objects into polygon objects (the representation of pedestrians on the map) allowed us to determine the area occupied by individual pedestrians in the research area. The area in question refers only to pedestrians recorded in individual images and does not determine the land use area. When obtaining images at specific intervals, it is impossible to obtain data on the location of individual pedestrians in the research area between the states recorded. However, the large number of recorded pedestrians in individual images of the same area and visibly repetitive pedestrian routes in the research area make it possible to obtain data on the location of pedestrians at the time between individual images. In the research, individual images recording the research area were obtained at 10-s intervals, which was related to the observation of the dynamic phenomenon. To determine the actual size of informal land use areas in the analyzed research area, we decided to aggregate (generalize) the polygons created. For this purpose, the *Aggregate Polygons* cartographic generalization tool available in the ArcMap software was used. During the process of defining aggregate variables we determined the aggregation distance representing individual pedestrians, which was 13.9 m. The value defined the average distance the pedestrian could cover in 10 s, i.e., during the interval between individual images. In our calculations, we assumed that the average walking speed was 5 km/h [28–30]. To avoid the aggregation of polygons located on the opposite sides of sidewalks, the dividing layer (i.e., the transport infrastructure) was determined. Because of this aggregation, we obtained the area of the informal land use by pedestrians in the research area (Figure 8). In addition, the repeatability of pedestrian traffic along the transport infrastructure, recorded on interval images from Low-Level Aerial Imagery and carrying out cartographic generalization consisting in the aggregation of surface objects presenting individual pedestrians on images made at different times, allowed for the estimation of their paths, and obtaining information about the location of pedestrians in the time between single images. For working out the map, we used the chorochromatic mapping method that presents qualitative data on land use. Additionally, aggregation allowed us to generalize areal data presenting the area occupied by individual pedestrians and to obtain the total area of informal land use (Figure 8).

The presentation of large sets of quantitative data makes it difficult or even impossible to interpret the represented phenomenon correctly. To effectively present the results of the quantitative data set study, we used the *heat map* [23]. In traditional cartographic methodology maps presenting quantitative data are considered cartograms if they cover the entire area [38]. However, if they show values of average intensity only for classified fields, they are considered dasymetric maps. In the research the dot method was used for working out the heat map that presented quantitative data by means of the area method (the dasymetric map). The heat map was created in the software using the *Kernel Density* tool. To work out the *heat map* [39], we used point objects presenting the location of pedestrians. According to the nomenclature of cartographic methodology, to design the heat map (Figure 9), we used a dasymetric mapping method, presenting the average phenomenon intensity only in classified fields, i.e., areas in which the presence of pedestrians was identified. To boost the effectiveness of data presentation, we showed the data in the background of the research area infrastructure, worked out with the use of the chorochromatic method. We calculated pedestrian density in the research area and provided it in the form of the number of people per 1 m^2. The heat map of the observed pedestrian movement allowed us to obtain information about the frequency of use of specific research areas by pedestrians (Figure 9). It showed that, despite the existing transport infrastructure, pedestrians use areas that are not meant to be used for walking. Moreover, the map makes it possible to draw a conclusion that such phenomenon occurs very frequently in some parts of the research area.

The use of the heat map compensated for the disadvantage of the dot map, i.e., the lack of differentiation when the dots touch each other or overlap. Therefore, we proposed to present quantitative data using a continuous-scale heat map to obtain information about the intensity of pedestrian traffic in the study area. Unlike the dot map, the heat map presents surfaces with a cumulative number of point features. In the case of this test, it is an area of 1 m^2. Additionally, we proposed a second version of the heat map (Figure 10), with a more interpretable step scale in four intervals based on Jenks' natural breaks classification method.

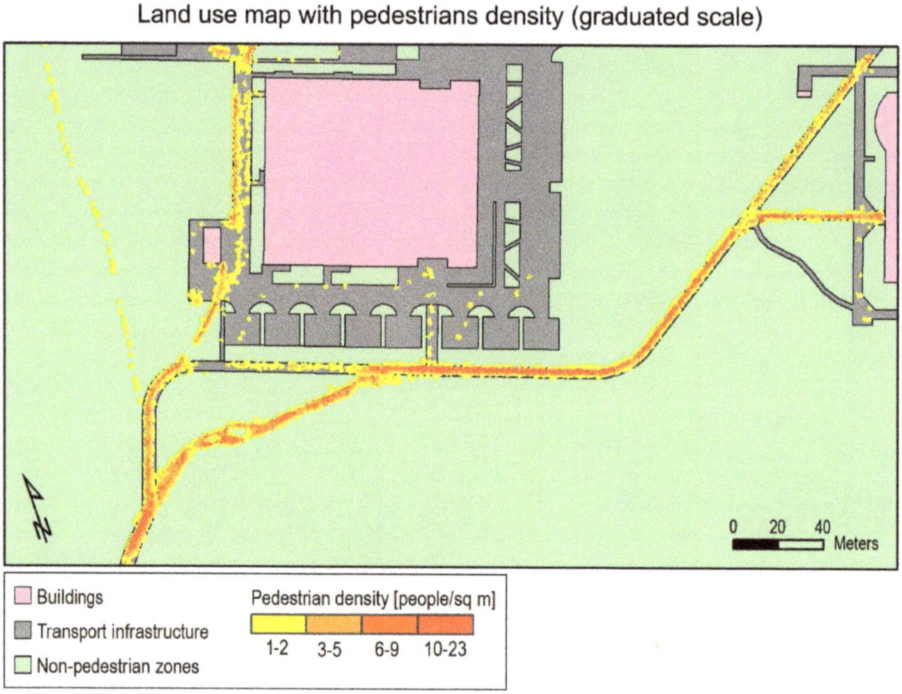

Figure 10. The map of pedestrian density in the research area (graduated scale): the chorochromatic method (qualitative data).

4. Conclusions and Discussion

In this article, we demonstrated a method of researching and visualizing actual land use. The research was conducted based on low-level aerial imagery obtained from UAVs. The UAV technology allowed us to observe and record pedestrian movement in the research area under analysis. As pedestrians were recorded, it was challenging to design a raid with the UAV platform to ensure maximum security for them. It was then necessary to meticulously observe the analyzed urbanized area in search of possible location of photogrammetric raid stations. Determining them with precision helped us accurately capture the entire research area. It mattered also in terms of opportunities to carry out the process of aerial triangulation of the imagery obtained and to achieve possibly accurate value of pedestrians' location in the research area. Modelling static objects by means of the UAV technology, as commonly described in the literature, allowed us to extend the methodology of observing pedestrian movement as a dynamic phenomenon. The methodology of designing maps presenting the actual land use and considering the areas used by pedestrians based on UAV images can be used in other areas while meeting two conditions. The first condition is that there should be relatively high pedestrian traffic in the transport infrastructure and in nonpedestrian zones. The second condition is the ability to use the UAV platform with safety for people and appropriate technical conditions.

Data obtained in this way made it possible to create thematic maps to present the actual use of a small, urbanized area, including areas where pedestrians changed the form of use.

To be able to obtain data on land use and the location of pedestrians through the observation of their movement, it was necessary to select the appropriate UAV platform. A study of the subject made it possible to choose a multirotor UAV as a highly effective tool, maintaining one position during a raid and constantly observing the entire research area. A selection of time intervals between images was also a highly significant factor. Taking into consideration the average speed of pedestrians during walking, we established that 10-s intervals are optimal for recording pedestrian movement.

It was possible to determine the area occupied by individual pedestrians in the research area thanks to the change of method, as the dot method (Figure 5) turned into the area method through the transformation of point objects into polygon objects based on the adopted basis of the buffer value (Figure 7). The value of the buffer diameter is a proposal based on the human stride length while walking [27]. It can be modified depending on the observed dynamic phenomenon; for example the visualization of runners. The cartographic generalization, consisting in the aggregation of polygon objects presenting individual pedestrians, allowed us to group individual polygon objects in informal land use areas (Figure 8). We presented a concept of the representation of the point-to-polygon transformation of pedestrians in Figure 3.

We worked out mapping methods adjusted to vectorial point-to-polygon transformation of pedestrians' representation and indicated validity of using qualitative methods (the range method) and quantitative methods (the dot method, the dasymetric version of the heat map, Figure 9) other than just the chorochromatic method.

The suggested method of new visualizations consisted in the point-to-polygon transformation of the representation of pedestrians [18]. We believe it is the way to enrich the process of designing land use maps that so far have focused on conveying spatial data in the traditional way in the form of chorochromatic maps. We are not trying to suggest the best method, as each of these mapping techniques depicts different features of the phenomenon [40]. Hence, the effectiveness of these maps depends on the aspect of the spatial phenomenon they analyze. The employment of the principles of static map design helps one to create a clear and transparent cartographic image [22].

Quantitative data, such as the location of individual pedestrians on the map, can be depicted by means of the dot method. However, if multiple point objects occur next to one another, the effectiveness of the map diminishes and it becomes less clear [41]. To boost the effectiveness of the map, such quantitative data can be presented by means of the heat map (specific variant of the choropleth map), which makes it possible to present the number of pedestrians per given area.

The point-to-polygon transformation of pedestrian representation constitutes a shift from quantitative to qualitative data. Thanks to the generalization of qualitative data in the area occupied by individual pedestrians, it is possible to depict the actual land use. However, such data presentation fails to provide information on magnitude of the occurring phenomenon, as opposed to heat maps. In our opinion, the actual use of land, considering the places for which pedestrians changed the form of use, is a good way to combine several mapping techniques to present the quantitative and qualitative aspects of this spatial phenomenon. We decided to add the second version of the heat map (Figure 10) with a more interpretable step scale [17,22,38,42].

The research conducted and the cartographic analysis designed constitute a suggestion how pedestrians recorded in the research area by UAVs can be represented. So far, researchers have attempted to develop a visualization of participants in a mass event based on low-level aerial imagery [18]. In their research, they proposed a set of animations that presented the distribution of participants in a mass event (dynamic objects) in the research area at specific times of the event. Additionally, the animations show the main points of interest of the participants of the mass event (static objects). Dynamic and static objects were presented using an orthophotomap, a dot map and a map of buffers then assessed by experts in terms of the effectiveness. Most of the pedestrians in the study cited sat, but were not in motion. Methods of presentation can be developed and modified, and then tested on users [18]. Sets of data resulting from such studies can be used for creating visualizations in 3D and 4D, which were not the subject matter of the research but have a great potential for perspective and temporal geovisualizations [7]. Broad sets of quantitative data, obtained by means of UAVs, can be also used for designing animated cartographic visualizations [43].

Currently, new methods of cartographic presentation can be analyzed more accurately in terms of effectively conveying data through eye-tracking studies carried out on users [44,45]. Recognition and tracking of human trajectories is a valuable issue due to many aspects of everyday life. A method commonly used for this purpose was satellite receivers [46]. In the past, attempts were made to determine human movement due to the epidemiological risk and the possibility of transmitting serious diseases through movement and contact with other people [47]. Data obtained with the use of basic GPS receivers, which were equipped with students moving around the campus. However, the basic GPS receiver does not allow to obtain accurate data on the location of individual people, unlike the UAV platform and the obtained images fitted into the coordinate system based on the GCPs. In addition, the proposed method of tracking pedestrians with the use of a GPS receiver requires that each tracked person be equipped with it, which may cause unnatural behavior. The method we propose using the UAV platform allows for the observation of people moving without affecting their behavior. As such, it was possible to determine the actual land use by pedestrians and to design a large-scale thematic map to present the phenomenon.

Author Contributions: Conceptualization, Maciej Smaczyński, Beata Medyńska-Gulij; Methodology, Maciej Smaczyński, Beata Medyńska-Gulij; Software, Maciej Smaczyński; Validation, Maciej Smaczyński and Łukasz Halik; Formal Analysis, Beata Medyńska-Gulij; Investigation, Maciej Smaczyński, Beata Medyńska-Gulij and Łukasz Halik; Resources, Maciej Smaczyński; Data Curation, Maciej Smaczyński; Writing-Original Draft Preparation, Maciej Smaczyński, Beata Medyńska-Gulij; Visualization, Maciej Smaczyński, Beata Medyńska-Gulij; Supervision, Beata Medyńska-Gulij; Project Administration, Maciej Smaczyński; Funding Acquisition. All authors have read and agreed to the published version of the manuscript.

Funding: This paper is the result of research on visualization methods carried out within statutory research in the Department of Cartography and Geomatics, Faculty of Geographical and Geological Sciences, Adam Mickiewicz University in Poznań, in Poland.

Conflicts of Interest: The authors declare no conflict of interest.

References

1. Niedzielski, T. Applications of Unmanned Aerial Vehicles in Geosciences: Introduction. *Pure Appl. Geophys.* **2018**, *175*, 3141–3144. [CrossRef]
2. Clapuyt, F.; Vanacker, V.; Schlunegger, F.; Van Oost, K. Unravelling earth flow dynamics with 3-D time series derived from UAV-SfM models. *Earth Surf. Dyn.* **2017**, *5*, 791–806. [CrossRef]
3. Colomina, I.; Molina, P. Unmanned aerial systems for photogrammetry and remote sensing: A review. *ISPRS J. Photogramm. Remote Sens.* **2014**, *92*, 79–97. [CrossRef]
4. Aicardi, I.; Garbarino, M.; Lingua, A.; Lingua, E.; Marzano, R.; Piras, M. Monitoring post-fire forest recovery using multi-temporal Digital Surface Models generated from different platforms. *EARSeL eProceedings* **2016**, *15*, 1–8.
5. Gerke, M.; Przybilla, H.-J. Accuracy analysis of photogrammetric UAV image blocks: Influence of onboard RTK-GNSS and cross flight patterns. *Photogramm. Fernerkund. Geoinf.* **2016**, *2016*, 17–30. [CrossRef]
6. Padró, J.C.; Muñoz, F.J.; Planas, J.; Pons, X. Comparison of four UAV georeferencing methods for environmentalmonitoring purposes focusing on the combined use with airborne andsatellite remote sensing platforms. *Int. J. Appl. Earth Obs. Geoinf.* **2019**, *75*, 130–140. [CrossRef]
7. Halik, Ł.; Smaczyński, M. Geovisualisation of relief in a virtual reality system on the basis of low-level aerial imagery. *Pure Appl. Geophys.* **2018**, *175*, 3209–3221. [CrossRef]
8. Anai, T.; Sasaki, T.; Osaragi, K.; Yamada, M.; Otomo, F.; Otani, H. Automatic Exterior Orientation Procedure for Low-Cost Uav Photogrammetry Using Video Image Tracking Technique and Gps Information. In Proceedings of the ISPRS—International Archives of the Photogrammetry, Remote Sensing and Spatial Information Sciences 2012, Melbourne, Australia, 25 August–1 September 2012; Volume XXXIX-B7, pp. 469–474. [CrossRef]

9. Barazzetti, L.; Remondino, F.; Scaioni, M.; Brumana, R. Fully automatic UAV image-based sensor orientation. In Proceedings of the International Archives of Photogrammetry Remote Sensing and Spatial Information Sciences, Commission V Symposium 2010, Calgary, AB, Canada, 15–18 June 2010; Volume XXXVIII, Part 5. pp. 1–6. Available online: http://www.isprs.org/proceedings/XXXVIII/part1/12/12_02_Paper_75.pdf (accessed on 30 October 2020).
10. Wang, J.; Garratt, M.; Lambert, A.; Wang, J.J.; Han, S.; Sinclair, D. Integration of Gps/Ins/Vision Sensors To Navigate Unmanned Aerial Vehicles. In Proceedings of the International Archives of the Photogrammetry, Remote Sensing and Spatial Information Sciences 2008, Beijing, China, 3–11 July 2008; Volume XXXVII, Part B1. pp. 963–970.
11. Eugster, H.; Nebiker, S. Uav-Based Augmented Monitoring—Real-Time Georeferencing and Integration of Video Imagery With Virtual Globes. *Archives* **2008**, *37*, 1229–1236.
12. Wielebski, Ł.; Medyńska-Gulij, B.; Halik, Ł.; Dickmann, F. Time, Spatial, and Descriptive Features of Pedestrian Tracks on Set of Visualizations. *ISPRS Int. J. Geo Inf.* **2020**, *9*, 348. [CrossRef]
13. Cybulski, P.; Wielebski, Ł.; Medyńska-Gulij, B.; Lorek, D.; Horbiński, T. Spatial visualization of quantitative landscape changes in an industrial region between 1827 and 1883. Case study Katowice, southern Poland. *J. Maps* **2020**, *16*, 77–85. [CrossRef]
14. Wielebski, Ł.; Medyńska-Gulij, B. Graphically supported evaluation of mapping techniques used in presenting spatial accessibility. *Cartogr. Geogr. Inf. Sci.* **2019**, *46*, 311–333. [CrossRef]
15. Żyszkowska, W.; Spallek, W.; Borowicz, D. *Kartografia Tematyczna*; Wydawnictwo Naukowe PWN: Warszawa, Poland, 2012.
16. MacEachren, A.M. The role of maps in spatial knowledge acquisition. *Cartogr. J.* **1991**, *28*, 152–162. [CrossRef]
17. Kraak, M.J.; Ormeling, F. *Cartography: Visualization of Geospatial Data*; Taylor & Francis Group: Abingdon, UK, 2020; ISBN 0429464193.
18. Medyńska-Gulij, B.; Wielebski, Ł.; Halik, Ł.; Smaczyński, M. Complexity Level of People Gathering Presentation on an Animated Map—Objective Effectiveness Versus Expert Opinion. *ISPRS Int. J. Geo Inf.* **2020**, *9*, 117. [CrossRef]
19. Cartwright, W.; Miller, S.; Pettit, C. Geographical Visualization: Past, Present and Feature Development. *J. Spat. Sci.* **2004**, *49*, 25–36. [CrossRef]
20. Forrest, D. Thematic Maps in Geography. In *International Encyclopedia of the Social & Behavioral Sciences*, 2nd ed.; Elsevier: Oxford, UK, 2015; Volume 24, pp. 260–267.
21. Robinson, A.H.; Morrison, J.L.; Muehrecke, P.C.; Kimerling, A.J.; Guptil, S.C. *Elements of Cartography*, 6th ed.; Wiley: New York, NY, USA, 1995.
22. Dent, B.D. *Cartography. Thematic Map Design*, 5th ed.; WCB/McGraw-Hill: Boston, MA, USA, 1999; ISBN 007231902X.
23. Netek, R.; Pour, T.; Slezakova, R. Implementation of Heat Maps in Geographical Information System—Exploratory Study on Traflc Accident Data. *Open Geosci.* **2018**, *10*, 367–384. [CrossRef]
24. Medyńska-Gulij, B. *Kartografia. Zasady i Zastosowania Geowizualizacji*; Wydawnictwo Naukowe PWN: Warszawa, Poland, 2015.
25. Yeap, E.; Uy, I. Marker Clustering and Heatmaps: New Features in the Google Maps Android API Utility Library. Google Geo Developers, 2014. Available online: https://mapsplatform.googleblog.com/2014/02/marker-clustering-and-heatmaps-new.html (accessed on 3 December 2020).
26. Trame, J.; Keßler, C. *Exploring the Lineage of Volunteered Geographic Information with Heat Maps*; GeoViz: Hamburg, Germany, 2011; Available online: http://carsten.io/trame-kessler-geoviz2011.pdf (accessed on 4 December 2020).
27. Neufert, E. *Neufert. Podręcznik Projektowania Architektoniczo–Budowlanego*; Wydawnictwo Arkady: Warszawa, Poland, 2011.
28. Mohler, B.J.; Thompson, W.B.; Creem-Regehr, S.H.; Pick, H.L., Jr.; Warren, W.H., Jr. Visual flow influences gait transition speed and preferred walking speed. *Exp. Brain Res.* **2007**, *181*, 221–228. [CrossRef]
29. Browning, R.C.; Baker, E.A.; Herron, J.A.; Kram, R. Effects of obesity and sex on the energetic cost and preferred speed of walking. *J. Appl. Physiol.* **2006**, *100*, 390–398. [CrossRef]
30. Levine, R.V.; Norenzayan, A. The Pace of Life in 31 Countries. *J. Cross Cult. Psychol.* **1999**, *30*, 178–205. [CrossRef]

31. Ruzgiene, B.; Berteška, T.; Gečyte, S.; Jakubauskiene, E.; Aksamitauskas, V.C. The surface modelling based on UAV Photogrammetry and qualitative estimation. *Measurement* **2015**, *73*, 619–627. [CrossRef]
32. Nex, F.; Remondino, F. UAV for 3D mapping applications: A review. *Appl. Geomat.* **2014**, *6*, 1–15. [CrossRef]
33. Clapuyt, F.; Vanacker, V.; Van Oost, K. Reproducibility of UAV-based earth topography reconstructions based on struc- ture-from-motion algorithms. *Geomorphology* **2016**, *260*, 4–15. [CrossRef]
34. Westoby, M.J.; Brasington, J.; Glasser, N.F.; Hambrey, M.J.; Reynolds, J.M. 'Structure-from-Motion' photogramme-try: A low-cost, effective tool for geoscience applications. *Geomorphology* **2012**, *179*, 300–314. [CrossRef]
35. Smaczyński, M.; Horbiński, T. Creating 3D Model of the Existing Historical Topographic Object Based on Low-Level Aerial Imagery. *KN J. Cartogr. Geogr. Inf.* **2020**. [CrossRef]
36. Medyńska-Gulij, B. How the Black Line, Dash and Dot Created the Rules of Cartographic Design 400 Years Ago. *Cartogr. J.* **2013**, *50*, 356–368. [CrossRef]
37. Medynska-Gulij, B. Point Symbols: Investigating Principles and Originality in Cartographic Design. *Cartogr. J.* **2008**, *45*, 62–67. [CrossRef]
38. Medyńska-Gulij, B. *Kartografia i Geowizualizacjia*; Wydawnictwo Naukowe PWN: Warszawa, Poland, 2011.
39. Slocum, T.A. *Thematic Cartography and Geovisualization*, 3rd ed.; Pearson Prentice Hall: Upper Saddle River, NJ, USA, 2009; ISBN 9780132298346.
40. Medynska-Gulij, B.; Cybulski, P. Spatio-temporal dependencies between hospital beds, physicians and health expenditure using visual variables and data classification in statistical table. *Geod. Cartogr.* **2016**, *65*, 67–80. Available online: http://journals.pan.pl/dlibra/publication/113301/edition/98381/content/spatio-temporal-dependencies-between-hospital-beds-physicians-and-health-expenditure-using-visual-variables-and-data-classification-in-statistical-table-cybulski-pawel-medynska-gulij-beata?language=pl (accessed on 30 November 2020). [CrossRef]
41. Forrest, D. Geographic Information: Its Nature, Classification, and Cartographic Representation. *Cartogr. Int. J. Geogr. Inf. Geovisualization* **1999**, *36*, 31–53. [CrossRef]
42. Freitag, U. Kartographische Massstäbe. In *Wiener Schriften zur Geograhie und Kartographie*; Institut für Geographie und Regionalforschung der Universität Wien: Vienna, Austria, 2004; pp. 159–173.
43. Cybulski, P.; Wielebski, Ł. Effectiveness of dynamic point symbols in quantitative mapping. *Cartogr. J.* **2019**, *56*, 146–160. [CrossRef]
44. Krassanakis, V.; Cybulski, P. A review on eye movement analysis in map reading process: The status of the last decade. *Geod. Cartogr.* **2019**, *68*, 17–35. [CrossRef]
45. Horbiński, T.; Cybulski, P.; Medyńska-Gulij, B. Graphic design and placement of buttons in mobile map application. *Cartogr. J.* **2020**. [CrossRef]
46. Isaacson, M.; Shoval, N. Application of Tracking Technologies to the Study of Pedestrian Spatial Behavior. *Prof. Geogr.* **2008**, *58*, 172–183. [CrossRef]
47. Qi, F.; Du, F. Trajectory Data Analyses for Pedestrian Space-time Activity Study. *JoVE J. Vis. Exp.* **2013**, *72*. [CrossRef] [PubMed]

Publisher's Note: MDPI stays neutral with regard to jurisdictional claims in published maps and institutional affiliations.

International Journal of *Geo-Information*

Article

TouchTerrain—3D Printable Terrain Models

Chris Harding [1,*], Franek Hasiuk [2] and Aaron Wood [3]

1 Department of Geological and Atmospheric Sciences and Human-Computer Interaction Program, Iowa State University, Ames, IA 50013, USA
2 Kansas Geological Survey, Lawrence, KS 66047, USA; franek@kgs.ku.edu
3 Department of Geological and Atmospheric Sciences Iowa State University, Ames, IA 50013, USA; awood@iastate.edu
* Correspondence: charding@iastate.edu

Abstract: TouchTerrain is a simple-to-use web application that makes creating 3D printable terrain models from anywhere on the globe accessible to a wide range of users, from people with no GIS expertise to power users. For coders, a Python-based standalone version is available from the open-source project's GitHub repository. Analyzing 18 months of web analytics gave us a preliminary look at who is using the TouchTerrain web application and what their models are used for; and to map out what terrains on the globe they chose to 3D print. From July 2019 to January 2021, more than 20,000 terrain models were downloaded. Models were created for many different use cases, including education, research, outdoor activities and crafting mementos. Most models were realized with 3D printers, but a sizable minority used CNC machines. Our own experiences with using 3D printed terrain in a university setting have been very positive so far. Anecdotal evidence points to the strong potential for 3D printed terrain models to provide significant help with specific map-related tasks. For the introductory geology laboratory, 3D printed models were used as a form of "training wheels" to aid beginning students in learning to read contour maps, which are still an important tool for geology.

Keywords: 3D printing; topography; terrain; elevation; geoscience education; web; Python; Earth Engine; open source

Citation: Harding, C.; Hasiuk, F.; Wood, A. TouchTerrain—3D Printable Terrain Models. *ISPRS Int. J. Geo-Inf.* **2021**, *10*, 108. https://doi.org/10.3390/ijgi10030108

Academic Editors: Wolfgang Kainz and Beata Medynska-Gulij

Received: 31 December 2020
Accepted: 19 February 2021
Published: 25 February 2021

Publisher's Note: MDPI stays neutral with regard to jurisdictional claims in published maps and institutional affiliations.

1. Introduction

One of the fundamental competencies of geoscientists is to describe the spatial nature of the solid Earth so the processes that modify it can be interpreted [1]. Indeed, the physical morphology of Earth has profound impacts on the evolution of species, the distribution of weather and climate, and provides the basis for millions of economic decisions made by governments, landowners, developers, architects and engineers every day.

Terrain, the surface of the Earth, is of fundamental importance for many aspects of the geosciences, be they education, research, engineering or resource management. Terrain is also a pivotal component for related efforts such as hydrology and climate modelling. Even when terrain does not play a large part in our daily lives, it is often part of the mental residue we collect from some places that we had a connection to where we grew up, where we went to school, where we traveled on vacation or where we went mountain biking. As such, "experiencing terrain" can afford a form of emotional connection to times and places that were important to us in some form and it can help to lead us back to them.

Despite this importance, the scale and nature of terrain is often difficult for people to comprehend because it is so much larger than the scale of a human being. Field trips, theoretically the most genuine way to experience terrain to the geology student, are often not feasible, which is why most of the time, terrain is only experienced intermittently through interacting with some form of cartographic artifact (e.g., a map). Terrain maps vary from 2D topographic maps, traditionally printed on paper, to various forms of 3D multimedia cartography viewed on a computer screen or projected in 3D "caves".

Such terrain maps are visualizations of data in the form of Digital Elevation Models (DEM), also known as a Digital Terrain Model (DTM) or Digital Surface Model (DSM). These digital models represent terrain as elevation offsets from sea level and are typically stored in a rasterized height field. Technology for creating such digital elevation models has proliferated rapidly over recent decades, both in terms of their coverage and their resolution. As of 2021, the globe has been fully covered by multiple datasets of various resolutions, such as the satellite derived ALOS World 3D–30m (AW3D30) DEM with a 30 m resolution at the equator [2]. A 1 km resolution DEM that includes the entire seafloor bathymetry also exists. In the US, the United States Geological Survey (USGS) maintains the nationwide 10 m resolution National Elevation Dataset (NED); its 3D Elevation Program (3DEP) is on track to create nationwide lidar coverage by 2023, affording a roughly 1 m resolution DEM.

Although the details are still debated [3], it seems clear that interacting with 3D terrain data "in 3D" should make it easier to perform map-related tasks as it imposes a lower cognitive burden than having to "translate" 2D information (such as 2D terrain contours) into a 3D mental model. Indeed, the difficulty students have when asked to make qualitative and quantitative measurements using 2D topographic maps has long been documented [4–7]. Beginner students often find it difficult to make the leap from reading a topographic map and "seeing" the 3D shape that it represents in their mind's eye. Using 3D computer graphics, even when displayed on a standard (i.e., non-stereo) display, affords the 3D cartographer a great deal of flexibility with regard to visualization (symbolization) of the terrain elevation, possible in conjunction with other terrain-based data. Increasingly sophisticated techniques are being developed to enable the user to interact with the data in ways that mimic the manipulation of everyday physical objects.

Indeed, the promise of "natural interaction" with 3D data is at the heart of value proposition for virtual reality (VR), augmented reality, and related technologies. VR cartography has been explored in recent years, for example by [8–10], and has shown promise. However, despite significant advances in recent VR technology, using VR to interact with 3D maps still requires a substantial amount of expertise in operating sophisticated hardware and software, imposing nontrivial hindrances to creating suitably natural user interactions. This makes VR costly and often impractical to deploy, especially to large audiences and in everyday "low-tech" scenarios. For most people, experiencing the majesty of the Grand Canyon from their living room or classroom is much more likely to involve Google Earth than donning a VR Head-Mounted Display (HMD).

3D printed terrain models (or physical, tangible models in general) fall on the opposite side of VR on the 3D technology spectrum. By their very nature, physical models are static. Unlike VR (or 3D computer graphics models in general), they cannot effortlessly change their colors or any other visual aspect. However, physical models bring something to the user that few, if any, VR systems can provide, an aspect that is usually neglected in cartography [11–13]—the sense of touch.

With tangibility, the complexity of smooth and efficient interaction with the model is reduced to triviality. From handling everyday objects, we already know how to perform a "rotate", "pan" or "zoom in" on a physical terrain model, either using our hands or, for very large models, walking around them. No need to look up which mouse buttons are used for zoom, pan and rotate. Given good lighting conditions, just handling the model reveals its surface details (plus, if available, painted colors or drawn annotations), without the need for complex 3D rendered visual effects like specular highlights, shaders or ray tracing. As a "no-tech" solution, a physical terrain model can be handled outdoors during glaring sunlight and during rain; it never stops working and has no batteries that can run out.

More importantly, the tactile aspect, i.e., the ability to feel the model's surface details (and its general macro shape), in concert with a visual exploration creates a much stronger sensory impression and potentially a much more vibrant and memorable mental model of the terrain. No wonder that we are intuitively tempted to touch an interesting looking physical object and run our fingers over it. Despite often being discouraged in children, actively touching a new object provides an underappreciated yet integral part of first exploring, and later understanding, complex objects.

A variety of approaches can be used to obtain elevation data and convert the DEM into a format suitable for 3D printing. Given sufficient expertise, a multi-stage workflow using 3D modelling software, such as Blender (with the BlenderGIS plugin [14], Meshlab, Meshmixer, Sketchup or threejs; and/or GIS software, such as ArcGIS, Global Mapper or QGIS (with its DEMto3D plugin [15]) can be developed to extrude the elevation values into a triangle surface that is part of a watertight mesh. Terrain2STL [16] and The Terrainator [17] are easy-to-use, free web applications with a approach similar to the TouchTerrain web application. However, they appear not to be in active development and lack many of the DEM sources and usability features that TouchTerrain offers.

Hobbyist-level 3D printers are now cheap enough and their software has matured enough to make filament-based 3D printing affordable to many. In addition to being operated as a hobby, 3D printers are now part of many makerspaces, schools, workshops and offices, where they can be used to print the 3D model files downloaded from TouchTerrain. Typical hobbyist-level 3D printers can print models with horizontal extents ranging from 10×10 cm to 40×40 cm, with very little material cost. With TouchTerrain, larger 3D terrain models can be assembled from multiple smaller tiles. The large model is split into equal sized tiles of a size small enough to be printed on a typical hobbyist 3D printer. After 3D printing each terrain tile separately, the tiles are assembled (e.g., via glue) into a much larger model. Depending on their use case, 3D terrain models can be painted or annotated with pencils.

3D printed terrain models have been used in several undergraduate geology courses at Iowa State University over the last 3 years, both in classroom exercise and to support geologic mapping in the field. In the classroom, 3D printed terrain models can be used in conjunction with paper contour maps of the same area to help students better understand the fundamental principles of contour maps and to provide a novel tool for supporting all learning styles. In the field, 3D terrain models were used to situate the students geographically in the area and to permit them to put annotations about their field work directly on the models.

TouchTerrain has grown in popularity, especially since version 3 was released in October 2020, which introduced a modern GUI and several features requested by users. Based on telemetry and feedback from TouchTerrain users, we attempted to answer the following research questions: *Who are our users? For what purpose are terrain models used for? What areas on the globe are most attractive to 3D terrain print?*

The rest of the paper is structured as follows: Section 2 provides a historical context for 3D physical terrain models. Section 3 gives an overview of the 3D printed process in general, followed by sections about how to use the TouchTerrain web application (server version), the standalone version and will describe some of its implementation details. The web application, which has been running for several years, has collected anonymous usage information via Google Analytics. Section 4 reports on this web usage and includes an analysis of feedback from 1500 users about how they intended to use their 3D terrain models. Section 5 provides more details on the use of 3D printed terrain models at Iowa State University geology curriculum followed by conclusions and future work in Section 6. Figures A1–A4 in Appendix A show images of 3D printed terrain models.

2. Background: Usage of Physical 3D Terrain Models in Geography

3D physical terrain models have long been used in geography and related fields. The following briefly covers some of these uses to provide context to the TouchTerrain project. Historically, physical 3D terrain models have been used primarily to support planning efforts in an engineering or a military context [18]. For an extensive description of such historical uses, cf. Chapter 2.1 of [19].

Priestnall [20] describes the partial reconstruction of a 4.57 m by 4.27 m plaster terrain model of the Lake District in the UK that was created in 1875. The model consisted of 142 tiles, each 1 × 1 foot. Each tile was created and painted by hand, based on Ordnance Survey of England's topographic maps. The model was created to promote the burgeoning tourism in the region. Viewers walked around the model to gain an appreciation of the landscape and to plan their own holiday excursions. In this way, they became familiar with the region without having to travel across it. Compared to smaller, handheld 3D models, such large 3D terrain models naturally provide a much fuller context but impose a different navigational paradigm. Users must move around the model rather than rotating them directly with their hands. Some interior parts of the model may be inaccessible to closer inspection.

Although recently produced terrain models are typically created in a purely machine-driven manufacturing context, there can still be room for the cartographic artistry of the human touch. Comparing two 3D models of the Eiger (Switzerland), Welter [21] pointed out that hand-made and hand-painted models can be cartographically superior to the technically precise, computer generated models because . . . ? A better, "hybrid" approach would be to use technology to quickly and precisely generate most of the geometry and to then "touch-up" or "post-process" the model by hand with paint to impart the artistic feel and cartographic quality of the traditional, hand-made models.

Kete [22] describes the creation of a 3D map which combines elevation data with infrastructure objects, such as roads, ski runs and ski lifts. Cartographic principles, such as generalization, were applied to the infrastructure data. Four tiles, each measuring 245 × 356 × 203 mm, were printed by a plaster-powder based 3D printer (Z510), which also applied color to the print surface. A final layer of epoxy resin was applied to stabilize the powder. While offering the advantage of direct color manufacturing, power-based prints are comparatively expensive and far more fragile than plastic-based 3D prints, even after being coated with epoxy resin.

2.1. Research in an Educational Context

Many aspects of the work involving physical terrain models are centered around their role in education. Oswald et al. [23] explored novel application of 3D printed terrain models in physical and urban geography education and outreach. Using GIS and other software, they created watershed models using a polygon as a boundary. These were then printed on a plastic 3D printer and a powder-based 3D printer. Both models used a z-scale of 10 to increase the details of the various river valleys within the watershed. On the 20 × 40 cm powder-based model, color was used to indicate major streams (blue), major roads (gray) and greens spaces (green). In a novel watershed puzzle (total size: 25 × 15 cm), a watershed was subdivided into 10 sub-watershed "puzzle pieces". The puzzle was used to demonstrate how landscapes can be partitioned into watersheds, these into sub-watersheds, etc. The puzzle was also used for comparing the area, relief and drainage density of the different sub-watersheds.

Furthermore, a 25 × 25 cm model (z-scale of 5) of the Toronto, Canada, waterfront including lake bathymetry was sprayed with water to demonstrate how water would drain downslope along the terrestrial part of the model into the lake part. These terrain models were used in outreach events and in a high school learning module on urban watersheds, where an increase in the level of student engagement when interacting with a 3D watershed model was noted. Other models, such as a two-tile, 50 × 25 cm model of a moraine (z-scale of 25) and cityscape models, where buildings are placed onto the topography, are also mentioned.

It should be noted that this is exactly the scenario TouchTerrain was designed to improve upon: to alleviate the need for GIS expertise and to thus make it possible for end users such as high school teachers or outreach personnel to create 3D terrain models with complex outlines (such as the polygon outlines defining the watersheds). With TouchTerrain, models can easily be given enough elevation exaggeration to show relevant terrain details even on low relief terrain. Models could also be potentially printed as multi-tile models.

In closing, the authors note that "Feedback from social media, the high school education sessions, expert workshops/conferences, and public events was positive without exception" and that "we believe there is a great potential to continue integrating 3D landscape prints (of watersheds and landforms) into our teaching and research outreach activities". Other publications, such as [24], offer a similar, but primarily anecdotal assessment of the potential educational benefits of using 3D printed terrain models.

In recent years, several educational studies were performed aimed at going beyond such anecdotal evidence and to quantitatively evaluate the role of 3D models in geoscience university teaching. The Topographic Map Assessment (TMA) instrument [25] is one way to objectively define a set of map reading skills. The TMA consists of 18 exercises each testing a certain type of topographic map-related task, such as drawing the water flow between two points in different geographical settings, comparing the steepness of two slopes or determining the visibility (uninterrupted line of sight) between two points. Points are awarded based on how well the tasks are performed.

Tests like the TMA can thus quantitatively determine how well groups of participants perform spatial tasks using one of two different types of maps, e.g., using a 3D terrain map and a traditional 2D contour map (as baseline). A statistical analysis of the results determines the efficacy (gain) for each type of teaching technique. As intrinsic motivation is an important factor in learning, such quantitative assessment may be complemented by an instrument such as the Intrinsic Motivation Inventory (IMI) [26], in which participants self-assess their experience of performing the spatial mapping tasks.

A study performed by Carbonell-Carrera et al. [27] compared 2D contour maps to 3D printed terrain (tangible 3D) and to a marker-based AR system in which a card of the same extent as the 3D print was held in one hand and viewed through a smartphone or tablet. Using those markers, the AR software deduces the card's orientation and creates a 3D computer graphic that shows the terrain in this orientation on the screen (digital 3D). The results suggest that using 3D, both tangible and digital, resulted in significantly better performance. Note, however, that no distinction was made between the tangible 3D and the digital 3D and it seems that members of the 3D group used both types together. Therefore, the effect of using the tangible, 3D printed terrain models we are interested in, is not quantifiable from this study.

Carbonell-Carrera et al. [28] performed a similar, larger study that also included use of the Intrinsic Motivation Inventory (IMI), but now compared 2D contour maps with AR-based 3D models only (i.e., it did not use 3D printed models). Again, 3D has some advantages, especially with regard to student motivation, excitement about the new technology and ease of use, but 2D showed greater efficacy in some aspects of spatial learning. Additionally, some students performed better with 3D while others performed the same task better in 2D. In the author's opinion, both 2D and 3D tools should be used in tandem to offer students different entry points to success, no matter their initial spatial abilities and preferred learning styles.

In a large study performed by Adams [19], the TMA was used in a pretest to establish a baseline of the participants' spatial skill. Participants would use either a 2D contour map, a roughly 30 × 30 cm 3D printed terrain model (3D group) or the 3D printed terrain model with the contours projected on it from above (3D/AR group). For the TMA pretest, none of the three groups was statistically different. For the actual experiment, participants were given a set of ten custom tailored spatial tasks that were equivalent to those in the TMA but involved different terrain not included in the TMA. All tasks were evaluated via multiple choice questions. For most, but not all, of the ten questions, both 3D groups performed statistically significantly better than the 2D group. In general, the two 3D groups themselves were not significantly different from each other, however, the 3D/AR group performed better than the 3D group in the photo interpretation relief tasks, perhaps hinting at the value of adding contour information on top of the 3D printed terrain model. For questions involving slope, 2D was found to be marginally better than 3D.

2.2. Research on Blind–Visually Impaired Map Users

The final area in which to put 3D printed terrain models into context is their use by blind and visually impaired (BVI) individuals. As touch is the primary or only sense a BVI person has to interact with and learn about terrain data, usage of 3D printed terrain models in areas such as wayfinding, mobility training and geography education would seem to be fertile ground for research and applications. However, we found few studies in this area of research.

Tactile maps, which traditionally use raised symbols embossed into paper, have long been employed in many BVI related mapping contexts [29]. Recently, 3D printing has been used to extend this tactile map concept [30]. Schwarzbach et al. [31] describe the cartographic process of a 25 × 35 cm "3D tactile map" of an outdoor hiking area. Instead of raised symbols, the map is based on a 3D landscape, created with a powder 3D printer. Its extended tactile symbology included raising roads to make them easy to follow and the use of tactile textures to differentiate types of land cover (e.g., forests from lakes). Results were anecdotally reported as successfully giving BVI individuals the ability to plan and prepare for hiking tours. However, no formal user study was performed.

Voženílek et al. [32] designed and 3D printed a tactile map of Europe for use in BVI education. The 3D model was printed with a 25 by 35 cm powder-based printer in color. The printed DEM was severely simplified to have only five elevation levels, at 0.5 mm intervals, which reportedly allowed BVI users to consistently perceive changes in elevation via their hands. To help low-vision users, each interval was painted in bright colors in a high contrast sequence: white (sea level), green, yellow, magenta and black. An observational user study was performed with BVI school children as participants. The reactions were positive with participants stating that this 3D terrain map gave them, for the first time, the ability to perceive and compare terrain features such as different European mountain ranges.

Perry [33] compared two methods of conveying topographic data for BVI individuals. In this study, the recognition of comparatively simple topographic landforms was used, based on a set of eight idealized variations of a simple hill. 3D printed terrain models were created from these eight hill variations which were used in the study's 3D mode. For its 2D mode, the eight hills were converted into contours which were braille printed as tactile paper maps. Using these two modes, the study performed a cross-modal experiment. Participants were given each of the hill variations either as a tactile map or as a 3D printed model, and were asked to find the matching model of the other modality. This was repeated for all models/maps. Learning time (i.e., the time needed to completely encode the map or model into a mental representation), selection time, (i.e., the amount of time needed to haptically explore all eight maps/models of the other modality to find a match), and the match's success/fail were recorded and analyzed.

Results indicated statistical significantly shorter learning times for the 2D mode, (i.e., for learning the terrain shape encoded in a 2D tactile map and then selecting the correct match from the eight 3D models), possibly because it is faster to explore each 3D model and to then decide to either declare it to be a match or to discard it and investigate the next model. In all other regards, the two modes did not differ significantly, which indicated that, at least for these very simplistic geomorphologies, tactile contour maps are just as effective in building up a mental representation as handling a 3D printed model.

3. Methods and Technology

3.1. 3D Printing Technology Overview

A 3D printer creates a physical 3D model based on a digital 3D model. 3D printers using Fused Deposition Modelling (FDM, also known as Fused Filament Fabrication, FFF) prints with plastic filament and are by far the most common method for printing medium

(10 × 10 cm) to large (40 × 40 cm) terrain models, combining cheap, yet reliable hardware, low material cost (1 kg of filament costs USD 15 to 30) with reasonably good detail.

Subtractive manufacturing, another method for creating 3D models, works by successively cutting material away from a solid block of material. For terrain models, typically a Computer Numerical Control (CNC) machine is used to mill away the model from a block of wood [34]. CNC-produced terrain models can be very large (e.g., 120 × 240 cm) and are commonly used for architectural dioramas or interior decorations.

In FDM printing a 3D digital model, usually contained in a STL file, is first processed by software known as a *slicer*. The slicer software creates horizontal profiles (slices) through the model at 0.1–0.3 mm intervals (the so-called "layer height"). Slices consist of many 2D lines (toolpaths). Lines designated as walls (perimeters), tops and bottoms are used to define the outside of the model. Lines forming a loose infill pattern (typical density: 10–20%) cover the inside of the model, resulting in a sturdy outer "shell" that envelopes a mostly hollow interior.

During the printing process the 3D print hardware moves a nozzle of typically 0.4 mm diameter in the horizontal (x/y) plane along the tool path defined for each slice/layer. After each slice is deposited, the nozzle moves up by the layer height (typically 0.1–0.3 mm) and deposits the next slice. The nozzle extrudes ~210 °C hot plastic filament, which quickly cools to a solid form and fuses with the layer below, thus creating a physical object (Figure 1).

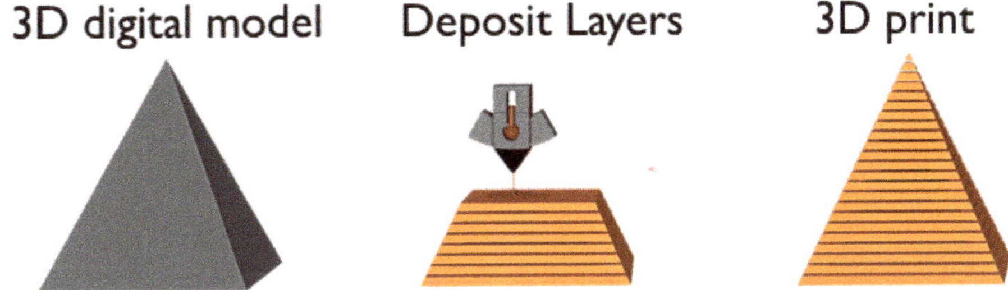

Figure 1. The Fused Deposition Model (FDM) 3D printing process: the digital model (left) is sliced into layers, layers are printed from bottom to top (center) and are fused into a 3D printed object (right).

3.2. Our Experience with 3D Printing Terrain Models (Guidelines)

It is important to understand some limitations inherent to FDM 3D printing. 3D Printers move the nozzle along the slice (x/y) plane with much higher precision than the smallest possible layer heights produced by TouchTerrain, which can be as small as 0.06 mm but is typically between 0.1 and 0.3 mm. 3D printed terrain therefore will always have a noticeable anisotropy along the horizontal plane, which manifests as horizontal "stripes". In our experience, noticeable stripes do not detract from the experience as they mimic contours found on topographic maps. However, low layer heights (0.15 mm or lower) result in more visually pleasing models and should be used when possible.

Although it is possible to print different layer heights with the same nozzle by extruding more or less filament, it is generally not possible to print details along the x/y plane that are smaller than nozzle diameter, typically 0.4 mm. In this respect, the degree of detail that is possible along the z-axis with a 0.1 mm layer height is considerably finer than what a 0.4 mm nozzle can create in the x/y plane. Smaller nozzle sizes (e.g., 0.25 mm nozzles) could be used to compensate for this but are generally more difficult to reliably print with.

When looking for suitable areas to print, many users seem to be naturally drawn to high relief terrain, such as the Grand Canyon or famous mountains, which, without z-scaling, prints out as a tall model with many layers (even using a coarse layer heigh of say 0.3 mm) and thus provides good detail and an aesthetic model. However, even areas in low relief regions (such as Kansas or Iowa) can be printed with good detail if the model's elevation is appropriately exaggerated (z-scaled) prior to printing. The z-scale factor should be large enough, so that the resulting print is at least 20–30 mm tall and thus has enough layers to show enough terrain detail.

Large-sized prints are more impressive and present the terrain in better detail to the viewer. The horizontal size of the print should therefore be chosen to be as large as can reliably be printed in good quality (given the size of the printer's build plate), but without becoming unwieldy for the intended use case. Going beyond the build plate limitations requires the printing of multiple smaller models (tiles), These can then be assembled into a larger print but will typically still show the internal tile boundaries, which are difficult to conceal well, even with extensive post processing.

Hillshading can help when selecting an area by geomorphological terrain features (e.g., by looking at ridges, valleys or drainage patterns). In TouchTerrain, the transparency, direction and severity of the hill shading can be configured to better bring out such features (even in low relief regions) and can help to inform the precise placement of the print area. It may also be desirable to define the print area with a polygon instead of a simple box, so that its boundary partially corresponds to a terrain feature, e.g., follow a river valley. For this, a polygon can be digitized in either Google Earth or a GIS and given to TouchTerrain as a .kml file. For further guidelines and heuristics for 3D printing terrain models, see Appendix B.

3.3. *TouchTerrain Project—Software*

TouchTerrain is an open source software project hosted in a Github repository [35]. It is written primarily in Python and is implemented both as a web application and a standalone version. The web application uses an interactive, Google Maps interface. After selecting the print area and setting options related to a user's 3D printer, online DEM data from Google Earth Engine is converted into digital terrain models and downloaded by the user for 3D printing. The web application is currently hosted by Iowa State at https://touchterrain.geol.iastate.edu/ (accessed on 20 February 2021) (or http://touchterrain.org (accessed on 20 February 2021)) but others are free to deploy the application on their servers.

In the standalone Python version of TouchTerrain, the input parameters (such as the coordinates of the print area) are given via code instead. The standalone version can be run via a jupyter notebook, which affords some interactive functionality. It can process DEM data from Earth Engine and also from local DEM geotiff files. Building on the 2017 version [36], TouchTerrain version 3 added several substantial upgrades and improvements, either in response to user feedback or contributed by others to the open source code (see http://blog.touchterrain.org/ for details (accessed on 20 February 2021)).

3.3.1. TouchTerrain Web Application

Print Area Selection

The first step for the user is to find the desired area on the Google Maps window (Figure 2) and then interactively outline it with the red area selection box. Standard Google Maps navigation is implemented, and a place search field (top) can be used to jump to the general area quickly. Coordinates for the corners of the print area can also be edited manually. Furthermore, a .kml file with a polygon can be uploaded and used as print area. Several online DEM sources, which are provided by Google Earth Engine, are available, providing different resolutions and coverages. For example, the USGS NED dataset, which has the highest resolution (10 m), is only available for printing areas within the contiguous US. A hill shade layer of the current DEM source is overlain over the Google Map. Users can adjust the transparency, and manipulate the hill shade sun direction and sun angle to create an effective relief effect for the area in question.

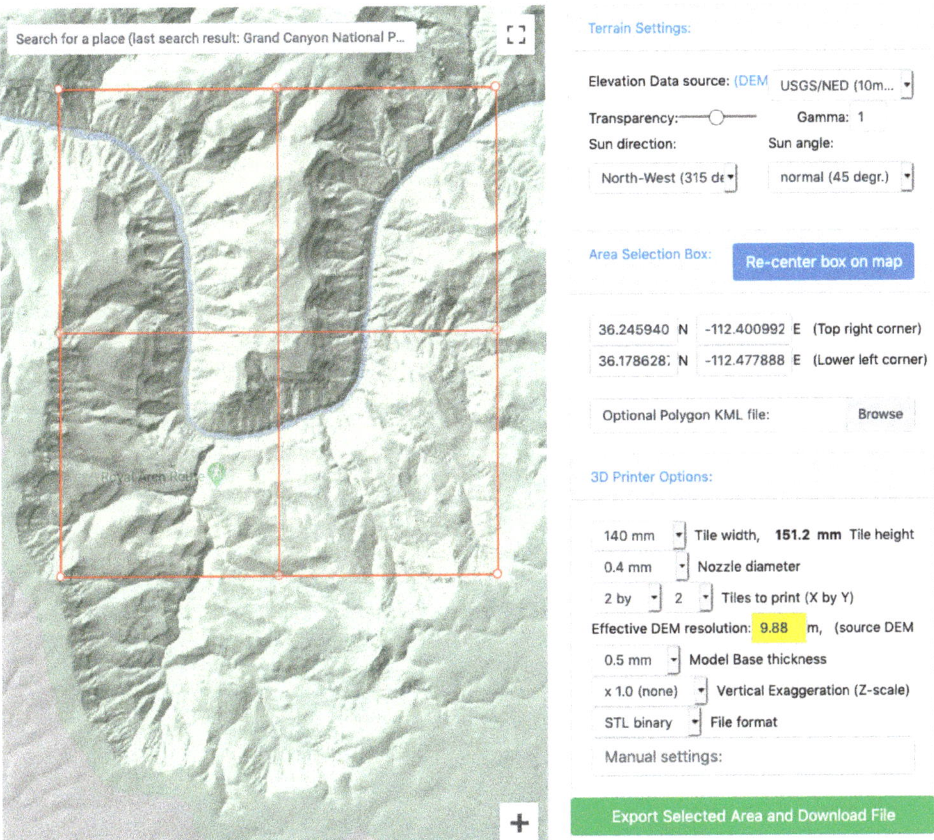

Figure 2. TouchTerrain Web Application with Google Maps Window at left and Terrain Settings, Area Selection Box and 3D Printer Options on the right.

Printer Settings

Once a print area has been selected, a few additional settings need to be given (3D printer Options). Tile width and height determine the real-world horizontal dimensions of the 3D print in mm. Only the width needs to be given by the user, the height is calculated as per the aspect of the area selection box. This requires the user to know the size of the build plate of their 3D printer. For larger models, TouchTerrain can instead divide the print area into multiple tiles of a size that fit the build plate.

Similarly, the user should know the nozzle diameter of their 3D printer. Although adjustable to values between 0.2 mm and 0.6 mm, the default value of 0.4 mm is an extremely common diameter for hobbyist and prosumer grade printers and should not need to be changed. This nozzle diameter plays a vital role in determining an internal metric called the 3D print resolution, which determines to what resolution the source DEM is down sampled before processing. The rationale for this down sampling is that a 0.4 mm printer nozzle can generally not express terrain details any smaller than 0.4 × 0.4 mm anyway. The DEM raster that is processed into a 3D digital model therefore needs to only be of a resolution that is equivalent to 0.4 × 0.4 mm raster cells instead of the source DEM, for which the resolution is typically considerably higher. This effective DEM resolution is shown to the user, together with the source DEM resolution.

For example, a tile that is as wide as the continental US that shall be printed out as 100 mm wide with a 0.4 mm nozzle, can be down sample from its 10 m resolution (for

USGS NED as source DEM) to a resolution of about 20,000 m. This down sampling results in a dramatically smaller digital 3D model file to be downloaded, which can be sliced at a fraction of the time and, in the end, will generate a 3D print that is indistinguishable from a 10 m resolution model. However, there are cases, where the combination of area geometry, 3D print size and nozzle diameter would result in an oversampling of the original source resolution from which nothing can be gained. In these cases, the field showing the effective DEM resolution will turn yellow to inform the user of this fact. While a small degree of oversampling is a non-issue, massive oversampling suggests that the user may want to change the area extent, number of tiles or nozzle size.

A model base thickness value (default: 0.5 mm) determines how thick of a horizontal slab will be added beneath the actual terrain model. This small downward extrusion, which is uniform across all tiles, can help improve the durability of otherwise very thin models and can improve model aesthetics.

The vertical exaggeration factor (or z-scale, default 1×) determines by how much the elevation will be multiplied when processed into a digital model. Given the comparatively coarse layer height used in typical FDM printing (0.1–0.3 mm) it is important that the z-scale is large enough to result in a reasonably large number of layers. Printing at least 50 layers ensures that the model will show a good level of terrain details. The value for a "good" z-scale will depend on the difference between the maximum and minimum of elevation within the requested area, with low-relief areas (with little differences) suggesting z-scale of 2× to 4×, sometimes (e.g., watershed models) as high as 15×. Prior to actually processing the DEM for the selected area, no measure of relief (i.e., elevation difference) can be given to the user, so trial and error must be used for now. However, as it is possible to preview the processed 3D model prior to download, the user may realize that a higher z-scale would be advantageous and re-process the model with it.

The file type for the digital 3D terrain model can be set to either to STL a simple, yet still extremely popular 3D print file format it defaults to, or to OBJ, which is more general and often used in 3D computer graphics and modelling.

The final set of potential 3D printer options are the manual settings, which are meant as a Command Line Interface (CLI) for power users who want to set options that are not directly covered by the Web application's User Interface (UI) elements. TouchTerrain supports a large number of these expert settings, which are documented on the GitHub page. Once the names of the settings are known, the user can set their values by typing them into the textbox using a JSON style notation. For example, to override the use of UTM as projection and to instead set it to a specific EPSG code, the user would type in "projection":12345. Multiple manual settings can be given by separating them with a comma.

Model Download

Export Selected Area will fetch a geotiff of the print area from Google Earth Engine and process it into a digital 3D terrain model according to the parameters given by the user. Processing times vary by requested tile number, size and server load, but are generally short: a single tile 3D model that results in a 50 Mb file is usually processed in less than 20 s. The user can now optionally preview the digital model in a WebGL based in-browser 3D renderer and, if needed, change setting and re-process until satisfied with the result.

The final step is to download a zip file, which contains the digital terrain model in the requested file format, the geotiff from which the model was created and a detailed log file indicating all the parameters used in generating the model. The model can now be opened in a slicer for 3D printing. Finally, the user is given a URL to "bookmark" the current print. If this URL is put in a browser, it will bring up the TouchTerrain web application with all the specific print settings for this model already filled in. This is useful for coming back later to repeat a print without having to manually re-enter the print settings. URL bookmarks can be emailed to others to enable them to recreate the same model.

3.3.2. TouchTerrain Processing Details

The processing step at the core of TouchTerrain involves the creation of a triangle mesh by extruding a DEM raster contained in a geotiff. This triangle mesh describes a watertight volume and is stored either in a STL or OBJ file, which can then be sliced and 3D printed. The geotiff, a *GIS raster*, consists of a regular 2D array of cells (or pixels), each cell containing an elevation value. All cells are of the same size and are typically, but not always, square. In a GIS raster, this 2D array is geo-referenced (i.e., draped onto the surface of the Earth). Each cell, therefore, covers a specific geographic location on the globe with its value.

Google Earth Engine, an online computing platform for raster data, combines over 200 public domain raster datasets (including several elevation datasets) with raster computation via an application programmer interface (API) in JavaScript or Python. TouchTerrain requests DEM rasters from Google Earth Engine, based on the selected DEM source, the lat/long extent of the print area and the down sampling resolution. In response Google Earth Engine generates a geotiff, which TouchTerrain receives and processes. Unless overridden, the coordinate system of the geotiff is set to the UTM zone that covers the print area with WGS84 as datum.

In some rasters, a special value is used for cells to represent the absence of an elevation value. Using such NoData (undefined, NaN) cells permits the definition of arbitrary shapes of terrain outlines (e.g., the terrain of an island surrounded by a sea of NoData values). This scenario still requires a regular 2D array of values; however, some of the cells are set to the NoData value and are, by convention, simply omitted when displaying (or processing) the elevation raster. This results in the appearance of a complex outline instead of a regular 2D array with orthogonal edges and 90° corners. For this, Google Earth Engine can be given a polygon, with which it will mask out any cells that are outside the polygon. The resulting geotiff will, thus, have no data for all cells outside the polygon, which TouchTerrain will ignore when creating the 3D printable model.

Mesh Creation

In the TouchTerrain triangle mesh, each square raster cell can be represented by a *quad* that covers the cell's x/y extent, which is then divided into two triangles. The z-value of these triangles depends on the elevation value of the raster cell it covers. Extruding all triangles upwards thus forms the top triangle mesh of the 3D model. In order to create a watertight, 3D printable model, a bottom mesh, consisting of a non-extruded version of the raster, and vertical sides (walls) must be added. Figure 3 illustrates the processing steps from the DEM raster to a watertight terrain mesh model.

The raster can also be divided into several equally sized sub-rasters (tiles), each of which can be processed into multiple 3D models that, when grouped together, are equivalent to the full 3D model. In Figure 4, using a polygon that sets the values of some cells to NoData resulted in a mesh with omitted cells. Wherever possible, TouchTerrain avoids producing such "stair-step" patterns and attempts to place only one triangle in a configuration that smoothes out the outline. For two cells (arrowed), it was not possible to remove one of the two triangles.

Figure 5 demonstrates the perfect fit of the walls at the tile boundaries when assembling a digital terrain model from four equal sized tiles.

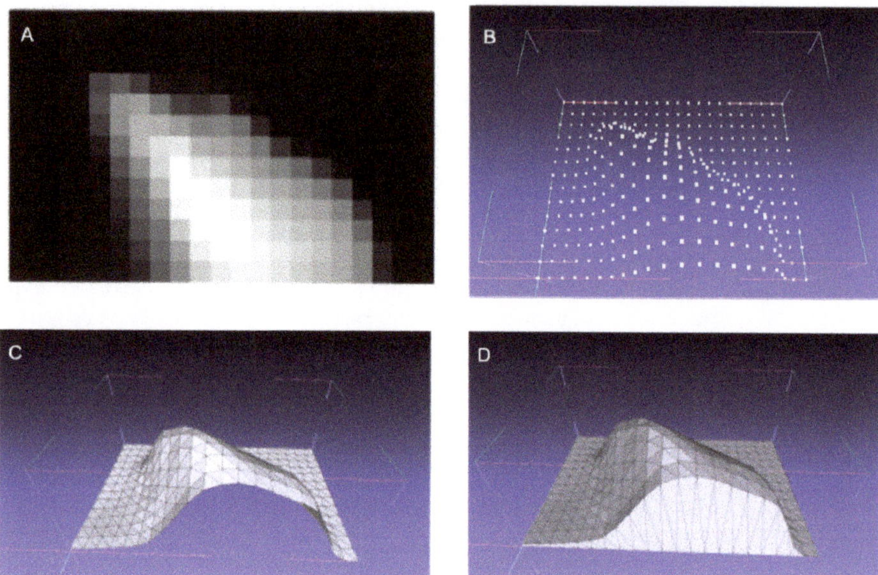

Figure 3. TouchTerrain processing steps for creating a watertight, 3D printable terrain model. (**A**) the Digital Elevation Models (DEM) raster (symbolized via a black to white increasing grayscale); (**B**) the positions of all the corners after they have been extruded from the ground upwards (viewed from above); (**C**) the top triangle mesh made from connecting the corners (oblique view from South); and (**D**) the complete model with bottom mesh and a vertical wall along the southern border.

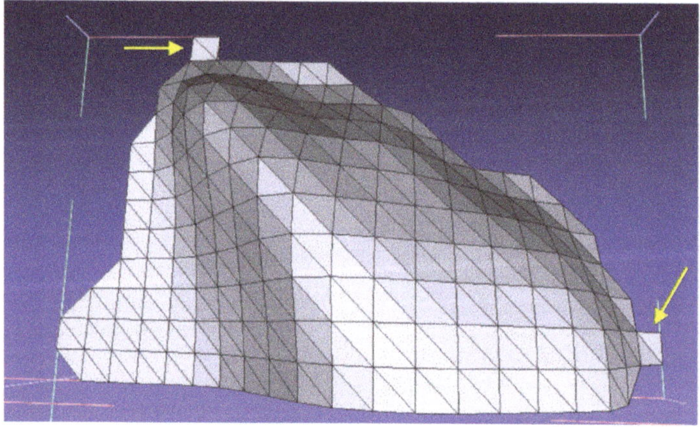

Figure 4. Optimized triangle structure when using a polygonal outline.

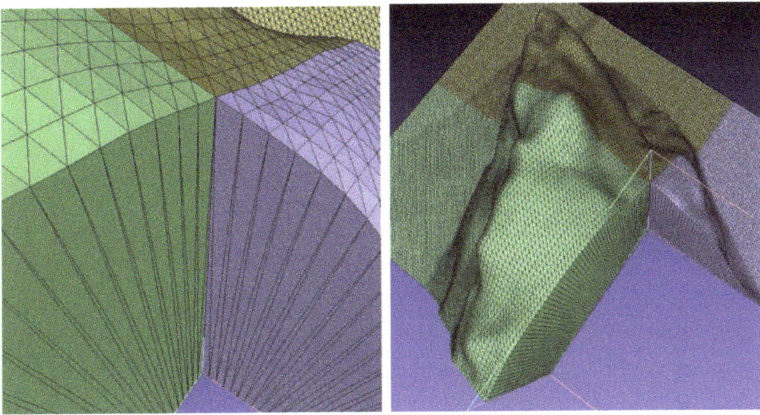

Figure 5. Assembly of a digital terrain model from four equal sized tiles.

Base Thickness

The base thickness setting determines the (vertical) thickness the print will have at areas where the elevation of the DEM is the lowest. In TouchTerrain, this is set by the base thickness parameter (in mm). 3D printing with the typical 0.2–0.3 mm layer heights requires at least two layers of filament to fuse into a reasonably solid connection. In Figure 6, a river bed occupies the lowest area of the model. Using a setting of 0 mm (Figure 6, left) may not look problematic, but will result in a very brittle result when printed. A setting of at least 0.5 mm (Figure 6, right) is advised.

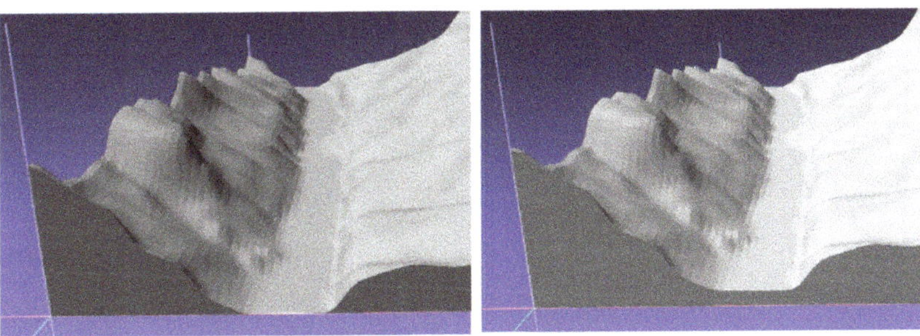

Figure 6. Effect of the base thickness setting. **Left**: With 0 mm, the riverbed will not be printed Scheme 0. mm (**right**) is recommended.

3.3.3. TouchTerrain Standalone Mode

Although using the web application's manual options expand TouchTerrain's functionality beyond the capabilities of the web user interface, there are still some operations that, thus far, can only be performed with the standalone (non-server) version of TouchTerrain. An important aspect of the standalone version is its ability to also process a *local geotiff file* (in addition to Google Earth Engine DEM sources), which opens the door to, e.g., printing high-resolution LIDAR terrain models (preprocessed into a geotiff raster).

Trails

The standalone version supports the ability to work with GPX (GPS Exchange Format) data files, which describe travel routes and are commonly produced by GPS devices. Thanks to TouchTerrain contributor Chris Kohlhardt, it is possible to "drape" these path lines over the terrain model and extrude them either up or down, thus creating trenches or crests within the model's top surface. This functionality creates visible imprints of line data on 3D printed terrain models, which is useful for 3D printed terrain models used to support, for example, outdoor activities.

Interactive Digitizing

Using the geemap package (geemap.org (accessed on 20 February 2021)), the notebook version gives the user the ability to interactively digitize the print area rather than defining it via python code settings (Figure 7). Geemap offers an in-notebook interactive map interface. The user can digitize a rectangle, a circle or even a polygon to define the print area. If GPX route files are used, these are also displayed on the gee map.

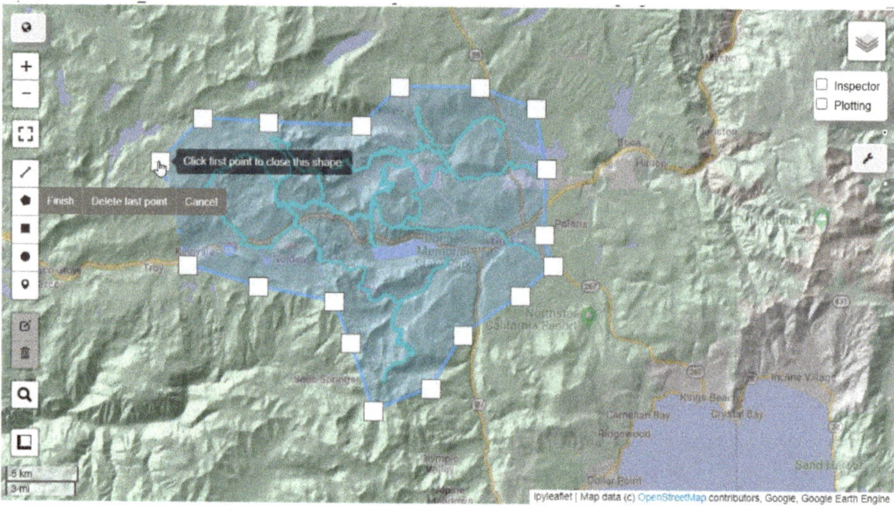

Figure 7. Digitizing a 3D print area polygon inside the interactive gee map jupyter cell around a set of GPX trails (cyan).

Bottom Mesh Reliefs

A final aspect of processing only possible with the standalone version is the use of a user-created grayscale (8-bit, 1 band) image for creating a relief on the bottom mesh of the model. The bottom mesh is then extruded upwards according to the value of its corresponding pixel value. White (value 255) pixels are not extruded, black (value 0) pixels are extruded upwards. Results are only legible at relatively large font sizes. Figure 8 shows the use of such a bottom relief to permanently encode metadata about the 3D terrain print. The image may also contain Braille text which could have potential application for terrain models in a BVI context.

Sheep Mountain
Shell, WY
0.2 mm layerh
x2 z-scale
Dec. 12, 2020
1:270,000

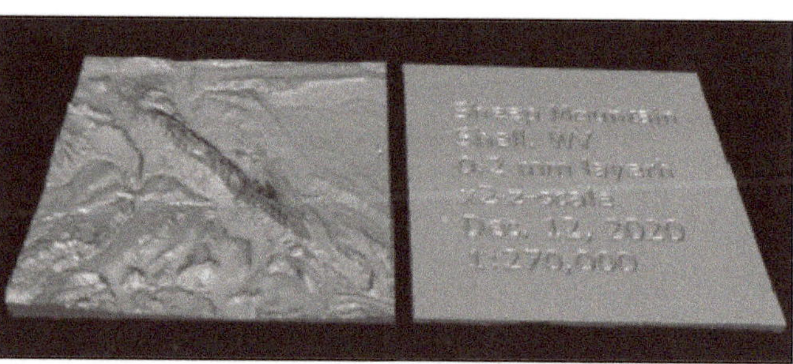

Figure 8. Effect of using a 8-bit grayscale image (**left**) to generate a relief effect on the bottom of a terrain model (**right**).

4. TouchTerrain Web Application Usage Analysis

TouchTerrain uses Google Analytics, a web analytics service offered by Google that tracks and reports anonymized website traffic, to gain insights into its use. Every time a browser connects to the TouchTerrain server, anonymous information is collected and stored at the Google Analytics server. This includes time of access, type of device used, how the user found the page (e.g., directly via URL or bookmark, via Facebook or via a Google search), and detailed geographic information about the location of the user (city, county, lat/long). It also stores the URL query string and thus all the print options selected.

Collecting this kind and amount of anonymous user information is very typical and completely standard for any web page that uses the Google Analytics service, which almost all commercial websites do nowadays. TouchTerrain does not store any personal information or cookies. Instead of using Google Analytics for its intended marketing purpose, TouchTerrain uses statistics to study usage patterns for academic purposes.

The application will also record certain events such as the user click on buttons: (1) the Export button (i.e., when processing commenced); (2) the Preview button (i.e., when the processed model was previewed inside the browser); and (3) the Download button (i.e., when a processed model was downloaded). Users have the option to give feedback (up to 150 characters) about the terrain model they are about to download ("Optional: tell us what you're using this model for"); this also generates an event that stores the feedback text. Beginning with TouchTerrain 3.0 (October 2020), the results of place searches are also stored via an event.

4.1. TouchTerrain User Analysis ("Who Are Our Users?")

The following characterizes the TouchTerrain users by analyzing Google Analytics web statistics. To exclude cases where a user did not actually process a model, we base our analysis on sessions. A valid session requires that the Export button, which processes the model, was clicked at least once. Feedback, Preview and Download can only be clicked after Export has been clicked.

4.1.1. User Characteristics

The following lists some Google Analytics results based on data from sessions between 1 July 2019 and 26 December 2020:

- Nearly 20,000 sessions were recorded, or around 38 sessions per day;
- The average user session duration was 8 min, during which 5.8 pages are typically viewed;
- Most sessions (79%) are from returning users (i.e., which used the site twice or more), only 21% of sessions are from new users. The top three repeat users logged 234, 177, and 170 TouchTerrain sessions, respectively, from December 2019 to December 2020;
- 47% of users are located in the USA, with California (12%), Texas (6%) and Colorado (5%) as the leading states. In addition, 7% of users are located in Germany, 5% in the UK, 4% in Canada, around 2% each in Italy, France, Australia or Spain. The IP-geolocated coordinates of their browsers were used to create an interactive ArcGIS Online dashboard (https://arcg.is/0L8fLz (accessed on 20 February 2021); see next section);
- 62% of users run Windows as their operating system, followed by 16% on MacOS and 2% on Linux. The remaining 20% of users apparently browse TouchTerrain from a mobile device (36% on Apple iPhones, 6% on Apple iPads, with the rest using an Android device);
- 53% of TouchTerrain sessions were initiated directly via a URL. This URL could come from the user's memory or from a browser bookmark, or perhaps the user's browser had stored the URL from a previous session. The full model URL that TouchTerrain shows the users at the end and permits them to bookmark specific model settings may also contribute to this unusually large percentage;

- TouchTerrain is also often found via a web search. 33% of sessions were the result of a Google search, with other search engines (Yahoo, Bing, duckduckgo) contributing a total of 2%;
- 2% of sessions were referred from blog.touchterrain.org (accessed on 20 February 2021), the TouchTerrain blog. Looking at social networking sites, around 5% of sessions come from Facebook, 1.5% from YouTube and 0.5% from reddit. Some traffic comes from community sites, such as geo.lmz-bw.de (1%) (accessed on 20 February 2021), which publish content about TouchTerrain.

4.1.2. User Workflow Analysis

Pressing a button generates events. Events can be used to analyze the workflow of a typical session. Table 1 shows the number and type of the different categories (types) of events recorded between 12 October 2020 to 26 December 2020. Unique events are counted only once per session, total events record all events during a session, typically multiple times. This multiplication factor is shown as Total Event Factor

Table 1. Web Usage statistics for version 3, 12 October 2020—26 December 2020 (75 days).

Event Category	Unique Events	Total Events	Total Event Factor	% of Unique Export Events
Export	5866	17,178	2.93	100%
Download	4826	9069	1.88	82%
Preview	2991	8008	2.68	51%
Feedback	324	616	1.9	6%
Place search	6703	8567	1.28	114%

Looking at this multiplication factor suggests that users typically refined their print settings after processing and reprocessing it again (possibly using preview to make this decision) before downloading the model. Using the number of unique exports events as 100% (as export is a prerequisite of the other events), only 82% of all user immediately downloaded their model. This indicates again that some users went back to the map, changed parameters and pressed Export once more. Around half of all sessions used the in-browser 3D visualization tool to preview the model prior to download. About 6% of users left some text as optional feedback about the model's usage.

The place search field is the only event that can be triggered on the map page, i.e., prior to processing Export. A large number of users seem to find this functionality useful. On average, each session performed 1.14 place searches. The 25 most popular places searched for can be found in Table A1.

4.1.3. User Locations

Google Analytics reports the user's location for 99.75% of all TouchTerrain sessions using IP-based geolocation. Note that there are many locations from which, over time, many downloads were made. As this leads to a "stack" of many points at the exact same location, a point data visualization cannot adequately represent this fact. However, a kernel density derived heatmap is able to account for such stacks. Figure 9 shows the heatmap, which again suggests the US and Western Europe as primary user locations for TouchTerrain. The full dataset, including the individual point data, can be explored at this ArcGIS Online Dashboard (https://arcg.is/0L8fLz (accessed on 20 February 2021)).

Figure 9 shows a heatmap of user geolocations (July 2019 to December 2020) at the time of hitting the download button. Each of the roughly 19,000 locations was converted into a point (using WGS84 as coordinate system) with the date and time, city, and county as data attributes. Using a geodesic gaussian kernel density operation with a search radius of 50 km, these points were used to create a global density raster (resolution 10 km), which was colored as a heatmap.

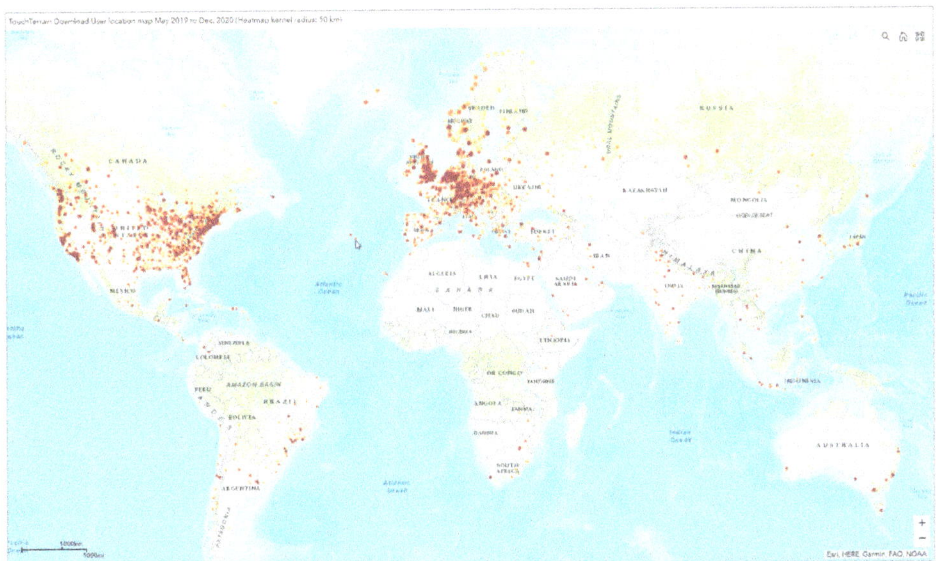

Figure 9. Map of TouchTerrain user locations generated via IP-based geolocation.

4.2. Area Selection Boxes ("What Areas on the Globe Are Most Attractive for 3D Terrain Printing?")

Perhaps more interesting is a look at the areas that users had selected for their 3D terrain prints (Figure 10). Due to changes in the format of the data transmitted to Google Analytics, only user data between November 2019 and December 2020 could be used for this analysis. Again, an ArcGIS Online Dashboard was created to allow the interactive exploration of this large data set here: https://arcg.is/0PO1nK (accessed on 20 February 2021).

From the coordinates of opposing corners of the area selection box, 9925 point pairs were extracted. Exact spatial duplicates were removed. In ArcGIS Pro, the Coordinate Table to two-point line created a diagonal line from each pair. The Feature Envelope-to-Polygon tool was then created a bounding box rectangle from each diagonal. Finally, box centroids used to create a heatmap with a 50 km density kernel. Boxes were divided into three categories by their area and color coded (large = purple, medium = green, small = red), which helps reduce visual clutter. Figure 10 shows a map of the print area boxes, colored by size category, suggesting that mountainous parts of Western Europe and the USA are particularly popular with users.

Figure 11 shows the area selection boxes in Europe. The cluster pattern observed here seems to suggest that users are drawn to mountainous areas, such as the Alps. Closer inspection also reveals "concentric" patterns, which may be the result of a user refining the extent of the print area over several steps. In the US, mountainous regions, such as the Rocky Mountains or the Cascades, monumental structures of the American Southwest, such as the Grand Canyon, and large volcanic structures (such as Mount Rainier in Washington State or Crater Lake in Oregon) appear to be popular. Looking at the most popular place search terms seems to confirm this (see Table A1).

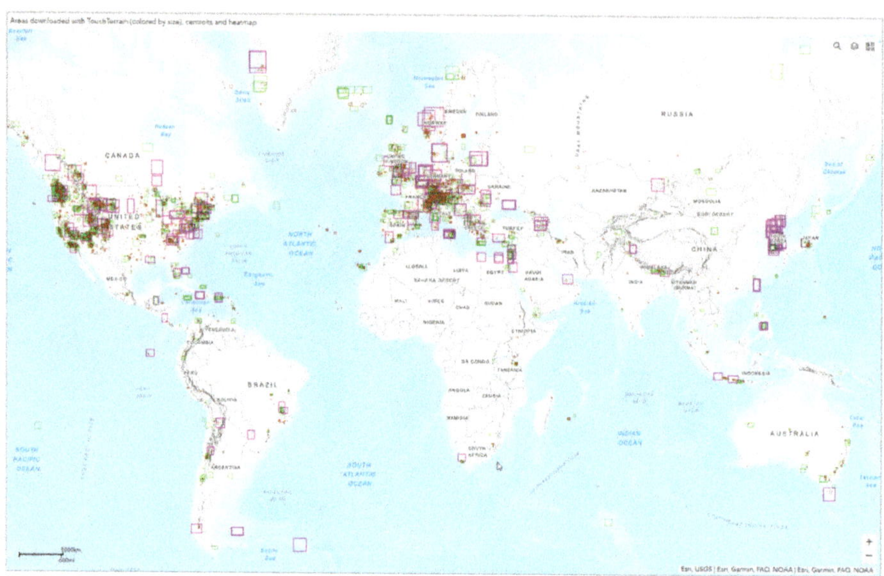

Figure 10. Map of the centroids of the area boxes that users had selected to print.

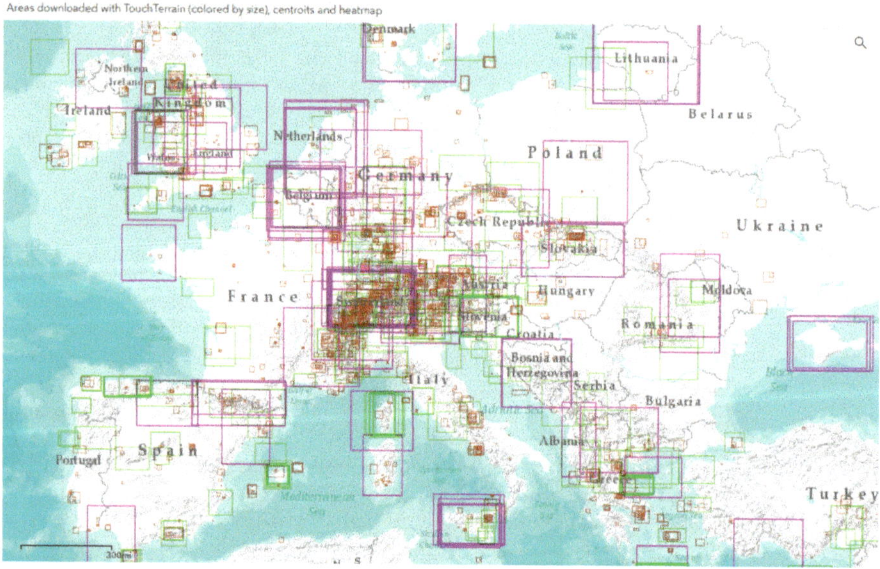

Figure 11. Bounding boxes of areas selected by TouchTerrain users for 3D printing in Europe.

Analysis of the Size of the Area Selection Boxes

Plotting a histogram of the areas of these 9925 selection boxes showed an extreme positively skewed distribution. When plotted without a transformation, the very smallest areas would seem to dominate. However, plotting it using a log10 x-axis (Figure 12) shows an approximately bell-shaped distribution with area for around 300 km^2 (corresponding to a square with 17 km long sides) as the most common choice.

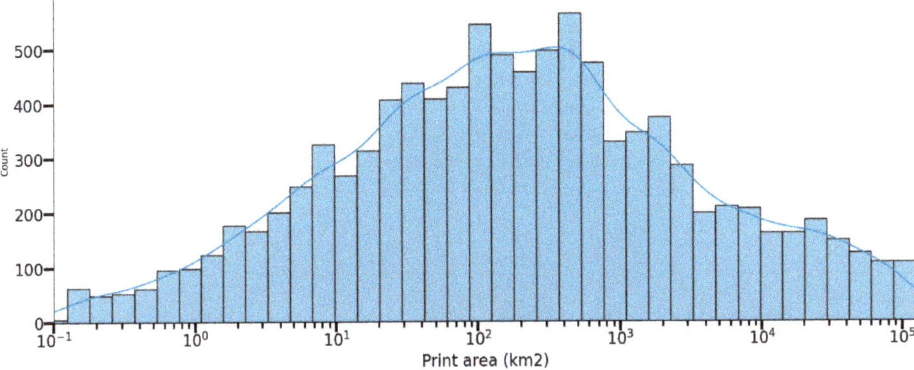

Figure 12. Histogram of the print areas in square kilometers, using a log10 x-axis.

4.3. Analysis of the Optional User Feedback

After processing is finished, users are given the option to comment on the purpose of the terrain model ("Optional: tell us what you're using this model for"). Between July 2019 and December 2020, 1947 users (~6% of all users) provided comments of varying length and quality. A table in csv format, containing these feedback comments, can be found in the Supplementary Material section. After discarding meaningless feedback (such as gibberish, n = 146, or users indicating they were just testing the software, n = 302), the remaining 1499 comments were divided into categories (n = 8) and subcategories (n = 77), which were meant to capture the essence of what the model is used for. If the information given in the comment was too sparse, no subcategory was assigned to it. The system of categories and subcategories used was subjective as was the interpretation of the comments. However, care was taken to be internally consistent. Table 2 shows the system of categories and subcategories used for the analysis. Cases where the category and subcategory are identical (e.g., Art–Art) indicate that a more specific subcategory could not be assigned.

Table 2. Complete listing of the categories and subcategories used to define model use cases.

Category	Subcategory	Category	Subcategory	Category	Subcategory	Category	Subcategory
Art	Art	Education	Art Project	Map	Country	Recreation	Aviation
Crafts	Carving	Education	BVI	Map	Farm	Recreation	Biking
Crafts	Crafts	Education	College	Map	Instructional	Recreation	Camping
Crafts	Furniture	Education	Documentary	Map	Land Use	Recreation	Climbing
Crafts	Gift	Education	Education	Map	Landscape	Recreation	Dirt Biking
Crafts	Hobby	Education	Elementary School	Map	Map	Recreation	Drone
Crafts	Memento	Education	Exhibition	Map	Mountains	Recreation	Gaming
Crafts	Puzzle	Education	Field Trip	Map	Neighborhood	Recreation	Hiking
Crafts	Railroad	Education	High School	Map	Outdoors	Recreation	Hunting
Crafts	Trophy	Education	Learning	Map	Park	Recreation	Leisure
Crafts	Wall Art	Education	Library	Map	Property	Recreation	Paraglider
Design	3D Design	Education	Middle School	Misc.	Commercial	Recreation	Radio
Design	Architecture	Education	Museum	Misc.	Curious	Recreation	Ski
Design	Art	Education	Personal	Misc.	Food	Recreation	Snowmobile
Design	Augmented Reality	Education	School Project	Misc.	Fun	Recreation	Sports
Design	Decoration	Education	Teaching	Misc.	Message	Recreation	Tourism
Design	Diorama			Misc.	Personal	Research	Archeology
Design	Modeling			Misc.	Private	Research	Geology
Design	Programming			Misc.	Special	Research	History
Design	Video Game			Misc.	Unknown	Research	Meteorology
						Research	Research

In addition, each comment was tagged with the type of 3D printing technology intended to be used. By default, that type of usage was assumed to be 3D printing (3DP), which seemed reasonable given the overall messaging of the entire web application as intended for 3D printing. However, a 11.7% of comments contained words like *CNC* or *carving*; indicating a subtractive, carving process which were flagged as "CNC." In three cases (0.2%), laser cutting was mentioned, which indicated that the model was to be assembled from laser cut slices of paper, wood or acrylic. The remaining 88.1% of all models was assumed to be intended for 3D printing.

Figure 13 shows the number of comments for each of the major categories. Most models are used for crafting, education or creating a 3D map of an area.

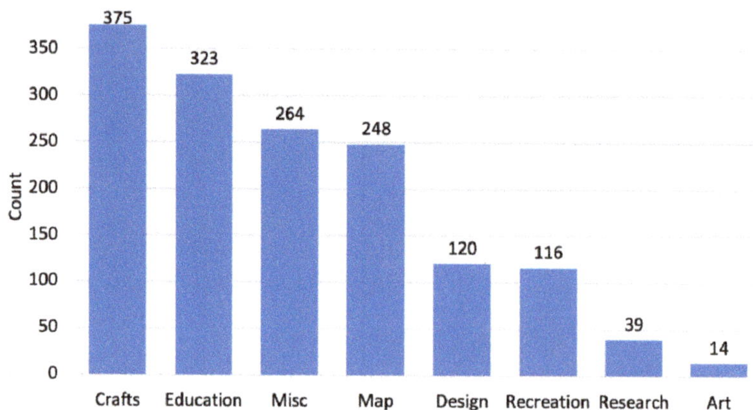

Figure 13. Number of comments classified by major category.

Tables 3 and 4 list the number of comments by category/subcategory and by type of technology used. n total shows the total number of comments assigned to each category/subcategory combination that was used. *n 3DP* shows how many of those comments were tagged as being created with 3D printing technology, *n CNC* those tagged as being created with CNC type technology. Table 3 also includes examples of the feedback text.

Table 3. Number of comments by category/subcategory and by type of technology used (with examples). Abbreviations: CNC = Computer Numerical Control; 3DP = 3D printing.

Category	Subcategory	n 3DP	n CNC	n Total	Examples
Misc	Personal	182	3	185	Model printing and painting as a hobby; 3D print for personal use
Education	Education	176	0	176	for educational purposes; school map; Class assignment!; ed printing
Crafts	Gift	141	12	153	Print a gift for my Dad - he was born nearby; Anniversary gift for my wife, we hiked here for our honeymoon.
Crafts	Carving	6	135	141	3D carving; CNC carving; CNC Machine; Wood relief; X carve;
Map	Map	94	2	96	3D printing Kauai island since we can't travel there currently; model of this place where we were with friends.
Map	Property	69	1	70	Bought new rural property. Wanted to evaluate and visualize landscape for trail planning and possible cabin locations; My property to teach my kids the boundaries and the land.
Misc	Fun	45	0	45	For Fun, conversation; PLAYING AROUND; Just for fun/ unsure yet
Education	College	42	0	42	I will teach geomorphology with this model; teaching oceanography course: section on coastal geology
Design	3D Design	39	1	40	Trial terrain in Blender 2.83; Topo Model in Rhino; fusion360 landscape austria
Design	Architecture	36	1	37	Understand land, easements, and possible building sites; Student architecture project
Recreation	Hiking	28	1	29	Trying to get a realistic model to plan for some climbing; trecking & climbing observation
Research	Research	25	0	25	Research of the Buffalo Creek Mining Disaster; Glacial modelling

Table 3. *Cont.*

Category	Subcategory	n 3DP	n CNC	n Total	Examples
Crafts	Wall Art	23	1	24	make a terrain map of Long Island where i am from to put on my wall; 3d print an area map for my wall
Map	Neighborhood	24	0	24	Teaching my kids their surrounds, way to school, beach, woods we visit; neighborhood exploring
Recreation	Ski	22	0	22	favorite ski area: Arapahoe Basin Colorado; 3D backcountry ski map
Education	School Project	21	0	21	For my Grandson School Project; I am using this for a school project.
Crafts	Crafts	14	6	20	3d printing key chain fob; Carving a door sign; CNC a topo for box lid
Education	High School	17	1	18	testing for education, geography course high school; teaching College prep geography students
Recreation	Gaming	17	0	17	Hobby for Gettysburg Campaign, personal use w my father; Model of Savo Island for WWII Naval Game
Design	Art	15	1	16	Fine Arts project dealing with geography and mapping; concept art illustrations; art show
Map	Mountains	15	1	16	I've always wanted so much 3D maps of my mountains, it feels like a dream coming true!; cnc carving of the smokies
Education	Learning	15	0	15	Learning to 3d Print; Learning
Education	Teaching	15	0	15	to teach some stuff to kids during Covid 19 confinment; classroom instruction / data visualization
Art	Art	14	0	14	Art
Map	Outdoors	13	1	14	to see a river that I kayak; want to cnc a model of Tablerock Lake
Misc	Curious	14	0	14	Curious to see what this does!; Satisfying my curiosity
Design	Modeling	13	0	13	Model making; modelling; model
Recreation	Camping	12	1	13	carving a model of the Philmont Scout Camp located in this area; Commemorate a camping trip
Education	Museum	12	0	12	For a museum to illustrate how increased tides will effect the area; display in a children's museum
Crafts	Hobby	8	3	11	A new project for creating collectibles; hobbie
Recreation	Tourism	10	0	10	Visualising holiday terrain; view our holiday rides; Tourism
Design	Decoration	9	0	9	probably a shelf decoration; Home decoration; Desk Décor
Map	Farm	8	0	8	My Dads old farm area; I want a printed 3d model of my farm; Houston Farm Maryville, Missouri
Map	Park	8	0	8	Mt. Pisgah State Park; Mammoth Cave National Park; Ohio Pyle SP
Research	Geology	8	0	8	attempting to detail area for gold prospecting; We want to use this to learn the country geological.
Crafts	Furniture	5	2	7	3D Wood Carving into a desk top and back filling with epoxy resin; designing a dining table
Crafts	Puzzle	6	0	6	For making a jigsaw puzzle. Thank you; puzzle
Crafts	Trophy	6	0	6	Make into a trophy for beer Olympics.; Trophy for a friend
Education	Elementary School	6	0	6	3D print for use in a hands-on, 4th grade science unit about flood and drought, looking at a local watershed.
Misc	Private	6	0	6	Private use; private

Sadly, the most common combination, Misc./Personal, is hardly indicative of its actual usage, indicating the difficulty of interpreting the often terse comments into anything more specific. However, the next most common usage is for education in general, followed by crafting terrain models intended as gifts. This is followed by models intended for CNC carving and by models for creating a 3D map of an area in general.

Table 4. Number of comments by category/subcategory and by type of technology used (w/o examples).

Category	Subcategory	n 3DP	n CNC	n Total	Category	Subcategory	n 3DP	n CNC	n Total
Education	Art Project	5	0	5	Recreation	Paraglider	2	0	2
Map	Landscape	5	0	5	Recreation	Snowmobile	2	0	2
Misc.	Message	5	0	5	Research	Meteorology	2	0	2
Education	Exhibition	4	0	4	Design	Augmented Reality	1	0	1
Misc.	Special	4	0	4	Design	Diorama	1	0	1
Recreation	Aviation	4	0	4	Design	Programming	1	0	1
Recreation	Leisure	4	0	4	Education	Bvi	1	0	1
Recreation	Radio	4	0	4	Education	Documentary	1	0	1
Crafts	Memento	3	0	3	Education	Field Trip	1	0	1
Crafts	Railroad	3	0	3	Education	Personal	1	0	1
Education	Middle School	2	1	3	Map	Country	1	0	1
Map	Land Use	3	0	3	Map	Instructional	1	0	1
Recreation	Hunting	3	0	3	Misc.	Food	1	0	1
Research	History	3	0	3	Recreation	Biking	1	0	1
Design	Video Game	2	0	2	Recreation	Climbing	1	0	1
Education	Library	2	0	2	Recreation	Drone	1	0	1
Misc.	Commercial	2	0	2	Recreation	Sports	0	1	1
Misc.	Unknown	2	0	2	Research	Archeology	1	0	1
Recreation	Dirt Biking	2	0	2					

5. Educational Use Cases of 3D Printed Terrain Models

At Iowa State University (Ames, IA, USA), 3D printed terrain models were first deployed in 2017 and continue to play a role in the undergraduate geology curriculum, primarily as teaching aids for mapping tasks.

5.1. Field Trip to Death Valley

A very large 3D printed terrain model (400 × 400 mm, z-scale 2×) was used during a 2018 field trip to study the geology of Death Valley (Figure 14). On location, the model (dubbed Mordor by the students) was used to teach them about the geomorphology of the valley (e.g., alluvial fans and other erosional features) and, in conjunction with a geologic map of the area, was used to put the terrain's various geologic units into spatial context. The model used more than 1 kg of red PLA filament (~USD 30 worth) and took ~50 h to print.

Figure 14. Large 3D Printed model of Death Valley, dubbed "Mordor" by students.

5.2. Geologic Mapping Exercise in Summer Field Camp

GEOL 302 is a 6-week summer field course required of all geology majors. The course focuses on applied geology and provides students with extensive experience in geologic mapping and performing field analyses related to structural deformation, stratigraphy, sedimentology, geomorphology, metamorphic processes and energy resources. The Alkali Anticline Mapping Project comprises a three-day excursion to an anticlinal structure in which student groups (3–4 individuals per group) map formational contacts and collect data on fold and fault geometries of this asymmetric, doubly plunging anticline in the Bighorn Basin, Wyoming, USA. Furthermore, considering the large size of the mapping area (13.5 km^2), students must plan their area traversals to efficiently collect enough field data to then be able to fully characterize this complex geological structure on their final 2D geologic map.

Fourteen 3D printed terrain models of Alkali Anticline were used during the last three field seasons to support the mapping project. The 14 cm $\times$ 19 cm terrain models represent the topography of a 13.5 km^2 area and were printed with a vertical exaggeration of $2\times$. The model size was chosen for ease of manipulation and transportation during field exercises. Models were printed in ABS material, which is sturdy enough not to suffer damage during the sometimes rough outdoor handling. They were sprayed with a light gray automotive primer paint to ensure good contrast, even under bright outdoor lighting, and to allow students and instructors to temporarily mark waypoints, data, and working hypotheses using soft pencil lead or colored pencils.

Each group of 3–4 students was provided with a 2D topographic base map, a printed satellite image, and a 3D printed terrain model at the start of the project (Figure 15). The instructors spatially oriented students on all representations at the beginning of each day in the field. Students were instructed on how to use the 2D topo map, satellite image, and the 3D model together to plan efficient and safe hiking paths that optimize data collection and make working hypotheses regarding the placement of formational contacts, fault traces, and fold axial traces and plunge. More specifically, the 3D terrain models were used to predict the contact locations by using the geomorphic expression of previously studied geologic formations (i.e., ridge-forming vs. valley-forming lithological units). Disjointed ridge segments and variable valley geometries on the 3D terrain models provide evidence of fault offset whereas slope steepness and orientation of dip slopes on the models delineates folds within the strata.

Figure 15. Iowa State University field camp students discuss their working hypotheses with an instructor using both a satellite image and 3D terrain model during the Alkali Anticline Project.

5.3. Topographic Map Exercise in Introductory Geology Laboratory

The 3D printed terrain models of Alkali Anticline are also integrated into an introductory physical geology laboratory course (GEOL 100L), allowing novice geology students to practice reading and interpreting complex maps of real topography. Topographic map exercises make up a two-week module of lab activities designed to increase student spatial thinking skills in the context of socially relevant issues, such as flood risks. Students are given a 2D topographic map and the 3D printed model and are asked to translate information from the 3D to the 2D representations. Valleys and ridges are marked on the 3D printed terrain models, and students find and label the same features on the 2D maps. After the students have oriented themselves with both the 2D and 3D representations, they are given tasks similar to those of the Topographic Map Assessment, including pathfinding, slope steepness, and visibility between points on the landscape.

5.4. Results

Judging by their feedback, students appeared to be more engaged when using the 3D terrain models than when using 2D topo maps and showed greater enthusiasm to explore the landscape on the 3D models. Introductory geology laboratory students often remarked on how the printed layers of the 3D terrain models helped them identify patterns in contour lines that represent ridges or valleys on the 2D maps. The ability to easily rotate the models and view them at different angles also helped them better understand concepts like the "rule of Vs" and how contour spacing reflects slope steepness.

In course evaluations (see Appendix D), 42 summer field camp students were asked for optional feedback on their use of 3D printed terrain models: *"Were the 3D models of Alkali Anticline useful during mapping? If yes, how so? If no, why not?"* Of the 42 students, twelve gave responses—all were positive. Eleven students found models beneficial to completing the mapping project, one student commented that they "liked" using them during mapping. Two other students made positive affective comments as well, writing "I also enjoyed the 3D models" and " . . . a 3D model of an area gets you settled, and you feel more comfortable." Eight students described the models as being useful in visualizing large-scale geometries, such as structural deformation or stratigraphic relationships, with four other comments regarding the benefit of the models in locating themselves on 2D topographic maps. Three students expressed a desire to expand the use of 3D models in the field course.

Although not a formal study, our results suggest that using 3D terrain models had at the very least a reassuring effect on the confidence of students, which would be in line with work mentioned earlier, such as [28]. Being able to locate the current position on the 3D model (or "getting settled") is a basic, yet underappreciated mapping task that is foundational to and provides necessary calibration for more complex tasks. The ability to visualize the large-scale tectonic structures, such as folded or faulted rock strata, is at the core of geologic mapping. Here, a key ability is to develop a model in the mind's eye of 3D geometries that extend both into the subsurface and above ground, visualizing the shape prior to the effects of erosion. Furthermore, the ability to project these geometries onto the real-world terrain surface is critical for forming working hypotheses and predicting the location of additional data. While sophisticated 3D computer graphics would be needed to truly "look inside" the terrain and see these 3D structures, 3D printed terrain models, in combination with the ability to annotate them, may help to bridge the gap between those two worlds.

Further, quantitative studies are certainly needed. However, these student responses, in combination with our other 3D model experiences, may point to the application of 3D terrain models to support geologic field work in general, and specifically for geologic mapping, as promising areas of future research.

6. Conclusions and Future Directions

Analyzing 18 months' worth of web analytics allowed us to determine who is using the TouchTerrain web application and what size of models they are making; and to map out what areas on the globe they chose to 3D print terrain models of. Since July 2019, more than 20,000 terrain models have been downloaded. Models are being created for many different use cases, including education, research, outdoor activities and crafting mementos. Users took around 8 min to create a model, much faster than with traditional GIS or 3D modeling software. Most models were realized with 3D printers, with a sizable minority using CNC machines.

Our own experiences with using 3D printed terrain in a university setting are so far very positive. Although a formal study has yet to be performed, anecdotal evidence points to the strong potential for 3D printed terrain models to provide significant help with specific map-related tasks. For the introductory geology laboratory, 3D printed models were used as a form of training wheels to aid beginning students in learning the "language" of contour maps, which are still an important tool for field geologists. Field camp students noted the benefit of 3D printed models in locating themselves on their 2D base maps and visualizing the large-scale structures.

In the future, we plan to continue examining the efficacy of 3D terrain models in the undergraduate curriculum at Iowa State University and possible branch out in adjacent disciplines, such as landscape architecture. We are also committed to further develop TouchTerrain's capabilities. Adding functionality to superimpose other map data (e.g., aerial photos) on generated 3D models would make terrain models more useful when used with 3D modeling software. In order to make it easy to create model that have enough layers to make the 3D printed terrain more interesting, an auto z-scale setting is planned for which the user would select the desired height of the 3D printed model (e.g., 3 cm from the topmost part to the bottom and TouchTerrain would automatically scale the model to this size.

We hope that the further proliferation of affordable 3D terrain models facilitated by TouchTerrain will find more and additional uses for them in science, education and engineering. In the field of supporting student education, we hope that TouchTerrain will stimulate efforts in the training and education of Blind and Visually Impaired (BVI) individuals. This is a difficult and underrepresented area of education research for which TouchTerrain could prove to be a key enabling technology. For example, using 3D printed TouchTerrain models with Braille annotations could prove useful for BVI STEM education or for mobility training.

In future, we expect to expand the use of Google Analytics for more insights. For example, we may be able to combine IP geolocation data of the users with what areas they download, which may reveal insights into concepts like place-based learning, which suggests that people benefit from learning concepts within the context of the landscape that surrounds them and that they are thus familiar with. It might also be interesting to delineate types of users, e.g., power users vs. standard users, in order to better support them.

Supplementary Materials: Supplementary materials are available online at https://www.mdpi.com/2220-9964/10/3/108/s1.

Author Contributions: Chris Harding is the principal contributor to the TouchTerrain code, contributed to the usage analysis and wrote the first draft of the manuscript. Franek Hasiuk contributed to the usage analysis and to the writing of the manuscript. Aaron Wood deployed 3D printed terrain models in geoscience education and contributed to the writing of the manuscript. All authors have read and agreed to the published version of the manuscript.

Funding: This research received no external funding.

Institutional Review Board Statement: Not applicable.

Informed Consent Statement: Not applicable.

Data Availability Statement: GIS Data is available in a publicly accessible repository. GIS data for Figure 9 can be found at https://arcg.is/0L8fLz (accessed on 20 February 2021), GIS data for Figures 10 and 11 can be found at https://arcg.is/0PO1nK (accessed on 20 February 2021). The user feedback data analyzed in Section 4.3 are available as supplementary material.

Acknowledgments: We'd like to acknowledge the contribution and assistance provided by Jaqueline Reber, Levi Baber and Nick Booher.

Conflicts of Interest: The authors declare no conflict of interest.

Appendix A. Gallery of 3D Printed Terrain Models

Figures A1–A4 show images of 3D printed terrain models created with TouchTerrain. Some were contributed by TouchTerrain users and are used with their permission. Otherwise, the print and images were made by the authors.

Figure A1. Grand Canyon Model with geologic formations painted on by Aaron Wood. Before painting, a coat of gray automotive primer filler was applied.

Figure A2. Print of the West coast of South America, including the bathymetry (z-scale x4) to demonstrate the enormous elevation change from the bottom of the Andes trench to the top of the Andes mountains.

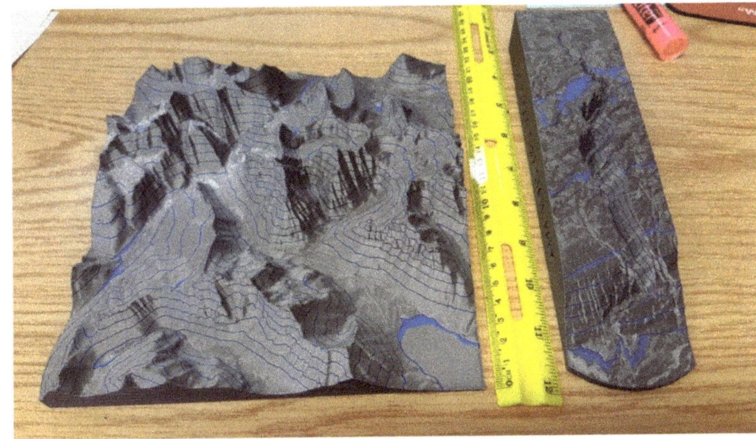

Figure A3. Two-color terrain print by Rob Lawry. The printer was paused at regular intervals and blue filament was used for one layer, after which gray filament was used again.

Figure A4. Print by Bryson Hicks, using a polygonal outline of a US county. Used in disaster management.

Appendix B. Additional Best Practices for 3D Printing Terrain Models

This appendix gives more details on best practices for 3D printing terrain models. Familiarity with filament materials and print parameter settings for the popular Cura slicer software is assumed, some of which are under "expert settings". However, most concepts will apply to other slicers (e.g., Prusa slicer), possibly with a different nomenclature.

Model robustness: To ensure a model is robust enough to withstand rough handline it should be printed with at least two walls (perimeters), at least two bottom layers and at least three top layers, assuming a layer height larger than 0.2 mm. For smaller layer heights, these numbers should be increased. Setting the topmost layer as "skin" in Cura can print it slower and increase print quality. Infill density can be as low as 10–15%. The Cura fill pattern zigzag is particularity economical with no ill effect.

Corner warping: the corners of a box-outlined terrain model may warp upwards as the filament cools and possible contracts. For PLA and PETG, using a heated bed and setting a 5–10 mm wide brim is usually sufficient to suppress corner warping. For ABS, which shrinks more, a larger brim and possible a heat-preserving enclosure may be needed.

Cooling and retractions: In general, terrain printing requires little to no part-cooling which also promotes layer adhesion. The exceptions are layer with are printed very quickly (e.g., small peaks) which require good cooling and/or a long minimum layer time (>15 s). Typical retraction settings should be used unless noticeable stringing occurs. If stringing occurs, brushing the top of the finished model with a wire brush is usually enough to remove it. A hot air gun is more effective but can easily warp the model when used for too long.

Improving the appearance of the top layers: For your top (skin) layer a concentric fill pattern should be used instead for the standard linear pattern. This will avoid artificial looking stripe patterns, which are especially obvious on flat terrain or lakes as the contrast sharply with the organic appearance that surrounds them. Ironing can be used to fill in and smooth the top layers, but is usually not worth the longer print duration.

Post-processing: 3D Prints are typically somewhat glossy. If a matte finish is preferred or is the model is to be drawn on or painted, it can be sprayed with a neutral gray automotive primer filler. To glue together multiple terrain tiles printed with PLA or PETG, superglue gel works well. For ABS, acetone can be used. To mount the print on a wooden plate, a hot glue gun works well; later, this can be removed with alcohol.

Appendix C. Top 25 Place Searches

Table A1 shows search results for place searches performed from 12 October 2020 to 26 December 2020 and ranked by the number of unique usage events. Unique denotes per-user session events, i.e., even if a user performed the same search multiple times during a session it was still counted as a single unique event. Note that these do not necessarily imply that this led to a later download of the searched for area, i.e., this is only an indication of the user's initial interest before selecting the area to print.

Table A1. Top 25 search results 12 October 2020 to 26 December 2020.

Search Results	Total Events	Unique Events
Grand Canyon National Park, Arizona, USA	90	57
Mt Everest	75	54
Yosemite National Park, California, USA	36	30
Mount Fuji, Kitayama, Fujinomiya, Shizuoka 418-0112, Japan	36	29
Mount Rainier, Washington 98304, USA	40	28
Matterhorn	31	25
Hawaii, USA	30	21
Denali, Alaska, USA	24	21
New York, NY, USA	23	21
Mt St Helens, Washington 98616, USA	26	19
Mount Everest	24	15
Half Dome, California, USA	20	15
Bryce Canyon National Park, Utah, USA	17	15
Santiago, Región Metropolitana, Chile	22	13
Norwich, VT, USA	18	12
Santiago de Chile, Región Metropolitana, Chile	18	12
El Capitan, California 95389, USA	16	12
Zion National Park, Utah, USA	15	12
Denver, CO, USA	14	12
Israel	14	12
Mt. Whitney, California, USA	17	11
Mount Rushmore National Memorial	11	11
Crater Lake, Oregon 97604, USA	25	10
Grand Teton, Wyoming 83414, USA	15	10
Yellowstone National Park, United States	15	10
Mt Hood, Oregon 97041, USA	14	10

Appendix D. Student Feedback on Using 3D Models for Geologic Mapping

In 2017, Iowa State geologic field camp students started using 3D printed models during the Alkali Anticline geologic mapping exercise. Working in the field, students were instructed to write annotations about their mapping activities directly on the model (with a soft pencil). The annotated models were used to help them construct the final geologic map. As part of the course evaluations filled out at the end of the field camp, students were asked:

Were the 3D models of Alkali Anticline useful during mapping? If yes, how so? If no, why not?

- The 3D model of Alkali Anticline was extremely useful in finding yourself on the topo maps and to determine if there was any offset on the ridge tops;
- We also got to use 3D models for one of our mapping exercises and if we could use those more that would help a lot. For some students it is really hard to think in 3D and the models fixed that. Contour maps are confusing at first and being able to look at a 3D model of an area really gets you settled and you feel more comfortable. I used the model for all of that project, it was a great learning tool that would be amazing to see implemented more;
- Yes it was very helpful to be able to see the mapping area on a small scale with the topography because it helped with seeing the big picture;
- The 3D models of Alkali Anticline were quite useful in correlating the topographical map and formations;
- I thought the 3D models of Alkali Anticline were useful because they aided us in mapping our position and allowed us to visualize the structure of the anticline by being able to write on them;
- I also enjoyed the 3D models used at alkali anticline and I learned a great deal from them. They were a great help to see displacement and helped us to understand the structure and geology of the location. I do think that an easier way to clean them needs to be found. I scrubbed my model for quite a while but was unable to get all of the graphite off;
- I also liked the 3D models for Alkali Anticline and I would have appreciated more models for other mapping projects;
- I felt that the 3D model was helpful when we remembered we had it, but could have utilized it more had we kept it out of our packs;
- The 3D models used for the Alkali Anticline were very helpful in visualizing what we were standing on and in locating ourselves on the map. I feel like this could be part of the future of geology and especially field camp. The more hands on, the better.
- The 3D model is very useful in mapping exercise in Alkali Anticline as it helps to visualize the structure more accurately;
- Not to me, but my groupmates used the model a lot. Making them available to students was very beneficial to the groups as a whole;
- The use of the 3D topographies of Alkali Anticline were extremely helpful in locating myself and determining what structures I was standing on.

References

1. Rapp, D.N.; Culpepper, S.A.; Kirkby, K.; Morin, P. Fostering Student's Comprehension of Topographic Maps. *J. Geosci. Educ.* **2007**, *55*, 5–16. [CrossRef]
2. Takaku, J.; Tadono, T.; Tsutsui, K.; Ichikawa, M. Validation of "AW3D" global dsm generated from Alos Prism. *ISPRS Ann. Photogramm. Remote Sens. Spat. Inf. Sci.* **2016**, *III–4*, 25–31. [CrossRef]
3. Juřík, V.; Herman, L.; Snopková, D.; Galang, A.J.; Stachoň, Z.; Chmelík, J.; Kubíček, P.; Šašinka, Č. The 3D Hype: Evaluating the Potential of Real 3D Visualization in Geo-Related Applications. *PLoS ONE* **2020**, *15*, e0233353. [CrossRef]
4. Ishikawa, T.; Kastens, K.A. Why Some Students Have Trouble with Maps and Other Spatial Representations. *J. Geosci. Educ.* **2005**, *53*, 184–197. [CrossRef]

5. Kastens, K.A.; Ishikawa, T. Spatial thinking in the geosciences and cognitive sciences: A cross-disciplinary look at the intersection of the two fields. In *Special Paper 413: Earth and Mind: How Geologists Think and Learn about the Earth*; Geological Society of America: Boulder, CO, USA, 2006; Volume 413, pp. 53–76, ISBN 978-0-8137-2413-3.
6. Manduca, C.A.; Kastens, K.A. Mapping the domain of spatial thinking in the geosciences. In *Geological Society of America Special Papers*; Geological Society of America: Boulder, CO, USA, 2012; Volume 486, pp. 45–49, ISBN 978-0-8137-2486-7.
7. Atit, K.; Weisberg, S.M.; Newcombe, N.S.; Shipley, T.F. Learning to Interpret Topographic Maps: Understanding Layered Spatial Information. *Cogn. Res. Princ. Implic.* **2016**, *1*. [CrossRef]
8. Lütjens, M.; Kersten, T.P.; Dorschel, B.; Tschirschwitz, F. Virtual Reality in Cartography: Immersive 3D Visualization of the Arctic Clyde Inlet (Canada) Using Digital Elevation Models and Bathymetric Data. *Multimodal Technol. Interact.* **2019**, *3*, 9. [CrossRef]
9. Hruby, F.; Ressl, R.; de la Borbolla Del Valle, G. Geovisualization with Immersive Virtual Environments in Theory and Practice. *Int. J. Digit. Earth* **2019**, *12*, 123–136. [CrossRef]
10. Hruby, F.; Castellanos, I.; Ressl, R. Cartographic Scale in Immersive Virtual Environments. *KN J. Cartogr. Geogr. Inf.* **2020**. [CrossRef]
11. View of The Tangible Map Exhibit | Cartographic Perspectives. Available online: https://cartographicperspectives.org/index.php/journal/article/view/cp82-dooley-et-al/1454 (accessed on 26 January 2021).
12. Heidkamp, C.P.; Slomba, J.T. GeoSpatial Sculpture: Nonverbal Communication through the Tactilization of Geospatial Data. *GeoHumanities* **2017**, *3*, 554–566. [CrossRef]
13. Rossetto, T. The Skin of the Map: Viewing Cartography through Tactile Empathy. *Environ. Plan. Soc. Space* **2019**, *37*, 83–103. [CrossRef]
14. domlysz BlenderGIS. Available online: https://github.com/domlysz/BlenderGIS (accessed on 8 February 2021).
15. Simon, F. DEMto3D. Available online: https://demto3d.com/en/ (accessed on 8 February 2021).
16. Hatch, J. Terrain2STL. Available online: http://jthatch.com/Terrain2STL/ (accessed on 8 February 2021).
17. Wilson, D. The Terrainator. Available online: https://terrainator.com/ (accessed on 8 February 2021).
18. Reed, H.P. The Development of the Terrain Model in the War. *Geogr. Rev.* **1946**, *36*, 632–652. [CrossRef]
19. Adams, A.M. A Comparative Usability Assessment of Augmented Reality 3-D Printed Terrain Models and 2-D Topographic Maps. Master's Thesis, New Mexico State University, Las Cruces, NM, USA, 2019.
20. Priestnall, G. Rediscovering the Power of Physical Relief Models: Mayson's Ordnance Model of the Lake District. *Cartogr. Int. J. Geogr. Inf. Geovis.* **2019**, *54*, 261–277. [CrossRef]
21. Welter, J. Solid Landscape Models in the Twenty-First Century—A Balanced Approach. *Cartogr. J.* **2013**, *50*, 300–304. [CrossRef]
22. Kete, P. Physical 3D Map of the Planica Nordic Center, Slovenia: Cartographic Principles and Techniques Used with 3D Printing. *Cartogr. Int. J. Geogr. Inf. Geovis.* **2016**, *51*, 1–11. [CrossRef]
23. Oswald, C.; Rinner, C.; Robinson, A. Applications of 3D Printing in Physical Geography Education and Urban Visualization. *Cartogr. Int. J. Geogr. Inf. Geovis.* **2019**, *54*, 278–287. [CrossRef]
24. Horowitz, S.S.; Schultz, P.H. Printing Space: Using 3D Printing of Digital Terrain Models in Geosciences Education and Research. *J. Geosci. Educ.* **2014**, *62*, 138–145. [CrossRef]
25. Jakovina, M.; Ormand, C.; Shipley, T.F.; Weisberg, S. Topographic Map Assessment (Instrument). 2014. Available online: http://www.silccenter.org/index.php/testsainstruments (accessed on 30 December 2020).
26. Plant, R.W.; Ryan, R.M. Intrinsic Motivation and the Effects of Self-Consciousness, Self-Awareness, and Ego-Involvement: An Investigation of Internally Controlling Styles. *J. Pers.* **1985**, *53*, 435–449. [CrossRef]
27. Carbonell Carrera, C.; Avarvarei, B.V.; Chelariu, E.L.; Draghia, L.; Avarvarei, S.C. Map-Reading Skill Development with 3D Technologies. *J. Geogr.* **2017**, *116*, 197–205. [CrossRef]
28. Carbonell-Carrera, C.; Jaeger, A.; Shipley, T. 2D Cartography Training: Has the Time Come for a Paradigm Shift? *ISPRS Int. J. Geo-Inf.* **2018**, *7*, 197. [CrossRef]
29. Perkins, C. Cartography: Progress in Tactile Mapping. *Prog. Hum. Geogr.* **2002**, *26*, 521–530. [CrossRef]
30. Götzelmann, T.; Pavkovic, A. Towards Automatically Generated Tactile Detail Maps by 3D Printers for Blind Persons. In *Proceedings of the Computers Helping People with Special Needs*; Miesenberger, K., Fels, D., Archambault, D., Peňáz, P., Zagler, W., Eds.; Springer International Publishing: Cham, Switzerland, 2014; pp. 1–7.
31. Schwarzbach, F.; Sarjakoski, T.; Oksanen, J.; Sarjakoski, L.T.; Weckman, S. Physical 3D models from LIDAR data as tactile maps for visually impaired persons. In *True-3D in Cartography*; Buchroithner, M., Ed.; Lecture Notes in Geoinformation and Cartography; Springer: Berlin/Heidelberg, Germany, 2011; pp. 169–183, ISBN 978-3-642-12271-2.
32. Voženílek, V.; Růžičková, V.; Finková, D.; Ludíková, L.; Němcová, Z.; Doležal, J.; Vondráková, A.; Kozáková, M.; Regec, V. Hypsometry in Tactile Maps. In *True-3D in Cartography*; Buchroithner, M., Ed.; Lecture Notes in Geoinformation and Cartography; Springer: Berlin/Heidelberg, Germany, 2011; pp. 153–168, ISBN 978-3-642-12271-2.
33. Perry, R. Converting Topographic Information with 3D Printed Models and Tactile Maps. Master's Thesis, School of Information Science, University of Maine, Orono, ME, USA, 2017; Unpublished.
34. Caldwell, D.R. Physical Terrain Modeling for Geographic Visualization. *Cartogr. Perspect.* **2001**, 66–72. [CrossRef]
35. Harding, Chris *TouchTerrain*. Available online: https://github.com/ChHarding/TouchTerrain_for_CAGEO (accessed on 30 December 2020).
36. Hasiuk, F.J.; Harding, C.; Renner, A.R.; Winer, E. TouchTerrain: A Simple Web-Tool for Creating 3D printable Topographic Models. *Comput. Geosci.* **2017**, *109*, 25–31. [CrossRef]

MDPI
St. Alban-Anlage 66
4052 Basel
Switzerland
Tel. +41 61 683 77 34
Fax +41 61 302 89 18
www.mdpi.com

ISPRS International Journal of Geo-Information Editorial Office
E-mail: ijgi@mdpi.com
www.mdpi.com/journal/ijgi

www.ingramcontent.com/pod-product-compliance
Lightning Source LLC
LaVergne TN
LVHW070157170726
843515LV00007B/1941

* 9 7 8 3 0 3 6 5 3 6 8 0 4 *